BALLYNAHATTY

Ballynahatty 6 during excavation. Billy Dunlop on left, Ken Pullin on right.

Dedicated to Ken Pullin and Billy Dunlop whose contributions to Ballynahatty and enjoyment of archaeology is an inspiration to us all.

BALLYNAHATTY

EXCAVATIONS IN A NEOLITHIC MONUMENTAL LANDSCAPE

Edited by

BARRIE HARTWELL, SARAH GORMLEY, CATRIONA BROGAN
and CAROLINE MALONE

With contributions from

*R.P. Barrett, Rowan McLaughlin, Eileen Murphy,
Eiméar Nelis, Gill Plunkett* and *J.L. Wilkinson*

Oxford & Philadelphia

Published in the United Kingdom in 2023 by
OXBOW BOOKS
The Old Music Hall, 106–108 Cowley Road, Oxford, OX4 1JE

and in the United States by
OXBOW BOOKS
1950 Lawrence Road, Havertown, PA 19083

© Oxbow Books and the individual authors 2023

Hardcover Edition: ISBN 978-1-78925-971-1
Digital Edition: ISBN 978-1-78925-972-8 (epub)

A CIP record for this book is available from the British Library

Library of Congress Control Number: 2023936752

All rights reserved. No part of this book may be reproduced or transmitted in any form or by any means, electronic or mechanical including photocopying, recording or by any information storage and retrieval system, without permission from the publisher in writing.

Printed and bound in Malta by Melita Press
Typeset in India by DiTech Publishing Services

For a complete list of Oxbow titles, please contact:

UNITED KINGDOM
Oxbow Books
Telephone (0)1226 734350
Email: oxbow@oxbowbooks.com
www.oxbowbooks.com

UNITED STATES OF AMERICA
Oxbow Books
Telephone (610) 853-9131, Fax (610) 853-9146
Email: queries@casemateacademic.com
www.casemateacademic.com/oxbow

Oxbow Books is part of the Casemate Group

Front cover: Reconstruction of the final built phase of timber enclosures BNH5 and 6 (Dr Robert Barratt).
Back (top photo): The inner 'temple' BNH6 during excavation (Barrie Hartwell).
Background image: Orthorectified drone image of the Giant's Ring (Image © David Craig | HeritageNI.com).

Contents

List of plates viii
List of figures and tables xi
Acknowledgements xv

PART 1: THE BALLYNAHATTY LANDSCAPE

1. The landscape and historical research 3
   1.1 Introduction, geology and topography of the complex 3
   1.2 Historical overview of Ballynahatty and the Giant's Ring 4
       1.2.1 Antiquarian study 5
       1.2.2 The 1855 tomb 7
       1.2.3 The later 19th-century landscape 8
   1.3 Archaeological investigations in the earlier 20th century 8
       1.3.1 H.C. Lawlor 1917: Giant's Ring bank, interior and megalith 8
       1.3.2 D.A. Chart 1929: cropmark within the Giant's Ring 10
       1.3.3 A.E.P. Collins 1954: Giant's Ring bank and megalith 10
   1.4 The later 20th century: BNH3 12
2. Archaeological surveys 14
   2.1 Aerial photography 14
       2.1.1 Historical imagery 14
       2.1.2 Archaeological aerial survey, 1984–2018 14
   2.2 Geophysical surveys 22
   2.3 Stray finds 24
       2.3.1 The Ulster Museum 25
       2.3.2 The National Museum of Ireland 25
       2.3.3 The excavations 25
       2.3.4 Field collection 25
3. Environmental history of the Ballynahatty area (*Gill Plunkett*) 28
   3.1 Introduction 28
   3.2 The Ballynahatty vegetation sequence 29
       3.2.1 The early post-glacial period 29
       3.2.2 Mesolithic 29
       3.2.3 Neolithic 30
       3.2.4 Bronze Age 30
   3.3 Discussion 30
4. Cumulative interpretation landscape map 31
   4.1 The wider Ballynahatty landscape 31

PART 2: THE EXCAVATIONS 1990–2000

5. Ballynahatty 5 and 6: excavating the enclosures 37
   5.1 Overview 37
   5.2 Passage Tomb related chamber and deposits of cremated remains 40
   5.3 Cremation deposits and pit 43
   5.4 The timber circle (BNH6) 44
       5.4.1 Four-Poster setting 44

		5.4.2	Central structure	46
		5.4.3	BNH6 Inner and Outer Ring	47
		5.4.4	The entrance to BNH6	53
		5.4.5	Eastern Setting (ES)	56
	5.5	Outer Enclosure (BNH5): feature analysis		58
		5.5.1	Overall shape and extent	58
		5.5.2	The entrance structure to BNH5	60
		5.5.3	North–South postholes (NS)	64
	5.6	The Annexe area		64
		5.6.1	Northern East–West postholes (NEW)	65
		5.6.2	West–North–West postholes (WNW)	66
		5.6.3	Outer Façade and entrance (IF, MF, OF)	69
		5.6.4	Southern East–West postholes (SEW)	71
		5.6.5	The Entrance Chamber	71
	5.7	Early medieval period		76
	5.8	A note on the archaeobotanical remains from excavated soil samples		78
		5.8.1	Phase 1a	78
		5.8.2	Phase 1b	79
		5.8.3	Phase 3	79
		5.8.4	Phase 4	80
		5.8.5	Early medieval period	80
		5.8.6	Discussion	81
6.	The pottery			82
	6.1	Introduction		82
	6.2	The distribution of the pottery at Ballynahatty		82
		6.2.1	Topsoil and ploughsoil (contexts C1 and C2)	82
		6.2.2	Features and contexts	83
		6.2.3	Interpretation of the pottery distribution	87
	6.3	Pottery traditions present at Ballynahatty		88
		6.3.1	Carinated Bowl pottery	88
		6.3.2	Early to Middle Neolithic 'Globular Bowl' pottery	89
		6.3.3	Middle Neolithic Coarse Ware	90
		6.3.4	Grooved Ware	91
		6.3.5	Early Bronze Age pottery	103
7.	The lithic assemblage: chipped stone (*Eiméar Nelis, abridged by Caroline Malone*)			106
	7.1	Introduction		106
	7.2	Composition: flint and non-flint		106
		7.2.1	The flint assemblage	107
		7.2.2	Primary technology	108
		7.2.3	Secondary technology	113
	7.3	The distribution of the lithics		123
		7.3.1	Topsoil and ploughsoil assemblage	123
		7.3.2	Feature and context assemblage	123
		7.3.3	Discussion of the distribution of the lithic assemblage	123
	7.4	Summary and discussion		127
8.	Other artefacts from the excavation			131
	8.1	Introduction		131
	8.2	Distribution		131
	8.3	Selected artefacts		131
		8.3.1	Stone axeheads	131
		8.3.2	Stone 'balls', possibly hammerstones, plus quartz pebble hammerstone	133
		8.3.3	Fragment of Arran pitchstone	134
		8.3.4	Possible chisel	134
		8.3.5	Beads	134
		8.3.6	Possible bead of lignite	135
9.	Human remains from excavations at Ballynahatty (*Eileen Murphy*)			136
	9.1	Cremated bone from excavations at Ballynahatty 5 and 6		136
		9.1.1	Deposits of cremated remains	136

		9.1.2	Identifiable burnt bone (including burnt animal bone) from other contexts	143
		9.1.3	Overall conclusions	144
	9.2	Human remains from the 1855 tomb: a re-evaluation		145
		9.2.1	Cremated bone	145
		9.2.2	Unburnt bone: Chamber D	146
10.	Dating and chronology (*Rowan McLaughlin*)			149
	10.1	Introduction		149
	10.2	Potential problems with radiocarbon dates		149
	10.3	Methods and modelling approach		150
	10.4	Ballynahatty radiocarbon dates		150
	10.5	Radiocarbon dates from comparable Irish sites		153
	10.6	Discussion		153
		10.6.1	Early activity	153
		10.6.2	Late Neolithic Ballynahatty	155
		10.6.3	Later activity	155
	10.7	Conclusions		156

PART 3: WHAT DOES IT ALL MEAN?

11.	Interpreting the excavation results in the wider context of prehistoric Ballynahatty			159
	11.1	Introduction		159
	11.2	Early and Early to Middle Neolithic activity		159
		11.2.1	Middle Neolithic activity	160
	11.3	Late Neolithic activity: the timber enclosures BNH5 and BNH6		162
	11.4	Early Bronze Age activity		164
	11.5	Early medieval activity		164
	11.6	Further consideration of the Late Neolithic timber monument complex		164
		11.6.1	The scale of construction	164
		11.6.2	Working to plan: Phasing the build	165
		11.6.3	Phase 1a BNH6: the planning and sequence of construction	166
		11.6.4	Phase 1b BNH5: enclosing the enclosure	169
		11.6.5	Phases 2 and 3, Annexe and entrance: impress and control	171
		11.6.6	Phase 4: destruction and removal or conversion and preservation?	173
		11.6.7	What do the artefacts tell us?	175
	11.7	Relationship between the timber monument complex and the Giant's Ring		177
	11.8	Relationship with other Irish timber circles, and with 'square-in-circle' structures in Britain		180
		11.8.1	The Brú na Bóinne connection	183
	11.9	The Ballynahatty timber monument complex: what was it all for?		185
12.	Digitally recreating Ballynahatty and simulating astronomical alignments in			
	Irish timber circles (*R.P. Barrett*)			187
	12.1	Introduction		187
	12.2	3D approximation: *in situ* elements		187
	12.3	3D approximation: hypothetical elements		188
	12.4	Astronomical simulation: from sites to data		189
	12.5	Astronomical simulation: results		191
	12.6	Conclusions		197
13.	The Ballynahatty landscape – past, present and future			199
	13.1	Introduction		199
	13.2	A short walk in the Neolithic		199
	13.3	Ballynahatty's archaeology in the more recent past		200
		13.3.1	No room for the ancients: the tenant farmer	200
		13.3.2	The Dungannons: owning the land. The intervention of antiquaries	202
		13.3.3	Cultural awakening: an intellectual puzzle	203
		13.3.4	20th century: excavation	205
		13.3.5	The 21st century: old problems, new solution	205
	13.4	Ballynahatty, the future: outstanding research questions		206

Bibliography 209
Index 219

List of plates

Plate 1.	Cropmarks photographed in 1989 reveal BNH5 with BNH6 visible within it on the ridge overlooking the Giant's Ring.
Plate 2.	Multiperiod cropmarks in Ballycarn townland taken in 1989.
Plate 3.	Southern Ballynahatty and the Giant's Ring at the end of the 1996 season of excavation.
Plate 4.	Orthorectified image of Ballynahatty td.
Plate 5.	View of Ballynahatty townland taken by a drone looking NW.
Plate 6.	View to the west of the Ring showing the position of the pond at the east end of the ridge.
Plate 7.	The BNH5 enclosure visible in 2018 with BNH6 at the near end.
Plate 8.	The BNH5/6 enclosures in 2018 from the west.
Plate 9.	The area of Bodel's fields in 2018 where so many sites had been destroyed by 1855.
Plate 10.	View NE across Ballynahatty Bog in 2018 towards Belfast Lough and the Irish Sea.
Plate 11.	The Giant's Ring in 2018, looking SE towards the Co. Down hills.
Plate 12.	Ring ditches BNH30–31, 10–12.
Plate 13.	Grooved stone balls, possibly recovered from the Giant's Ring.
Plate 14.	Selection of worked tools found in Edenderry and Ballynahatty areas.
Plate 15.	Selected pollen percentage data replotted against revised chronology.
Plate 16.	Early modern field ditch.
Plate 17.	Chamber C591 floor.
Plate 18.	Chamber C591 during excavation.
Plate 19.	Chamber C591 pit with stones removed.
Plate 20.	Cremation deposits C1014, C1016 and C1018.
Plate 21.	Cremation deposit, C588.
Plate 22.	C588 after removal of cremation deposit showing the carefully arranged split stone setting.
Plate 23.	Pit containing split stone and pebbles.
Plate 24.	Surface identification of postholes.
Plate 25.	Stones supporting the fill of OR27.
Plate 26.	OR25 during excavation.
Plate 27.	Postholes on the north side of the entrance to BNH6.
Plate 28.	Slot between OR1 and OR2, revealed after removal of C835.
Plate 29.	Charred plank from the secondary fill of OR32–33.
Plate 30.	Cobbled floor of northern Annexe.
Plate 31.	Grooved Ware pot BNH4 from northern Annexe.
Plate 32.	NEW postholes between the Inner Façade and BNH5 OR/IR at rear.
Plate 33.	NEW postholes during excavation showing the stone-filled secondary deposit.
Plate 34.	Section through a NEW posthole showing the stone column created in the secondary fill.
Plate 35.	The Annexe Façade from the entrance looking N.
Plate 36.	Looking S along the gently curving Façade to the Giant's Ring.
Plate 37.	The southern end of the Façade.
Plate 38–39.	The Outer Façade, looking N from the entrance.
Plate 40.	Façade, looking S at the lowest level.

List of plates

Plate 41. Showing leeched ring and central hump of OF5.
Plate 42. Truncated postholes of the SEW group.
Plate 43–44. Section through the Burning Pit, C1453.
Plate 45. Distribution of the Grooved Ware pottery by sherd type.
Plate 46. Distribution of Coarse Ware pottery.
Plate 47. Sherds from a Globular Bowl found in postholes EC15–17 in the Entrance Chamber.
Plate 48. Reconstruction of Vessel **1**.
Plate 49. Reconstruction of Vessel **2**.
Plate 50. Grooved Ware sherds with sooty accretions.
Plate 51. Distribution of the Grooved Ware pottery by type.
Plate 52. Composition of the primary technology assemblage.
Plate 53. Length by breadth of cores.
Plate 54. Length by breadth of flakes and blades produced by platform percussion flaking and bipolar techniques.
Plate 55. Platforms found on complete flakes and blades.
Plate 56. Length by breadth of flakes and blades: type of platform present.
Plate 57. Types of terminations found on complete flakes and blades.
Plate 58. Length by breadth of flakes and blades: types of terminations.
Plate 59. Types of platforms found on proximal flake and blade shatter.
Plate 60. Types of terminations found on complete platform flakes and blades.
Plate 61. Basic composition of the tool assemblage.
Plate 62. Length by breadth of modified tools.
Plate 63. Composition of the scraper assemblage.
Plate 64–66. Length by breadth of complete scrapers, hollow scrapers and possible blanks.
Plate 67. Types of 'miscellaneous' tools.
Plate 68. Types of functioning edges present on edge retouched tools.
Plate 69. Length by breadth of edge retouched tools: types of edges present.
Plate 70. Edge morphology of utilised tools.
Plate 71. Length by breadth of utilised tools.
Plate 72. Length by breadth of core tools: possible function.
Plate 73–74. Distribution of lithics in the primary and secondary fills of postholes in each feature group.
Plate 75. Burnt flakes recovered from Ballynahatty.
Plate 76. Two intact stone axeheads recovered during excavations.
Plate 77. Fragments of three porcellanite axesheads.
Plate 78. Flint and stone 'balls'.
Plate 79. Fragment of pitchstone artefact.
Plate 80. Beads.
Plate 81. Possible bead of lignite.
Plate 82. Contents of the 1855 tomb in the collection of the Ulster Museum.
Plate 83. Skull A64, one of the two skulls published in 1855.
Plate 84. Animal bones from the 1855 tomb.
Plate 85. The passage tomb (BNH2) looking E.
Plate 86. A modern example of two oaks growing in a densely wooded environment at The Argory in Co Armagh.
Plate 87. The entrance to Avebury.
Plate 88. Ramming the primary fill around the post.
Plate 89. Levering out the post.
Plate 90. The butt of the post topples as it is removed, causing considerable disruption to the surface.
Plate 91. The hollow post mould shown by the ranging rod.
Plate 92. Post extraction experiment: dropping stones into the void to create the secondary fill.
Plate 93–94. Newgrange 'Four Poster Enclosure' and encircling bank.
Plate 95. Giant's Ring: NE entrance and straight east section.
Plate 96. Newgrange Henge A.
Plate 97. 3D approximation of the timber circle in Ballynahatty.
Plate 98. Four different versions of the model, showing choronology of the site.
Plate 99. View of the left annexe with possible arrangment of skulls.
Plate 100. Platform structure portrayed as an excarnation platform with hypothetical elements used to convey the experience of visiting the site in use.

Plate 101. Early development of the model showing testing of different types of fills.
Plate 102. The timber 'Temple' from the south.
Plate 103. The Grand Façade.
Plate 104. The Entrance Chamber, looking into the interior of BNH5.
Plate 105. The tall timbers of the Entrance Chamber draws focus on the sky.
Plate 106. Inside the Annexe.
Plate 107. Racks of skulls in the Eastern Settings: the ancestors.
Plate 108. View into the Temple (BNH6).
Plate 109. Decaying cadavers on the central Platform open to the skies.

List of figures and tables

List of figures

Figure 1.1.	Part of OS 1:10,000 map 147 showing Ballynahatty td and the Giant's Ring.	4
Figure 1.2.	Four views of the passage tomb within the Giant's Ring.	6
Figure 1.3.	Charles Ligar's plan of the Giant's Ring 1837.	6
Figure 1.4.	Plan of the subterranean cist investigated in 1855.	7
Figure 1.5.	Location of the 1917, 1929 and 1954 excavations within the Giant's Ring.	10
Figure 1.6.	The RAF aerial photograph, which prompted Chart's excavation of 1929.	11
Figure 1.7.	Section through Giant's Ring bank and section through the bank.	11
Figure 1.8.	Distribution of flint artefacts from field collection, magnetic susceptibility survey and spot phosphate survey.	13
Figure 2.1.	Giant's Ring from the northeast in 1927.	15
Figure 2.2.	Cropmarks in the field north of the Giant's Ring photographed in 1984.	16
Figure 2.3.	Ballynahatty townland in July 1989 looking south towards the Giant's Ring.	17
Figure 2.4.	Enhanced monochrome orthorectified image of Ballynahatty td.	18
Figure 2.5–2.10.	Drone survey photographs of archaeological landscape features.	19–23
Figure 2.11.	1984 aerial photo of the Giant's Ring, showing internal concentric rings in the grass and the central circular feature identified in the gradiometer image.	24
Figure 2.12.	Resistivity profiles east across the Giant's Ring from the passage tomb to the embankment.	24
Figure 3.1.	View north across the kettle lake basin from which the Ballynahatty core was taken.	28
Figure 3.2.	Age-depth model for the Ballynahatty core based on radiocarbon samples and tephra.	29
Figure 4.1.	Cumulative map of sites identified through aerial photography and geophysical survey.	33
Figure 5.1.	Area numbers allocated during the excavation based on a 4 m grid.	38
Figure 5.2.	Site plan of excavation showing main feature groups.	39
Figure 5.3.	Chamber with line of Inner Façade to right and northern side of Entrance Chamber to left.	40
Figure 5.4.	Plan and section through stone lined chamber C591. Entrance at B.	42
Figure 5.5.	Plan of BNH6 and ancillary structures, showing excavated features and original context numbers and renumbered postholes.	43
Figure 5.6.	Estimated height of posts above ground.	45
Figure 5.7.	Excavation plans of 4P2.	45
Figure 5.8.	East–west section through 4P2.	46
Figure 5.9.	North–south section through IR10 and 4P2.	47
Figure 5.10.	Excavated postholes of the central platform and Inner Ring of BNH6, entrance at the far side.	48
Figure 5.11.	Postholes EP9 (C738) and EP13 (C830) of the Central Platform.	48
Figure 5.12.	BNH6 Outer Ring: radial posthole profiles, right side towards the centre.	49
Figure 5.13.	OR24–27: plan and longitudinal section.	50
Figure 5.14.	OR24–27: cross-sections.	51
Figure 5.15.	BNH6 Inner Ring: radial posthole profiles. Right side towards the centre.	52
Figure 5.16.	BNH6 south terminal postholes from north.	53
Figure 5.17.	BNH6 entrance: south terminal.	54
Figure 5.18.	BNH6 entrance profiles: OR1, 2, 4 and IR1, 2.	55

Figure 5.19.	Posthole IR3 at centre, which cuts two of the intermediate postholes.	56
Figure 5.20.	The Eastern Setting, adjacent to the BNH5 enclosure.	57
Figure 5.21.	The IR and OR postholes of the outer enclosure (BNH5) looking south.	57
Figure 5.22.	Plan of BNH5 excavated features.	59
Figure 5.23.	Longitudinal section through BNH5 IR and OR, and NS postholes.	60
Figure 5.24.	Plan of BNH5 entrance complex and associated features.	61
Figure 5.25.	Section drawings of BNH5 postholes.	62
Figure 5.26.	Profile and section through BNH5 OR12.	63
Figure 5.27.	Section showing the relationship of NS1 (C303) to OR1 (C203).	63
Figure 5.28.	Section showing the relationship of NS1 (C303) to OR1 (C203).	65
Figure 5.29.	Annexe area to the north of the Entrance Chamber showing feature numbers.	66
Figure 5.30.	Southern area of the Entrance Chamber, the WNW posts and SEW posts.	67
Figure 5.31.	Selection of sections from the Northern East–West (NEW) posts which define the north of the Annexe.	68
Figure 5.32.	Plans and sections of WNW4 and 13, 2, 9 and 17.	68
Figure 5.33.	Plans and sections of IF 5, 6 and 7.	70
Figure 5.34.	Outer Façade at 0.20 m and 0.60 m depth below C2.	72
Figure 5.35.	Outer Façade at lowest level.	73
Figure 5.36.	Sections across postholes of the Outer Façade.	74
Figure 5.37.	Southern East–West postholes (SEW)	75
Figure 5.38.	Plan of central Annexe with Entrance Chamber.	75
Figure 5.39.	Section through EC5.	76
Figure 5.40.	Sections through Four-Posters EC37 and 38 showing distinct secondary fill.	76
Figure 5.41.	Postholes EC15–17 of the Entrance Chamber.	77
Figure 5.42.	Slot on north side of entrance connecting main façade to Entrance Chamber.	78
Figure 5.43.	Section through Burning Pit C1453.	78
Figure 6.1.	Distribution of the pottery recovered from the topsoil and ploughsoil contexts.	83
Figure 6.2.	Distribution of pottery by type in the topsoil and ploughsoil layers of each area square.	84
Figure 6.3.	Number of pottery sherds recovered from each 'feature group'.	84
Figure 6.4.	Percentage of pottery sherds recovered from the primary and secondary fills of postholes.	85
Figure 6.5.	Number of pottery sherds from the postholes of each 'feature group'.	88
Figure 6.6.	The Carinated Bowl sherds from Ballynahatty.	89
Figure 6.7.	Coarse Ware pottery from Millin Bay and Ballynoe, Co. Down.	91
Figure 6.8–6.13.	Knowth Style 1 vessels	92–97
Figure 6.14.	Knowth Style 2 pot.	97
Figure 6.15.	Knowth Style 3 vessels.	98
Figure 6.16.	Reconstruction of the Vase Food Vessel from Ballynahatty.	104
Figure 6.17.	Rim sherd from a Bronze Age Collared Urn from Ballynahatty.	105
Figure 7.1.	Lithics from Ballynahatty, scrapers.	115
Figure 7.2.	Lithics from Ballynahatty, scrapers and arrowheads.	116
Figure 7.3.	Distribution of the lithics recovered from topsoil and ploughsoil contexts.	124
Figure 7.4.	The distribution of the lithics by feature group.	125
Figure 7.5.	Number of lithic artefacts from each feature group.	126
Figure 7.6.	Number of lithics recovered from the postholes of BNH6 and the Entrance Chamber.	126
Figure 7.7.	Percentage of lithic artefacts recovered from primary and secondary fills of postholes.	127
Figure 7.8.	The petit tranchet derivatives from Newgrange, Co. Meath.	129
Figure 8.1.	Percentage of the coarse stone artefacts recovered from the primary and secondary fills of the postholes.	131
Figure 8.2.	Broken tip of a possible carved stone chisel.	134
Figure 9.1.	Distribution of bone fragments and cremated deposits from the excavations.	137
Figure 9.2.	Proportion of bone fragments recovered from the primary and secondary fills of the postholes.	137
Figure 10.1.	Radiocarbon dates from the Ballynahatty complex.	151
Figure 10.2.	Kernel density models of activity levels at Ballynahatty and comparisons from Ireland.	155
Figure 10.3.	Posterior probability density of the date of the phases of activity at BNH6 and BNH5.	155

List of figures and tables xiii

Figure 11.1.	Comparative plans and sections through passage tomb BNH2 (1), and related subsurface chambers.	161
Figure 11.2.	Site plan of excavation showing main feature groups.	163
Figure 11.3.	Construction plan of BNH6 and 5.	166
Figure 11.4.	Reconstruction of the central post setting as a platform.	167
Figure 11.5.	Suggested construction sequence of 4P2 and IR8–11.	167
Figure 11.6.	Possible construction order of postholes BNH6 4P3 and the Inner Ring postholes IR15–18.	167
Figure 11.7.	Postholes of BNH6 south entrance showing a possible construction order.	168
Figure 11.8.	Reconstructions of Knowth timber circle.	169
Figure 11.9.	An early reconstruction of BNH6 showing the effect of infilling with wattlework panels and planking.	170
Figure 11.10.	Reconstruction of Sarn-y-bryn-caled.	170
Figure 11.11.	Timber wall: An impenetrable barrier created by infilling branches and trimmings between upright posts.	171
Figure 11.12.	Durrington Walls, Southern Circle, Phase 1.	172
Figure 11.13.	Durrington Walls, Northern Circle.	173
Figure 11.14.	Conjectural reconstruction of the Giant's Ring bank.	178
Figure 11.15.	Definite and possible Irish timber circle sites in Ireland.	181
Figure 11.16.	Plans of Neolithic Irish timber circles.	182
Figure 11.17.	Comparison between the Newgrange Four Poster Enclosure and BNH6.	184
Figure 11.18.	Giant's Ring, showing distance between the passage tomb and the embankment as radii and diameters.	184
Figure 11.19.	Giant's Ring and Newgrange Henge A: proportional relationship.	185
Figure 12.1.	Orientation chart of Irish timber circles.	191
Figure 13.1.	The house, formerly belonging to the Bodels, before demolition.	202
Figure 13.2.	The Dungannon plaque at the entrance to the Ring.	203
Figure 13.3.	Arthur Hill-Trevor, 3rd Viscount Dungannon (1798–1862).	203
Figure 13.4.	East–west section through the field fence on top of the ridge at the eastern end of the excavation.	205
Figure 13.5.	The standing stone (BNH8) with the ground levelled around it and an accumulation of field stone.	206
Figure 13.6.	Diagonal lines show damage by subsoiling.	206
Figure 13.7.	Erosion of the bank at the entrance to the Giant's Ring as a result of mountain bike use.	206

List of tables

Table 1.1.	List of sites discovered by the Bodel family during the 18th and 19th centuries.	9
Table 4.1.	List of BNH sites cross-referenced with NISMR numbers.	31
Table 5.1–5.5.	Archaeobotanical remains.	79–81
Table 6.1.	Breakdown of the different pottery traditions recovered during the excavation at Ballynahatty.	82
Table 6.2–6.3.	Details of the distribution of the pottery sherds.	85, 86–87
Table 6.4.	Summary of the characteristics of the Ballynahatty Grooved Ware pottery.	99–100
Table 7.1.	Classification of flint and non-flint assemblage and types of material present.	106
Table 7.2.	Inferred source of flint utilised.	107
Table 7.3–7.4.	Basic composition of flint assemblage.	107
Table 7.5.	The condition of the flint assemblage by count and mass.	108
Table 7.6.	Basic composition of unworked flint.	109
Table 7.7.	Composition of core assemblage.	109
Table 7.8.	Complete flakes and blades, indicating extent of cortex and burning.	110
Table 7.9.	Length, breadth, thickness and mass range for platform and bipolar flakes and blades.	110
Table 7.10.	Types of complete flakes and blades.	111
Table 7.11.	Types of bipolar flakes and blades.	111
Table 7.12.	Classification of flake shatter, indicating extent of cortex and burning.	112
Table 7.13–7.15.	Aspects of flake and blade shatter.	112–113

Table 7.16–7.17.	Aspects of angular shatter.	113–114
Table 7.18.	Basic composition of retouched flint assemblage, showing extent of cortex and burning present.	114
Table 7.19–7.22.	Details of scrapers.	117–118
Table 7.23.	Types of knives, petit-tranchet and derivative types, extent of cortex and burning.	119
Table 7.24.	Types of projectiles, and extent of cortex and burning.	119
Table 7.25–7.26.	Hollow scrapers and blanks.	119–120
Table 7.27–7.33.	Details of miscellaneous, edge-retouched and utilised tools.	120–122
Table 7.34.	Types of core tools, indicating types of cores utilised.	123
Table 7.35.	Location of the flint artefacts at Ballynahatty by feature group.	128
Table 8.1.	Miscellaneous stone objects from Ballynahatty, by feature group.	132
Table 9.1.	Distribution of bone fragments recovered from the primary and secondary fills of the postholes and from the early medieval 'Burning Pit'.	138
Table 9.2.	Burnt bone from other contexts.	144
Table 9.3–9.6.	Dentition.	147–148
Table 10.1.	Radiocarbon dates from the Ballynahatty excavations.	152
Table 10.2.	Dates from the Ballynahatty '1855' tomb.	153
Table 10.3.	Summary of events present in the Bayesian model of the site's chronology.	153
Table 10.4.	Radiocarbon dates from Late Neolithic sites in Ireland with timber structures and postholes.	154
Table 11.1.	Species identifications of charcoal.	165
Table 11.2.	Construction phases represented at Ballynahatty.	165
Table 12.1.	List of sites used for this study, with orientation calculated from plans.	190
Table 12.2.	Results of alignment study using *TarxienCore*.	192–197

Acknowledgements

The Ballynahatty excavations began in 1990 and have taken over 30 years to bring to publication. During this time the project has been supported by many people and organisations, each of which have been essential to the its continued development. The Department of Archaeology and Palaeoecology under the umbrella of a number of different Schools in Queen's University Belfast has provided continuous academic, financial and logistical support including a room set aside specifically for this project over the last year. For this I must thank Profs Caroline Malone and Eileen Murphy and the Director of the School of Natural and Built Environment, Prof Gerry Hamill. An annual grant and guidance during the excavation was readily given through Dr Chris Lynn, Claire Foley and Prof Brian Williams of the Historic Monuments Division of the Northern Ireland Environment Agency, Department of the Environment. Rhonda Robinson facilitated financial support for the publication from the Historic Environment Fund of its successor organisation, the Historic Environment Division of the Department for Communities. The Queen's University AHRC Impact Acceleration Account has supported the additional cost of colour printing in the volume.

Thanks go to Jessica Hawxwell and colleagues at Oxbow for their patient and constructive support of this project and finally bringing it to publication.

Permission to excavate was tentatively given then cheerfully renewed for a decade by the landowner, the late Mr Jim Thompson of Ballylesson. The other essential for excavation, water, was endlessly provided by the Dunlop family of Ballynahatty. The project formed one of the undergraduate teaching excavations at Queen's and most of the diggers were drawn from this admirable pool of enthusiastic students, many of whom continued to volunteer beyond their course requirement, becoming site assistants in later years, and the excavation team greatly benefitted from the help of local 6th formers. The smooth running of the site relied on several supervisors over the years especially Cia McConway, Linda Canning, Richard Lamb and Robert Chapple. John Davison supervised the laboratory processing and sorting of tonnes of soil samples. Ken Pullin and Billy Dunlop, both members of the Ulster Archaeological Society, deserve a special mention for returning year after year and enthusiastically sharing their expertise and accumulative knowledge of the site with each new intake of diggers.

My editorial colleagues Sarah Gormley, Catriona Brogan and Caroline Malone have equally made substantial contributions to this publication by organising the disparate elements of the excavation record, pulling the pieces together for publication, constant encouragement and creating a working environment to allow all this to happen. Without them the hiatus between excavation and publication would have been even longer. Thanks also go to Dr Gill Plunkett, Dr Eiméar Nelis, Prof Eileen Murphy, Dr Rowan McLaughlin, Dr Robert Barratt for their specialist chapters and further contribution by Dr Helen Roche. Our readers, Prof Jim Mallory and Dr Alison Sheridan, have provided sterling advice and manuscript corrections and Alison has been instrumental in the interpretation of artefacts, the cultural contexts and the final shaping of the chapters. The views expressed, though, are the responsibility of the editors.

Finally, any long running field project has repercussions on family life and so I am thankful for the continued forbearance of Chris, Amy, Alix and their families.

Barrie Hartwell

Part 1

The Ballynahatty landscape

1

The landscape and historical research

1.1 Introduction, geology and topography of the complex

This book is principally a report of the excavations, carried out between 1990 and 2000, in the townland of Ballynahatty (IG ref. J326677) in Co. Down, Northern Ireland, within the lowland area of the Lagan Valley (Fig. 1.1 OS 1:10000 map). Townlands are the smallest areal divisions in the Irish landscape and, as such, are commonly used to record the location of archaeological sites. Ballynahatty td contains one of the largest prehistoric sites in the north of Ireland – the Giant's Ring henge – which sits on the edge of a plateau above the River Lagan just 6 km south of the centre of Belfast. A prehistoric cemetery spreads north from it, covering nearly a quarter of the townland. The excavation was centred on a large cropmark site found by aerial photography consisting of an oval, pit-defined enclosure about 100 m long (BNH5), containing a smaller enclosure at one end (BNH6). It proved to be a remarkable timber ritual structure dedicated to the processing of the Late Neolithic dead.

The results of the excavation form the meat in a three-part sandwich. The area has attracted the attention of antiquaries and archaeologists for nearly three centuries, and field walking continues to produce lithic finds. Part 1 provides an historical and environmental introduction to Ballynahatty and the Giant's Ring. This is followed by the results of aerial and ground surveys, which have done so much to record the prehistoric cemetery, and concludes with a summary map of known sites. Part 2 covers the excavation itself – the features, lithics, pottery and dating. Part 3 attempts to interpret these remains and reconstruct both the building, how it functioned and its relationship to the Giant's Ring and other Irish sites. The final chapter presents an overview of Ballynahatty through time from prehistory, through the depredations of 19th-century farming to the present day.

The valley is enclosed by the higher ground surrounding it: to the north, the landscape is dominated by the termination of the Antrim basalt plateau, as the summits of Divis, Black Mountain, Squire's Hill and Cave Hill overlook the Lagan; to the south, the landscape is characterised by the rolling hills of the Castlereagh plateau. The townland of Ballynahatty itself extends 100 ha in area and lies on the basal Permian sandstones and Brockram. Its boundary is formed by the River Lagan and the Purdy's Burn to the west and north and the Giant's Ring enclosure to the south. Much of the townland makes up a discrete undulating plateau, 40 m high, and while the land falls away steeply to the south and east, there is a more gradual decline towards the Lagan at 5 m AOD. The soil covering is glacially derived clay, sands and gravels, typically laid down by meltwater close to or beneath an ice sheet. Rivers and streams washed glacial deposits into the pro-glacial Lake Lagan, then mixed and resorted as the ice-dammed lake discharged. The drying bed was then dissected by streams and rivers, such as the Lagan, with periods of downcutting energised by a fluctuating but falling sea level (O'Reilly 2010, 21–24). This bed can still be appreciated when looking southwest from the head of Belfast Lough. From this distance the hinterland of Belfast appears uniformly raised and flat between the Antrim and Down hills. Basaltic rocks of all dimensions, torn from the Antrim Plateau, together with rolled and battered flint from the underlying chalk, were incorporated in this disorderly, semi-sorted, fluvioglacial matrix. The weight of the ice caused considerable surface compaction in Ballynahatty, resulting in an uneven, dense, indurate lodgement till. The result is a subsoil in which human and natural activity are sometimes difficult to differentiate and which is predictable only in its unpredictability.

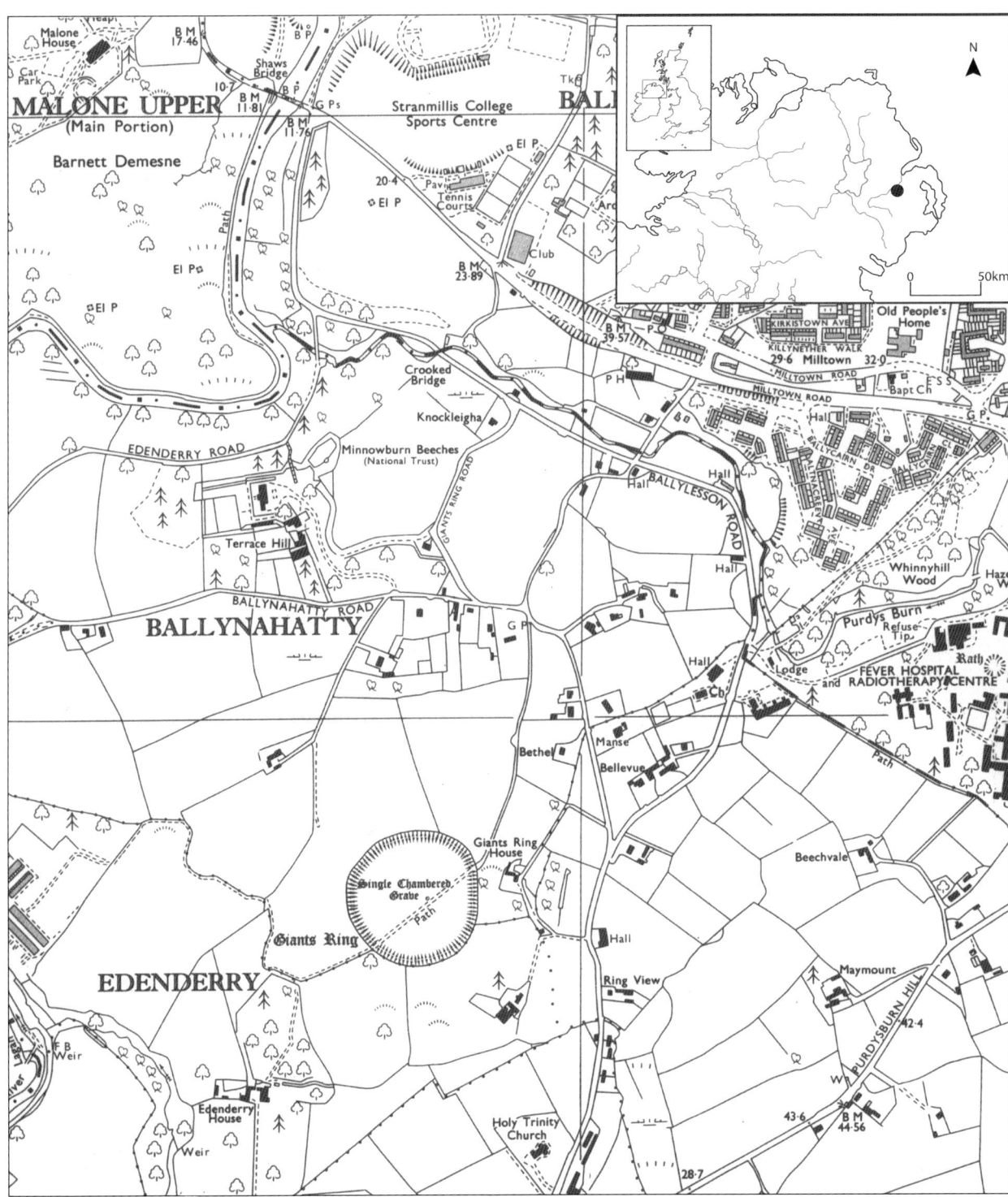

Figure 1.1 Part of OS 1:10000 map 147 showing Ballynahatty td and the Giant's Ring, the location of which is marked with a black circle in the insert map. The townland is defined by the R. Lagan to the northwest and west, the village of Edenderry to the west and the Purdy's Burn to the north. Edenderry House and Giant's Ring House lie at the bottom of the slope to the south and east. Reproduced from the 1971 Land and Property Services/Ordnance Survey of Northern Ireland ® map.

1.2 Historical overview of Ballynahatty and the Giant's Ring

Ballynahatty probably means the townland of the site of a house (Kay Lambkin, pers. comm., 29 October 2001). The visible prehistoric remains in Ballynahatty consist of the Giant's Ring, a large Neolithic embanked enclosure variously described as a henge or a hengiform enclosure; an earlier single-chambered grave or passage tomb within it; and a standing stone 100 m to the northwest. The 'house' in question could refer to the henge, which dominates the landscape. Details of the archaeology of Ballynahatty were first recorded in print in 1744 as part of Harris' survey of *The Ancient and Present State of the County of Down* (Harris 1744, 200), and prior to the latest programme of work, four excavations had been undertaken in the area. The first, in 1855, was at the site of an 'ancient sepulchral chamber' in the passage-grave tradition, which had been uncovered while digging potatoes in a field to the north of the 'Ring' (MacAdam and Getty 1855). Three more excavations investigated the Giant's Ring in 1917, 1929 and 1954 (Lawlor 1918; Richmond 1929; Collins 1957).

1.2.1 Antiquarian study

In an Inquisition of Charles I in 1605, Ballylarry (later Ballynahatty) is recorded as having 21 oaks of 6 ft (1.8 m) girth, and in the adjoining townlands of Ballynavally, Ballydollaghan, Ballycowan and Edenderry, a further 487. These were part of the ancient woodland that carpeted the Lagan Valley but which Lord Donegal later used for building timber to expand the growing towns of Knockfergus (Carrickfergus) and Belfast. The earliest record of the Giant's Ring itself is in a deed of 1672 from the Earl of Clanbrassil, included in the grant of various properties to St John Web (*Northern Ireland Land Act* 1925, 2). At this stage, the townland was known by several aliases, including Ballynehatty, Larey, Hatty McLarey and the Giant's Den. The land is noted as being in the occupation of 'Michael Dunne, William Beers or ... their ... Cottiers or tenants'. However, an earlier occupation of the area is shown by the 12th century AD Anglo-Norman mottes which occur 0.90 km to the northwest at Edenderry and 2.75 km northeast at Belvoir – both on the River Lagan.

Nothing more is known until the visit of Walter Harris in 1744, when he records two visible sites within the complex. He describes a passage tomb within the Giant's Ring as having an almost circular capstone measuring 7 ft 1 in × 6 ft 11 in (2.38 × 2.1 m) supported by two ranges of stones consisting of seven stones each. Four feet (1.22 m) from this are located several fixed stones, which stand no higher than 2 ft (0.61 m) (Harris 1744, 200). This is the only time when the tomb is described as having an outlying kerb of stones. Interestingly, he mentions a third site containing large quantities of bone, probably a cemetery mound, which was destroyed when some of the tenant farmers dug through it to obtain building stone.

The Giant's Ring was brought to the attention of General Vallency, Vice President of the Dublin Society and Secretary to the Society of Antiquaries of Ireland, in 1802, when the naturalist John Templeton wrote an account of the antiquities of the North of Ireland for him (Hinks 1825, 26–27). Templeton also included detailed measurements of the site in this account. Just over a decade later, Anne Plumptre, an English writer and translator, visited the Giant's Ring and Drumbo on 27 August 1814, during an excursion to Lisburn whilst staying in Belfast. She describes the site as having a cromlech enclosed within a high earthen bank, upon which were the remains of a round tower standing around 15 ft (4.6 m) foot tall (Plumptre 1817). The identification of a round tower on the bank is clearly erroneous and, although she may have conflated the description of the Ring and the Drumbo round tower, it is possible that she was describing a bleach-green watchtower (Macdonald and Hartwell 2009, 152). The field to the west of the embankment was owned by John Russell, who ran a linen manufactory at St Ellen, just downslope at Edenderry.

In 1823, George Benn produced the first illustration of a site within the complex (Fig. 1.2a). It showed a picture of the cromlech, now termed a 'druidical altar', drawn from the northeast in a romantic setting. He says:

> ... the first which shall be described, is that stupendous work, called the Giants Ring, in the Parish of Drumbo...It consists of an enormous circle, perfectly level [it is not] ...nearly one third of an Irish mile circumference. This vast ring is enclosed by an immense mound or parapet of earth, upwards of eighty feet in breadth at that the base, and though in the lapse, it is probable, of nearly 2000 years, the height of this bank must have much decreased, it is still so great as to hide the surrounding country, except the tops of the mountains, entirely from view, and in its original state there is not a doubt but they also were invisible. (Benn 1823, 256)

Benn also questions the accuracy of Harris's account. His illustration and description of the passage tomb equate more readily with the remains we see today than with Harris's account of 79 years before. Benn's description records a capstone of similar size to that which Harris details; however, he describes the capstone as resting on four and not 14 uprights, and he could locate only one of the several kerb stones described by Harris. Benn states that the tomb consists of ten stones in all (Benn 1823, 258; Benn 1834). His account was duplicated in the *Dublin Penny Journal* (1834, 77) but this time accompanied by a new view of the cromlech from the southwest and, for the first time, with an indication of the Giant's Ring behind it (Fig. 1.2c).

In 1833 the Ordnance Survey cartographers visited the area. A site as substantial as the Giant's Ring did not escape their attention (Bordes and Scott 1833). A trig point was placed on the Ring bank and Lieutenant G.F.W. Bordes duly described the Ring and sketched the cromlech, this time from the southwest. In his rather free interpretation, a recumbent stone was 'elevated' to fill an erroneously drawn gap between the surviving capstone and an upright (Fig. 1.2b). Charles Ligar produced a generalised plan of the Giant's Ring for the Ordnance Survey in 1837 (Fig. 1.3) and commented on the destruction caused by agricultural enterprises (Day 2014, 38–41). As a result, the 3rd Viscount Dungannon built a protective stone wall around the Ring in 1837.

Figure 1.2 Four views of the passage tomb within the Giant's Ring: a) Benn 1823; b) Bordes and Scott 1833; c) The Dublin Penny Journal, 1834; d) Borlase 1897.

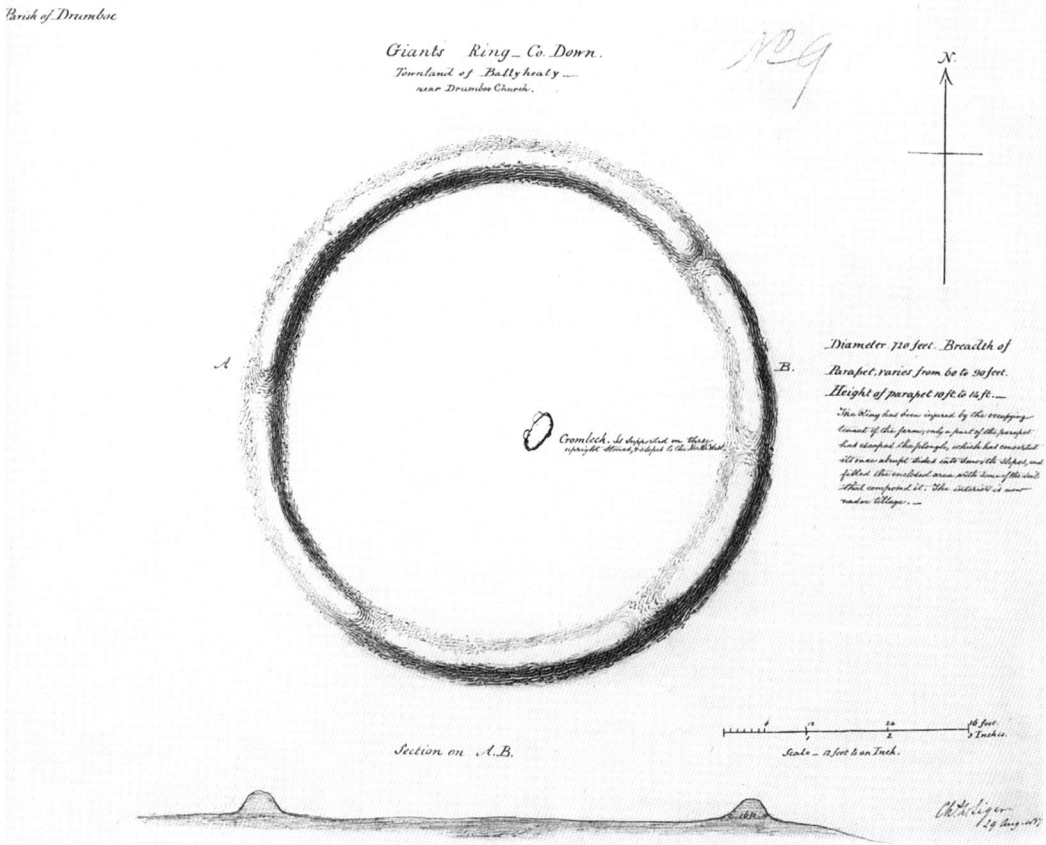

Figure 1.3 Charles Ligar's plan of the Giant's Ring 1837 (Day 2014, 138–141).

1.2.2 The 1855 tomb

The most interesting account, both as the first scientific study of an archaeological site in the prehistoric environs of the Giant's Ring and as a frank description of what had already been destroyed by the mid-19th century, is MacAdam and Getty's 1855 paper – 'Discovery of an ancient sepulchral chamber', published in the first series of the *Ulster Journal of Archaeology* (MacAdam and Getty 1855, 358–365). The article's subject was not the Giant's Ring, but an 'ancient tomb' located to the northwest and reported in the *Belfast Newsletter* on Wednesday, 21 November 1855, under the title 'Discovery of an ancient tomb'. The event is described thus:

> ... they came upon a broad, flat stone, which, upon being removed, proved to be the entrance to a tomb, most probably, the period of the ancient Druids. Two boys who were in the field at the time, immediately descended into the place and examined it. It is about 6 feet in diameter, and four in depth, and is nearly round at the base. (*Belfast Newsletter*, 21 Nov 1855, 1)

The article goes on to describe in some detail the construction of the tomb and its contents:

> ... In one of the compartments formed by them, was discovered an urn filled with bones, and three skulls, two of which are perfect; the third was, by accident, broken. From the appearance of the bones, it is evident that they had been burned previous to being deposited in the urn; but the skulls had been placed in the sand in their natural condition... The sides of the urn are curiously carved, but it cannot be removed in its perfect state. (*Belfast Newsletter*, 21 Nov 1855, 1)

How the *Newsletter* came to hear of the discovery is unknown but it was clearly deemed to be of sufficient local interest to dispatch a reporter to the scene. Though not written by an antiquarian, this proved to be an accurate and perceptive description. By another stroke of luck, the article was read by Mr James MacAdam, the first editor of the *Ulster Journal of Archaeology*, who immediately visited David Bodel's farm with Mr Getty. With the help of some members of the Belfast Natural History and Philosophical Society, they carried out an investigation of the tomb. MacAdam and Getty's visit resulted in the 1855 publication of an eight-page scientific description and analysis of the tomb. They describe how a circular cist measuring 7 ft (2.13 m) in diameter by 2 ft (0.6 m) in height was found to have been cut into the natural soil (Fig. 1.4). An entrance opening faced east, and the interior had been radially divided into six chambers. The tomb had a corbelled roof.

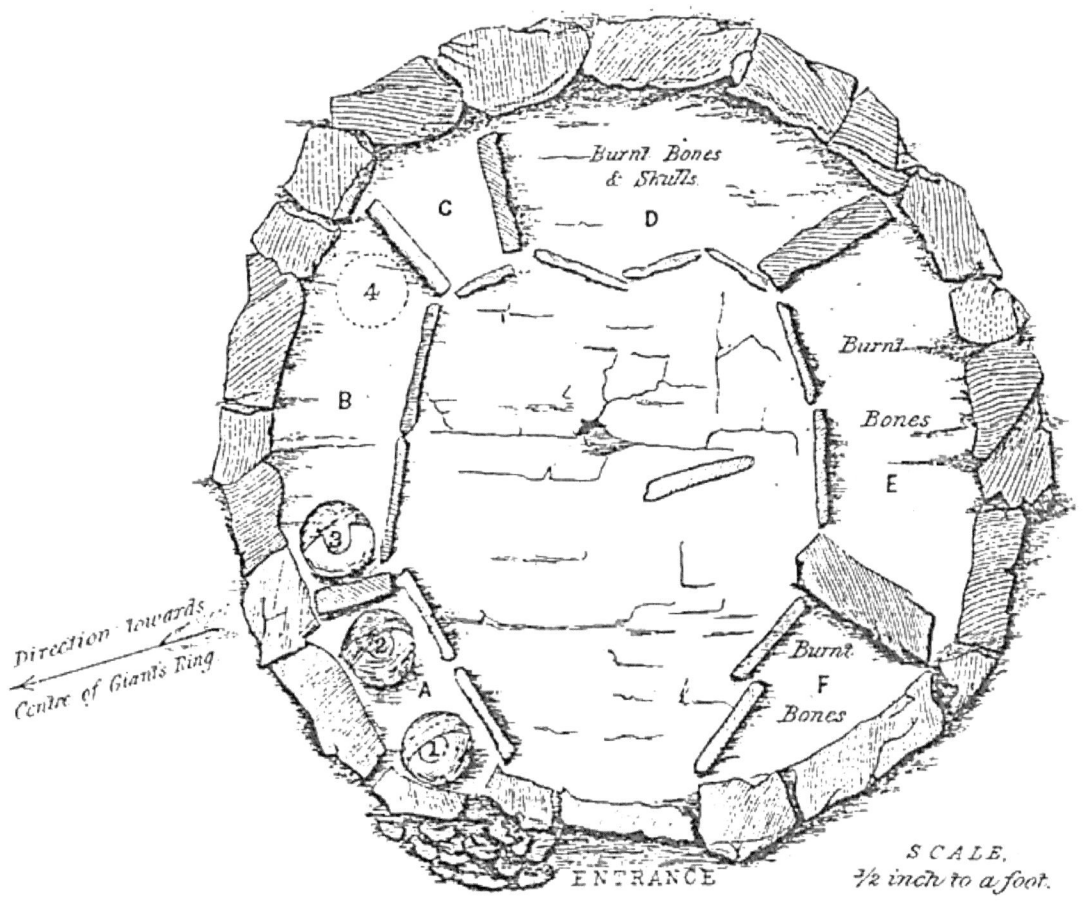

Figure 1.4 Plan of the subterranean chamber investigated in 1855 (MacAdam and Getty 1855, 358).

1.2.2.1 The tomb contents

Cremated human remains in pots, separate parcels of cremated bone and disarticulated unburnt human and animal bone were recovered from the chambers. The human bones included ribs and two skulls (one without a mandible, a cranium and three other mandibles) (MacAdam and Getty 1855, 360–362; see Plates 82 and 83). The tomb seems to have been reused on a number of occasions. The form of the tomb is unparalleled, but the coarse decorated pottery inside it is of the same type as that found in passage tombs, and the presence of cremated remains also connects it to that kind of monument; it could be argued that the compartments echo the chambers of some passage tombs. However, its round shape, lack of a passage and covering cairn and the fact that it is a subterranean chamber distinguish it from passage tombs (Hartwell 1998, 35).

The report on the excavation only provides a brief description of the pottery vessels found in the tomb, and no specific illustrations are included. The excavators describe 'in the compartments ... were found three urns of burnt clay (1, 2 and 3) filled with burnt bones' with dimensions of 12 inches (30 cm) high and 10 in (25 cm) broad (MacAdam and Getty 1855, 360). The excavators state that the vessels were fragmentary when removed, and another (4) was too fragile to be recovered. Wilde, in his Royal Irish Academy Catalogue of 1857 (193, 358–365), described the vessels in more detail. All three (Nos 33, 36 and 42), donated by Lord Dungannon, came from the same stone chamber in Ballynahatty and are similarly decorated with impressed 'scratches or indentations'. No. 42 is described as 'a large globular urn' with all-over stabbed decoration and is indisputably Middle Neolithic Coarse Ware. No. 33 is described as a 'rudely formed urn, scratched roughly all over' and is probably of the same type. No. 36 is simply 'a large and rudely decorated urn' which, because of its association with Nos 33 and 42 is probably the third Coarse Ware pot illustrated by MacAdam and Getty. Their report, otherwise so thorough, did not differentiate between the vessels suggesting that they were of the same type though the decoration was each of a different character. The plan of the tomb chamber supports this, with three identically shaped (Neolithic) 'globular' bowls being sketched in compartments A and B. They appear to be lying on their sides but were simply drawn obliquely while viewed beyond compartment D. Another pot, No. 32 may have been found in a different tomb, though still from Ballynahatty.

1.2.3 The later 19th-century landscape

The seeming paucity of sites and stray finds recovered in the townland over the years is in conflict with the account given by Mr David Bodel, the farmer at Ballynahatty in 1855 at the time of MacAdam and Getty's investigation. Mr Bodel stated that, on several occasions, he and his predecessors had uncovered the remains of human interments, cists, pottery, stone axeheads and flint tools, and he indicated that ploughing had resulted in the discovery of 'vast quantities' of human bones. During the construction of his house, a cemetery mound was removed containing several short stone coffins, earthen urns and burnt bones and many cart-loads of unaccompanied bones were removed, presumably from a number of cist graves in Bodel's field (MacAdam and Getty 1855, 364). Similar discoveries had been made on adjacent farms (Table 1.1), and in one case, a subterranean chamber like the one excavated in 1855 was discovered (MacAdam and Getty 1855, 365).

The next notable mention of the Giant's Ring occurs in Alexander Knox's *History of County Down* of 1875 (174–183). In this account, he confused Giant's Graves with stone circles and saw the Ring as a defensive earthen rath despite the absence of a ditch. At the end of the 19th century, Borlase published another view of the passage tomb (Fig. 1.2d) and returned to the question of discrepancies in Harris's statement. Unlike Benn, Borlase did not see a problem with Harris's account and argued that it was likely that stones would have been removed for other uses, leaving the tomb in its present condition (Borlase 1897, 276).

1.3 Archaeological investigations in the earlier 20th century

By the early 20th century, the Giant's Ring was firmly on the antiquarian map, and its presence in the landscape could not be ignored by tenant farmer, landowner or academic. The site remained a perplexing subject for the next 40 years; even though much was known, little of substance was understood. As archaeology began to supplant the antiquarian's pre-occupation with Druidism as a solution to any site without a history, so the Giant's Ring complex became ripe for re-interpretation. Excavation was seen as a way to answer some fundamental questions about this ancient landscape, and three excavators made the attempt (Fig. 1.5).

1.3.1 H.C. Lawlor 1917: Giant's Ring bank, interior and megalith

The first archaeological excavation of the Giant's Ring took place in 1917 when Henry Lawlor opened a trench around and below the off-centre passage tomb and another in the measured centre. Six radial 18 in (0.46 m) wide trenches, totalling 554 yards (507 m), were opened across the southeastern interior, with one trench extending through the southeast bank: trenches in total, measuring around 277 yd^2 (231 m^2). Because of the negative results, trenching was discontinued in the northern section (Lawlor 1918, 20). He found undisturbed till at a depth of 15–18 in (0.38–0.46 m) at the centre, which increased to 5–6 ft (1.50–1.80 m) depth at the bank. The centre of

Table 1.1 List of sites discovered by the Bodel family during the 18th and 19th centuries.

Site no.	Site type	Source description	Source location	Actual location	Page nos in MacAdam and Getty (1855)
1	Cist, compartmentalised & subterranean	Circular 'buried sepulchral monument' with 4 'urns' (incl Coarse Ware), burnt/unburnt human & animal bone	Field almost adjoining N side of Giant's Ring[2]	Not known	358–364
2	Cist, compartmentalised & subterranean	At least 3 artificial chambers similar to (1)	Lands owned by George Thompson, Mr McKeown, Frederick Russell	Not known	365
3	Cists (several)	Stone slab at base & another as lid 'shorter than a man' Mostly with 'urns'	Bodel's field[2]	Not known	364
4	Cist	As (3) Contains skull found by David Bodel	Bodel's field[2]	Not known	364
5	Cist	As (3) Contains bones & stone implement	Bodel's field[2]	Not known	364
6	Cist	As (3) Contained an 'urn' with bone & stone implement	Bodel's field[2]	Not known	364
7	Cist	As (3) Contained burnt bones & 2 flint arrowheads	Bodel's field[2]	Not known	364
8	Cists	At least 3 stone cists & 'urns' 'exactly similar' to (3)	As (2)	Not known	365
9	Cemetery, flat [1]	'vast quantities of human bones'	Vicinity of Bodel's house at plough depth	Not known	364
10	Cemetery, unmarked flat [2]	'many cartloads of human bones'… 'a little below the surface'	Near stone (13)	Not known	364
11	Cemetery mound	Contained 'several short stone coffins' & 'earthen urns and burnt bones'	Site of Bodel's house	House replaced *c.* 1965	364
12	Megalithic tomb	Small mound with 'three very large stones, placed on end' slopping inwards at top. Contained 'urn & small quantity of burnt bones'	Extremity of field furthest from Bodel's house in 'somewhat elevated' position Removed *c.* 1795	Not known, opened by William Bodel *c.* 1800	364
13	Standing stone or megalith	'… enormous stone …', suspected cromlech which Bodel 'intends to have uncovered'	In boundary fence[1] facing Giant's Ring	Survives in field 90 m NW of Giant's Ring bank	364
14	Standing stone	Pillar stone buried during field clearance	Bodel's field[2]	Not known	365
15	Pit	'several feet deep' & filled with 'a peculiar, dark-coloured and soapy mould' & stones 'discoloured by strong heat' … 'several perches[3] in extent' where vegetation 'always bad'	Extremity of field[2] furthest from Bodel's house Near (14)	Not known, destroyed *c.* 1800.	364

[1] Removed in 2000; [2] Part of a farm held by the Bodel family for several generations; [3] 1 perch = 21 ft/6.4 m

the Ring was 10 ft (3.00 m) higher than at the base of the bank, giving the appearance of an upturned saucer.

A 4 ft (1.2 m) wide extension of the southeast trench cut two-thirds of the way across the bank, presumably as far as the Dungannon stone wall (Fig. 1.5). The bank consisted of small, rounded boulders and earth, though larger stones were found elsewhere in the excavation. Lawlor hypothesised that the bank was formed by people

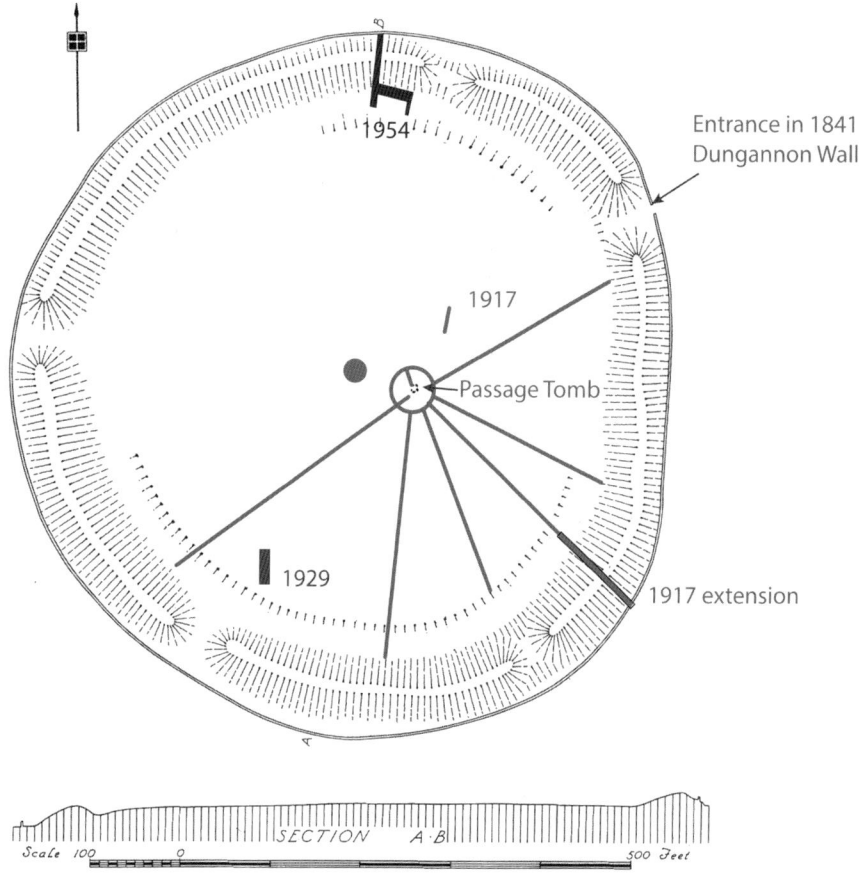

Figure 1.5 Location of the 1917, 1929 and 1954 excavations within the Giant's Ring (based on Collins 1957).

carrying single stones to the site in memory of the deceased interred within the passage tomb. He believed that, over time, this practice led to the enclosure reaching its immense proportions and assumed that the quantity of topsoil at the base had weathered downslope from the bank, which was originally considerably higher (*ibid.*, 21).

Unfortunately, the excavations of the passage tomb proved fruitless since the site had been looted by treasure hunters. At a depth of 4 ft (1.2 m), the tomb produced only lemonade and porter bottles, although the fill did contain 'not a teacup full' of burnt human bone (*ibid.*, 19). The trench around the passage tomb produced nothing except for the remains of a small fire at the northeast side and a hammer stone, but Lawlor established that supporting stones for the megalith were set on hard till (*ibid.*, 16). He argued that Harris and Borlase were incorrect in assuming an outer ring of stones (*ibid.*, 17).

1.3.2 D.A. Chart 1929: cropmark within the Giant's Ring

A second investigation of the Giant's Ring was undertaken in 1929 led by D.A. Chart (see Section 2.1.1), following the observation of a circular cropmark within the enclosure on an aerial photograph (Fig. 1.6). The excavation team opened a trench 25 × 3 ft (7.6 × 0.9 m) over the south side of the cropmark, but nothing of archaeological significance was discovered apart from a few flint flakes and farmhouse crocks (Richmond 1929). However, the location of the mark on the photograph and the position of the trench given in the report do not tally, and the excavation would have missed any possible site by a few metres.

1.3.3 A.E.P. Collins 1954: Giant's Ring bank and megalith

The only competent archaeological investigation of the Giant's Ring was the three-week excavation in 1954 in preparation for the *Archaeological Survey of County Down* (Jope 1966). A.E.P. 'Pat' Collins, a highly experienced archaeologist, investigated two main areas (Fig. 1.7). The first, aligned along the inner lip of the bank, essentially confirmed that there was no inner ditch but, rather, a shallow, flat-bottomed quarry trench, up to 2 m deep, from which the bank was constructed. Unlike Lawlor, Collins recovered numerous worked flint flakes, including a leaf-shaped point and some cores, from the layer above the natural soil in the quarry ditch, which was 'darkened by numerous flecks of charcoal' (Collins 1957, 44).

1. The landscape and historical research

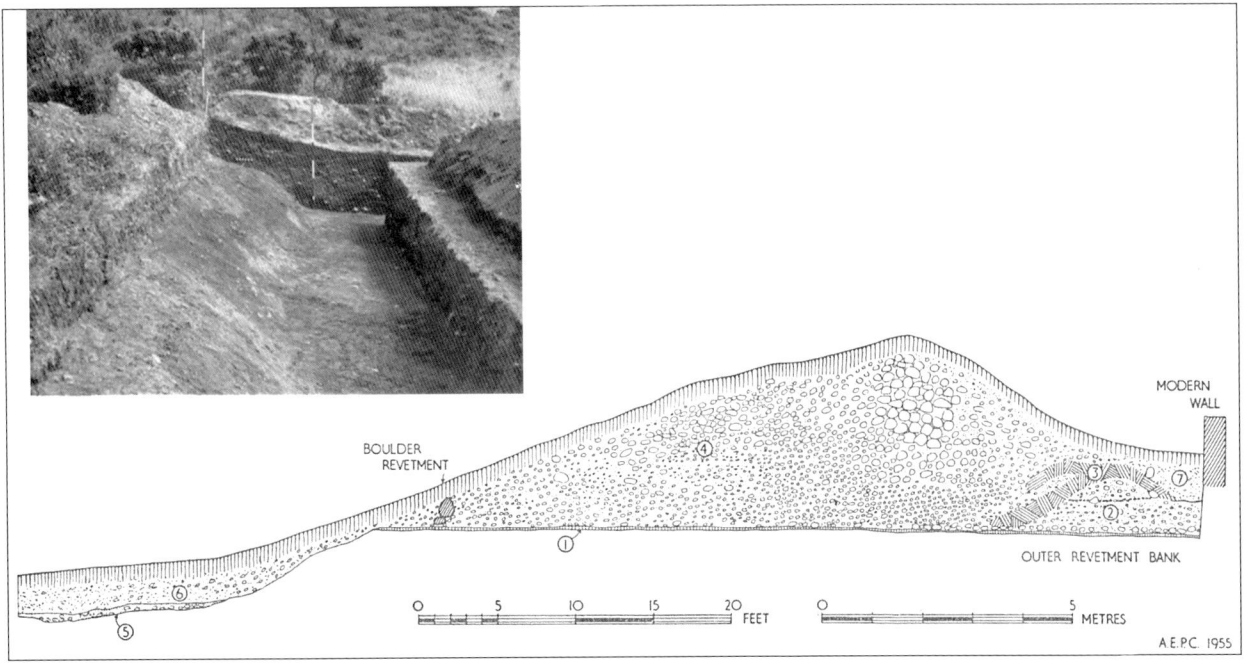

Figure 1.6 The RAF aerial photograph, which prompted Chart's excavation of 1929. The circular feature (marked by the arrow) is at the lower left side of the enclosure against the bank.

Figure 1.7 Section through Giant's Ring bank, from Collins 1957, pl. II, above, and section through the bank (47, fig. 4).

A 1.8 m wide trench was excavated through the outer bank as far as the Dungannon wall (Fig. 1.5). Collins reported that the bank had an inner revetment of boulders, with the outer edge demarcated by a small gravel bank. He suggested that this may have been a revetment but could also have acted as a marker or guide for adding the main bank material (*ibid.*, 46). In the core were 'massive boulders' dipping at a 45° angle, suggesting a dump construction. He recovered 24 worked flint artefacts, including flakes and cores of poor-quality glacial drift material, and described the recovery of a rough out for a leaf-shaped point (*ibid.*, 48). The Ulster Museum now holds these lithics. This material was examined by Nelis (2003, 681–684), and the assemblage was found to comprise 15 complete flakes, three blades, four shattered pieces and two cores, as well as a single modified tool (*ibid.*, 681). The modified piece is bifacially thinned and may be an unfinished laurel leaf – probably Collins' leaf-shaped point (*ibid.*, 683). Nelis proposes that the material excavated by Collins seems to have an Early/Middle Neolithic character, but the context of this material cannot be assigned to either the passage tomb in the centre of the Giant's Ring or to the Giant's Ring henge itself. The finds do, however, lend support to the assertion that there was activity in the Ballynahatty area prior to Coarse Ware activity at the timber circle site. An investigation of the southeast edge of the passage tomb confirmed complete disturbance, and although burnt bone was noted, the 'farmyard' pottery recovered dates to the 19th/20th centuries.

Collins discussed the contemporaneity, or otherwise, of the passage tomb and the Giant's Ring but came to no firm conclusion. The lack of linking stratigraphy and the off-centre position of the dolmen within the enclosure were problematic. A covering mound could have pushed the tomb closer to the centre, but the passage was on the wrong side to allow this (Collins 1957, 49).

All three excavations had a common outcome – little was found in the way of diagnostic, datable artefacts although, ultimately, it was discovered how the enclosure was constructed. The excavators' hopes of discovering the date and function of a monument that 'may well be classed among the most remarkable in Ireland' (Lawlor 1918, 13) were soon dampened, as is evident from their reports. Lawlor states '... the results obtained in the excavations already made were so completely negative ...' (*ibid.*, 20), similarly, Collins writes '... the excavations here described have done little to clear away the mystery' (1957, 49).

1.4 The later 20th century: BNH3

Air photography work undertaken in 1984 and 1989 identified new features concentrated in the south of the townland, giving some visual confirmation of the historic accounts identifying Ballynahatty as a focus of ritual activity from at least the Middle Neolithic to the Early Bronze Age. This prompted fieldwork at the site of cropmark BNH3 (Hartwell 1988) in 1991, which was visible in the oblique aerial photographs of 1984 (see Fig. 2.2), and in various OS vertical aerial photographs, as a pale, ragged, circular feature indicating a bank. More irregular darker marks were produced by lateral exposures of the underlying natural silty bands on the sides of the ridge. On the ground, BNH3 appeared as a shallow depression on the north slope of the ridge between the standing stone (BNH8) and the timber enclosure (BNH5). A 50 × 60 m grid was established over the northern part of the site at 10 m intervals, and a total surface collection was made (see Section 2.3.2) (Hartwell 1988). Apart from four isolated flakes, all the lithics, including four scrapers, came from beyond the interior of BNH3 (Fig. 1.8a).

Magnetic susceptibility tests were also carried out in the grid area and primarily indicated that there was substantial soil erosion in the central and southern areas (Fig. 1.8b). There was also no specific enhancement above field strength, which, along with the small artefact recovery, led to the initial belief that the site was unlikely to have had a domestic or industrial function. High levels were recorded during the spot phosphate tests (Fig. 1.8c), which could indicate the presence of animal or human excreta or the presence of bone material, but again these tended to avoid the central area.

The first trial trench opened at the north side of the cropmark, located a ditch, which was filled with coarse stones, gravel and a small amount of charcoal. It was generally difficult to distinguish between natural gulleys and constructed ditches. A trench opened on the east of the cropmark uncovered a thick stony layer, which may have been the remains of a bank. A third trench opened in the middle of the cropmark showed that this area had been disturbed to a depth of 3 m and gradually backfilled over the last two centuries. The reason for digging out the centre of the feature is unclear. Harris records that: 'Contiguous to this rath [the Giant's Ring], there was a small mount, that some of the neighbours dug through, in order to get stones for building; and in the middle thereof a great quantity of bones was found' (Harris 1744, 218). However, this could equally have been referring to the mound on the site of Bodel's house. Alternatively, it could have been the unrecorded site of a sand quarry – at this point, the natural subsoil contains exceptionally fine bands of sand. The excavation concluded with the site remaining an enigma, particularly as no artefacts were found associated with the bank and ditch. Subsequently, a deposit of charred grain found at the bottom of the presumed perimeter ditch was radiocarbon dated to the 5th–6th centuries AD. The site remains perplexing, but the most likely interpretation is an early medieval ringfort,

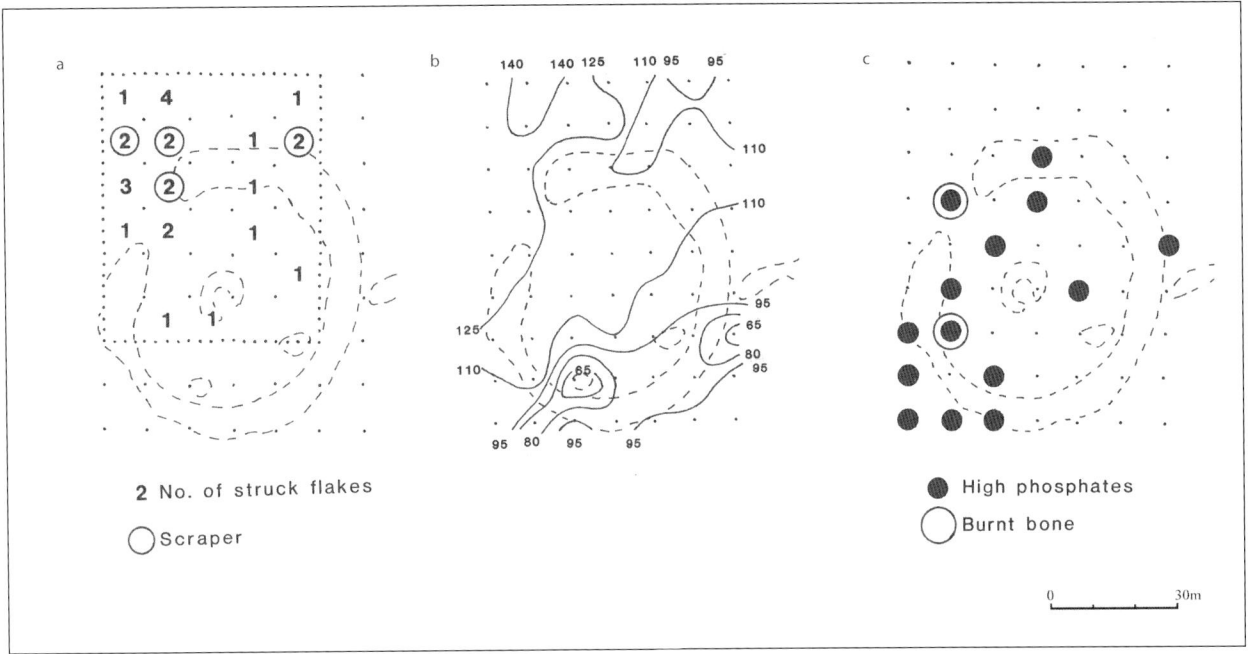

Figure 1.8 a) Distribution of flint artefacts from field collection; b) magnetic susceptibility survey; c) spot phosphate survey.

badly damaged by quarrying and the bank levelled by agricultural activity. There are a number of ringforts in the immediate area, and another early medieval pit was discovered 100 m east along the ridge during the excavation of the BNH5 Annexe. The charred grain shows that the area was now under cultivation.

2

Archaeological surveys

2.1 Aerial photography

2.1.1 Historical imagery

Although the importance of the Giant's Ring had long been appreciated, and its association with the funerary and ritual aspects of Neolithic life realised, the programme of survey, research and excavation undertaken by the Archaeology Department of Queen's University Belfast since the 1980s has demonstrated more clearly the scale of activity in the area. The nature of the soil and subsoil initially dictated that an aerial survey would be the ideal tool to locate any remains of these levelled sites. The first aerial photograph of the Giant's Ring was taken in 1927 (Hartwell 1988, 22) from a Vickers Vimy biplane operated by the Ulster Bombing Squadron, which resulted in the 1929 excavation (see Section 1.3.2., Figs 1.6 and 2.1). Cambridge University Committee for Aerial Photography photographed the Ring in 1952 and Jonathon Pilcher took more photographs for the visit of the Prehistoric Society to Ulster in 1967. However, it was not until the 1980s and 1990s that a series of sorties identified a concentration of new sites in the south of the townland (Chapter 4 outlines the principal sites), confirming Ballynahatty as a site of considerable archaeological significance and a focus of ritual activity and mortuary practice from at least the Mid-Neolithic to the Early Bronze Age.

2.1.2 Archaeological aerial survey, 1984–2018

The dry summers of 1984 and 1989 revealed a number of cropmarks, and because many of the sites remained to be confirmed and had not yet been given NISMR unique identifiers, an arbitrary numbering system was employed in the townland which incorporated the standing monuments and these new sites. Thus, the Giant's Ring became BNH (BallyNaHatty)1 and the enclosed passage tomb, BNH2. A standing stone, BNH4, sits in a significant position at the western end of the low ridge immediately to the north of BNH1. For convenience, and because this numbering system has appeared in several publications, it is being retained, whilst a conversion table of BNH to NISMR numbers appears at the end of Chapter 4.

2.1.2.1 The 1984 aerial survey

Oblique photographs were taken from a light aeroplane in May after a particularly dry spring (Fig. 2.2). Four concentric, darker bands appeared in the grass in the interior of the Giant's Ring. The widest, outermost one, which runs against the interior of the bank, corresponds to the quarry trench identified by Collins (see Section 1.3.3). The remaining three could represent ditches or conjoined postholes, and a preliminary soil resistivity survey at the Giant's Ring site showed that the three concentric cropmarks represent actual subsoil features and that their lack of conformity with the passage tomb suggests a stronger link to the later bank of the Giant's Ring and may represent a later phase of use of the enclosure or its reuse (see Section 2.2, below).

The broad, pale strip crossing the field outside the Ring from east to west represents the plough-shaved top of a natural ridge. The circular cropmark, BNH3, measuring approximately 50 m in overall diameter, at the western end of the ridge can be seen with an opening facing northwest. This was confirmed by Ordnance Survey vertical stereo photographs taken in June 1962 and August 1974. The excavation in 1991 (see Section 1.4) was inconclusive, but BNH3 can probably be interpreted as an early medieval ringfort, of which there are several in the area.

2.1.2.2 The 1989 aerial survey

An aerial survey was undertaken in the dry summer of 1989 as soil moisture deficit levels peaked, inducing differential ripening in a field of barley (Fig. 2.3). To the north of the Ring, a large enclosure (BNH5), 100 m long,

Figure 2.1 Giant's Ring from the northeast, taken by the RAF on 6/9/1927. There is a pathway along the bank top, but the sides are covered in whins, and the centre is under grass. From the original (and only) entrance in the Dungannon wall (left), well-worn paths lead past the passage tomb to two low points in the opposite bank.

appeared with a smaller enclosure (BNH6) at the east end. Postholes materialised as dark blotches in the pale, ripening crop (Plate 1). The holes had been driven through an obdurate impacted layer into light, mixed glacial soils allowing the cereal roots to penetrate to the groundwater below, keeping the crop greener for longer. The field beside it had an unresponsive potato crop. The remains of possible barrows and cists were also found in the fields surrounding what was Bodel's farm (MacAdam and Getty 1855; Hartwell 1998, 39), and a line of three barrows can be seen in Figure 2.3. This view encompasses a number of locations introduced in the text from the Ballynahatty Bog in the north to the Church of Holy Trinity in the south. A clearer resolution of many of these sites was obtained in the 2018 survey and are discussed below in more detail.

As well as identifying archaeological features in Ballynahatty townland, the 1989 sortie recorded possible sites nearby. Plate 2 shows a series of cropmarks in a field 1.1 km to the south-south-west, in Ballycarn townland (I.G. Ref. J321665), probably of Early Bronze Age to early medieval date. These features emphasise the evident archaeological richness of the Lagan Valley landscape.

2.1.2.3 The 1996 aerial survey

A further aerial survey was undertaken midway through the excavation in 1996 and revealed the emerging BNH5/6 sites located on top of the ridge, but it does not add more cropmark features (Plate 3). The aerial imagery had proven to be an excellent survey and location tool, although it was subsequently found to have missed 40% of the postholes. By 1997, all the available evidence had been summarised in a map (Hartwell 1998, 33, fig. 3.1) however, the position of many of the sites were estimates – a problem of converting site locations from oblique photographs when points of ground control were very limited.

2.1.2.4 The 2018 drone survey

The dry conditions of 1989 occurred again in 2018 but, on this occasion, the survey platform was a drone, taking advantage of its ability to respond rapidly to transient cropmarks and enabling the imagery to be enhanced by advances in digital processing. Without any modification, drones can take georeferenced vertical photographs and oblique imagery. All the Ballynahatty drone photography was taken and processed by David Craig of HeritageNI.

Figure 2.2 Cropmarks in the field north of the Giant's Ring photographed in 1984 showing the ploughed-out circular bank of BNH3 in the right foreground and the ridge extending to the left. The standing stone shows as a white dot in the bend of the field bank on the right. Four darker concentric bands show in the interior of the Ring (QAD/13-5-84/II/8).

com and shows many of the advantages of using a drone platform for aerial survey when operated by an experienced drone pilot. Because Ballynahatty is on the approach to Belfast City Airport, the flight was undertaken with a CAA Permit for Commercial Operations (PfCO). It is now classed as a Flight Restricted Zone (FRZ) requiring CAA Operational Authorisation (OA), and it is illegal for amateurs to fly drones here. Initially, under pilot control, the drone was moved around the site to take oblique photographs, with the images being sent live to the operator.

The area to be surveyed was defined in the field with reference to a Google Earth photograph on an iPad. The drone was then programmed to fly autonomously with the correct side and forward overlap in parallel rows to cover the designated area. The drone then followed this predetermined route to give sufficient overlap of the photographs to enable them to be later stitched together. At all times, the drone is guided by the georeferencing system, which gives the exact spatial reference for each photograph. Altogether, 51 ha were covered and 16 km flown, with 29 minutes of

Figure 2.3 Ballynahatty townland in July 1989 looking south towards the Giant's Ring. At the bottom right is the glacial kettle lake (Chap. 3), and above it, the modern house sits on the site of tenant farmer Bodel's house (Chaps 1 and 14). In the field beyond it are the cropmarks of three ring ditches, and to the left, in a bend in the field bank, sits a standing stone (Chap. 2) on the end of a low, pale-coloured ridge which extends left across the field. Spread over the near slope is the cropmark of BNH5 and 6 (Part 2). Beyond this is the Giant's Ring enclosing the passage tomb (Chaps 1 and 13), and at the top left is the Churchyard of Holy Trinity, Ballylesson, where many of the Bodels are buried (Chap. 13) (QAD17-7-89/V/25) (photo: Barrie Hartwell).

photo time producing 304 images. These images were then processed into a single orthorectified image (GeoTIFF file), georeferenced and output in a spatial Coordinate Reference System (CRS) that was suitable for importing into Geographical Information System (GIS) for accurate measurement and visualisation (Plate 4).

An orthophotograph can be used to measure the true distances of features within the image. A three-dimensional

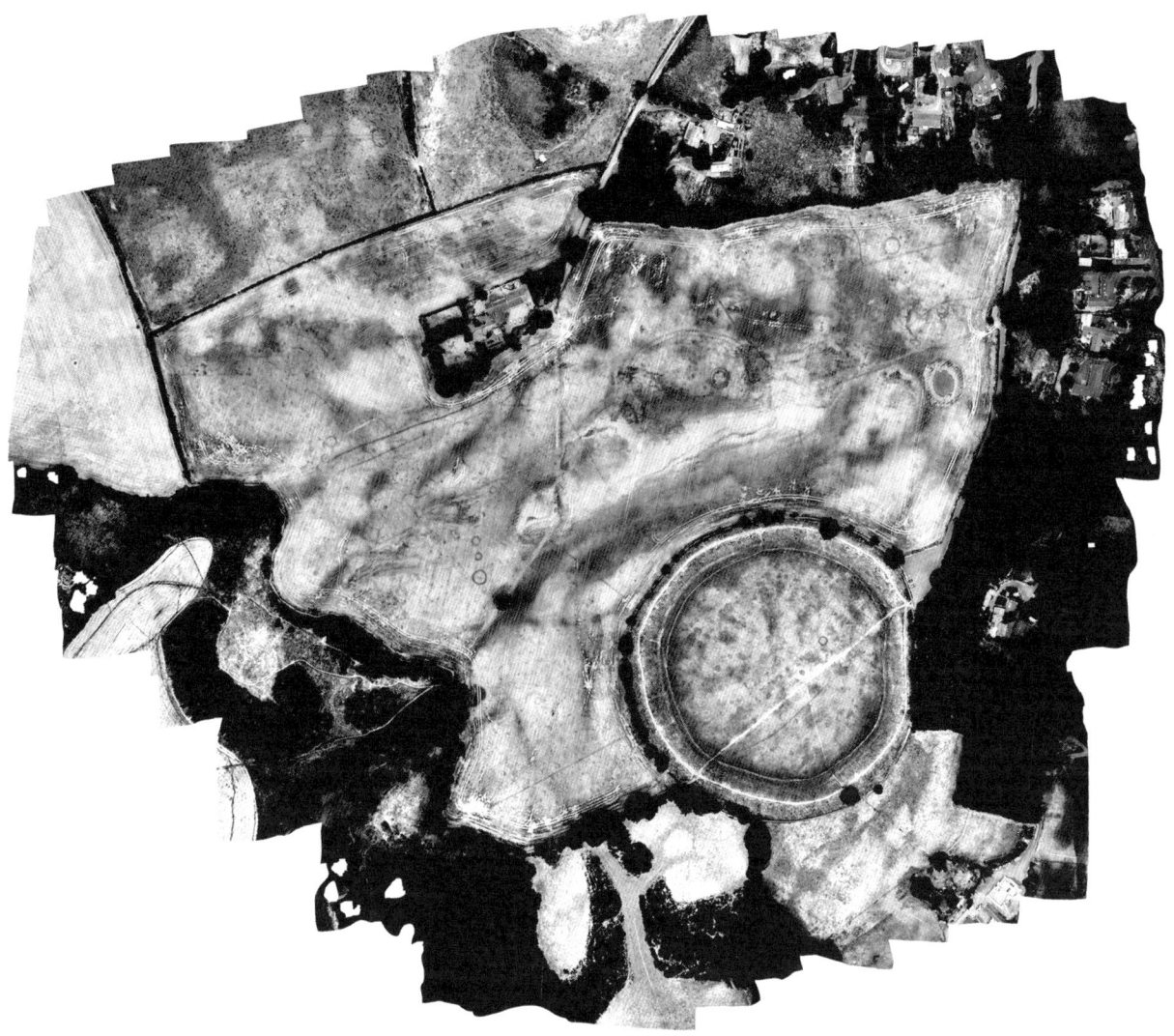

Figure 2.4 Enhanced monochrome orthorectified image of Ballynahatty td (image © David Craig/Heritage NI.com).

surface was created in *Agisoft Metashape Pro* and uploaded to Sketchfab.com. This can currently be explored at: https://sketchfab.com/models/7ea1a399314746c-092c528fcb791b1a3/embed. By selective use of the colour bandwidth, the Ballynahatty data was processed as an enhanced monochrome image (Fig. 2.4). For clarity, the individual sites identified in Figures 2.5–2.8 are shown in this format. North is at the top in all images.

The excavation of BNH5/6 confirms that regular cropmarks can be matched to excavated features. Where these marks create a pattern, we can be assured of human agency, but where they appear randomly or are more diffuse, interpretation becomes much more difficult. The cropmarks in a punctuated line running from the northwest of the Giant's Ring west towards the isolated tree would be an example of the latter (Fig. 2.4). However, the implication is that every dark dot on the images represents a hole deliberately punched through the hard subsoil. These could well represent the individual cist burials described by Bodel in 1855. BNH13 (Fig. 2.6), a dark patch, could be the site of the 1855 tomb and is the highest point on the plateau. The exceptions are the two fields shown in the north of the coloured orthophoto and the Giant's Ring, which are both under grass and give a different response than the ripening cereal (Plate 4).

The principal sites are discussed elsewhere, but this leaves 17 small circular marks (Figs 2.6 and 2.7; Pls 5–11), often with a linear or group arrangement (BNH10–12, 14, 25–28, 30–34, 38–40). These are probably Bronze Age ring ditches (Plate 12), possibly Iron Age, but the isolated and larger ones – BNH40 (Fig. 2.7: 22) and BNH42 (Fig. 2.5: 4) could also belong to the Neolithic. The nature of this landscape suggests they are mortuary structures rather than domestic. BNH25 (Fig. 2.6: 8) is set within an arc or possibly an oval setting of postholes and the arc of BNH25 may be associated with a square structure. Of particular interest is BNH36 (Fig. 2.7: 17), a circular setting of 40 postholes, closely set in a single 20 m diameter circle. It has a double posthole either side of a clear southwest entrance, possibly reflecting an earlier phase of BNH6.

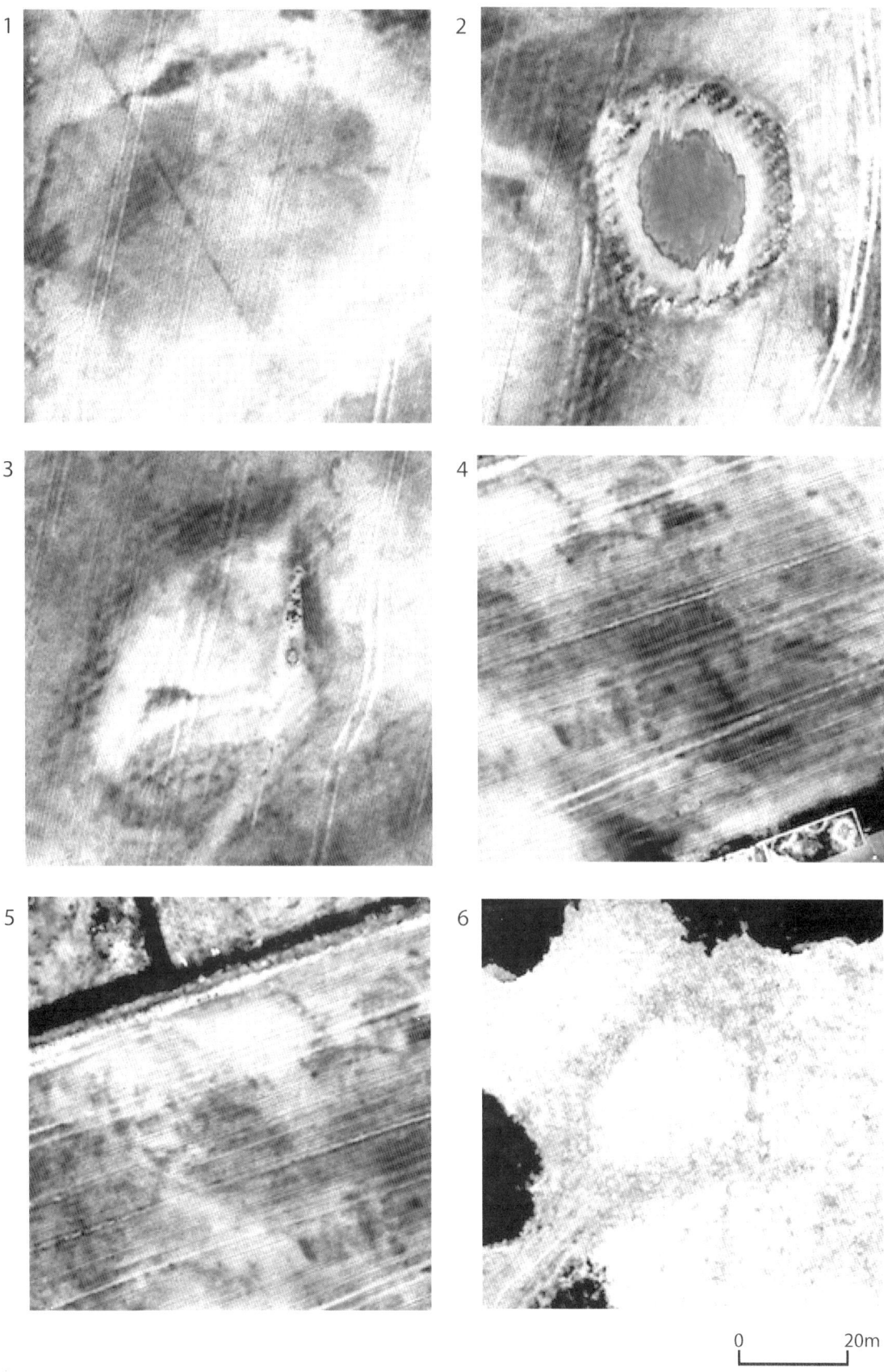

Figure 2.5 1: BNH3; 2: BNH24; 3: BNH8; 4: BNH42; 5: BNH41; 6: BNH22. See Fig. 4.1 for location (image © David Craig/ HeritageNI.com).

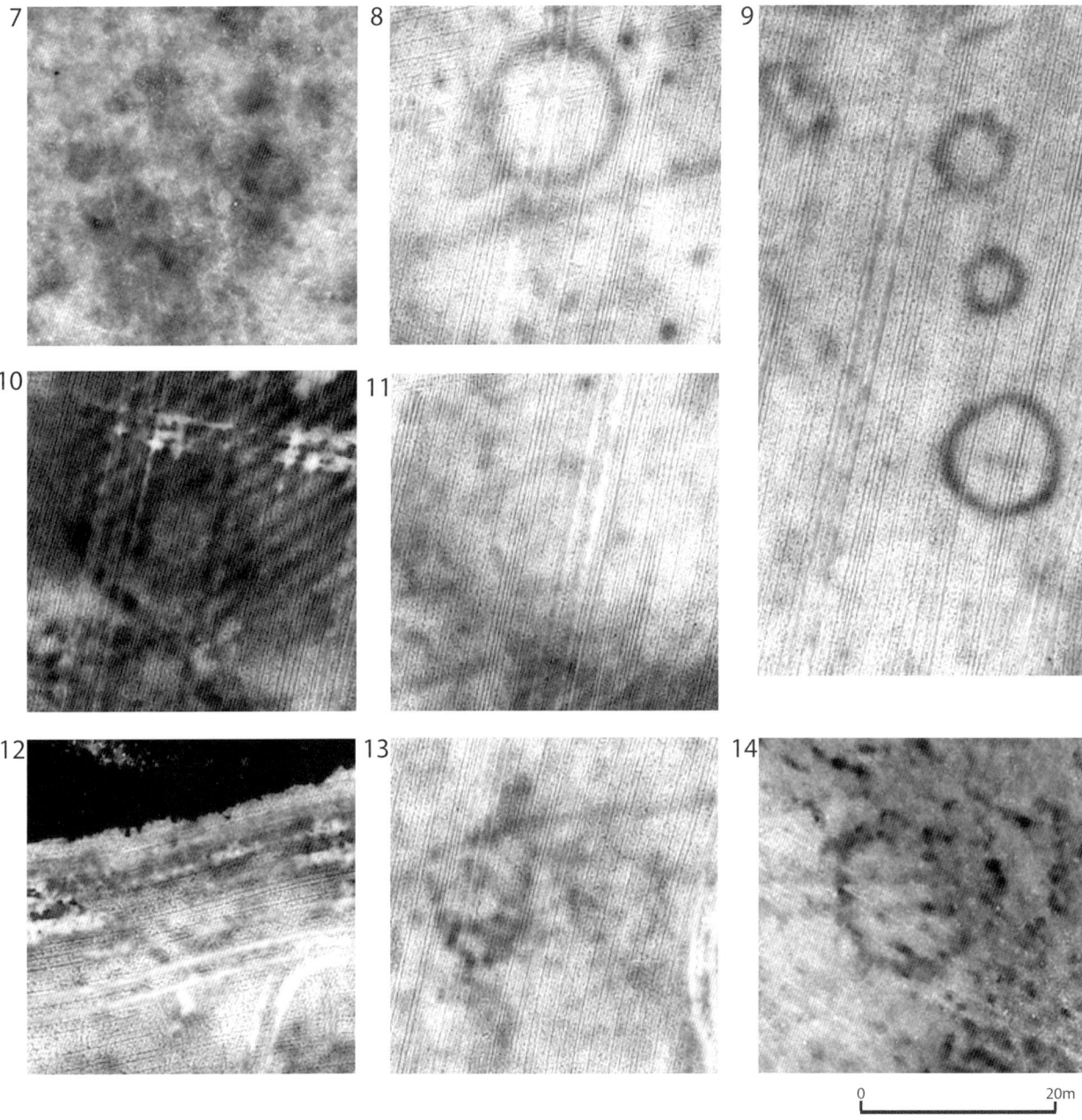

Figure 2.6 7: BNH23; 8: BNH25; 9: BNH9-12; 10: BNH26; 11: BNH27; 12: BNH28; 13: BNH29; 14: BNH14. See Figure 4.1 for location (image © David Craig | HeritageNI.com).

There are also linear features, principle of which is BNH16 (Fig. 2.8: 29). It runs east from near a cleft in the plateau above the village of Edenderry, and the River Lagan then bends northeast and appears to be heading towards the Purdy's Burn before being lost in modern housing developments. It could be argued that it avoids the old Bodel property and so is recent, but as their farmstead was built on the site of a prehistoric cemetery mound, the linear feature could be much older. The parallel ditches are slightly irregular in width, wider at the west end, and do not display the greater precision of post-medieval and later land management. Neither do they feature on the earliest Ordnance Survey County Series maps. If prehistoric, this could be the rare example in Ireland of a cursus.

A more obvious example of a trackway is BNH15 (Fig. 2.8: 28), which crosses the ring ditch BNH14 (Fig. 2.6; 14) to run east into Ballynahatty Bog, BNH4 (Fig. 2.7; 24). The parallel lines are regular and much closer together but do not appear on the OS 1st Edition 1834 map, although trackways to the lake are recorded from the north and east. It precedes a 2nd Edition field wall and may have connected the kettle lake to Edenderry House or an Anglo-Norman motte, 500 m west.

BNH17 (Fig. 2.8: 27) is a more substantial field ditch than others in the image and conforms at least in part to one shown on the 2nd Edition OS map. The image shows a staggered entrance at the north end, which is not seen

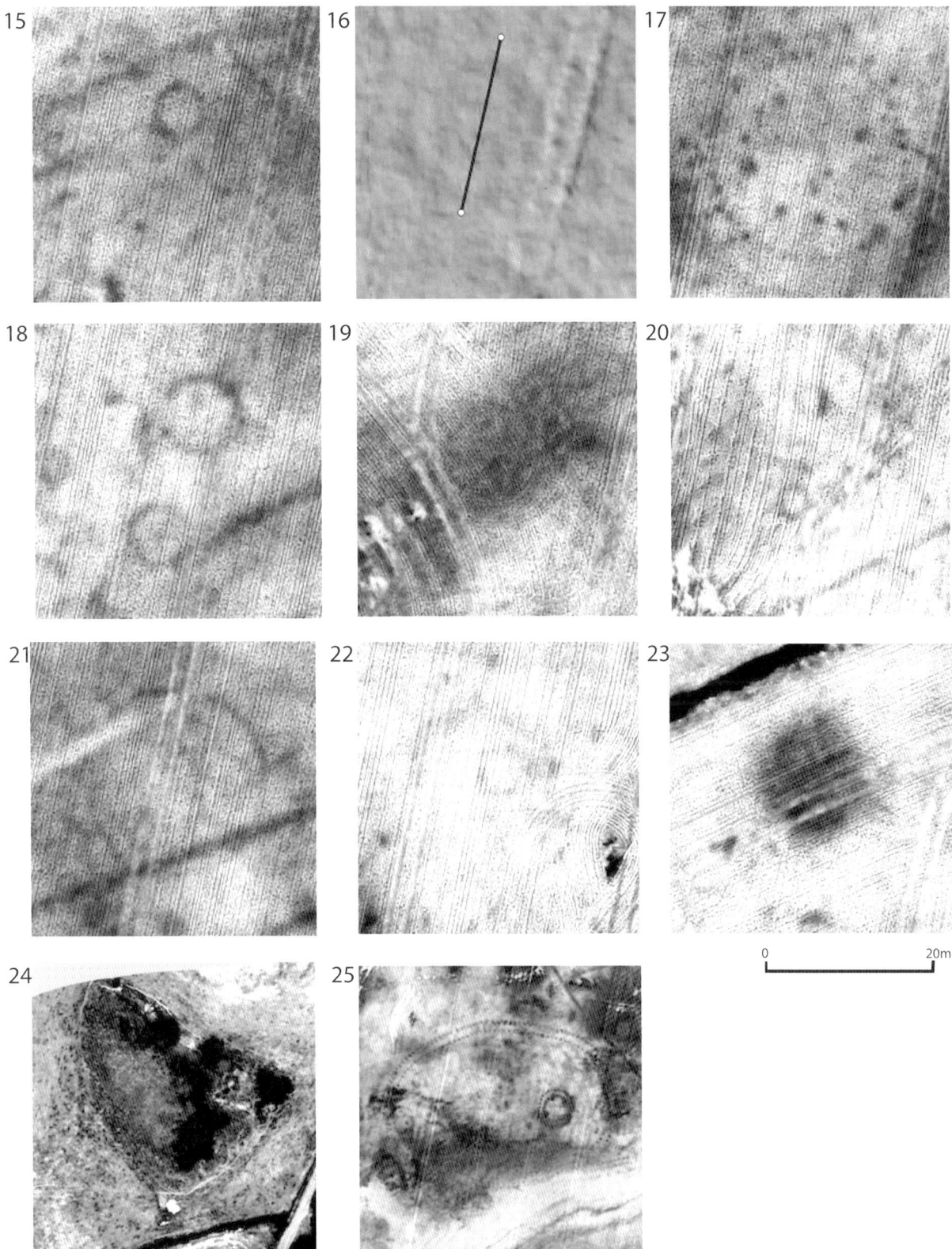

Figure 2.7 15: BNH30; 16: BNH31; 17: BNH36; 18: BNH32 and 33; 19: BNH37; 20: BNH38 and 39; 21: BNH34; 22: BNH40; 23: BNH13; 24: BNH4; 25: BNH5/6/7. See Figure 4.1 for location (image © David Craig/HeritageNI.com).

on the map, and the south end deviates from the southerly turn on the map to trend southeast.

BNH23 (Fig. 2.6: 7) is a roughly circular cluster of diffuse pits or postholes in the grass within the interior of the Giant's Ring. This is in a significant position on the north side, close to the bank and immediately opposite the BNH5 enclosure.

The final example is BNH43 (Fig. 2.8: 26), a faint complex consisting of two wide apart, east–west parallel

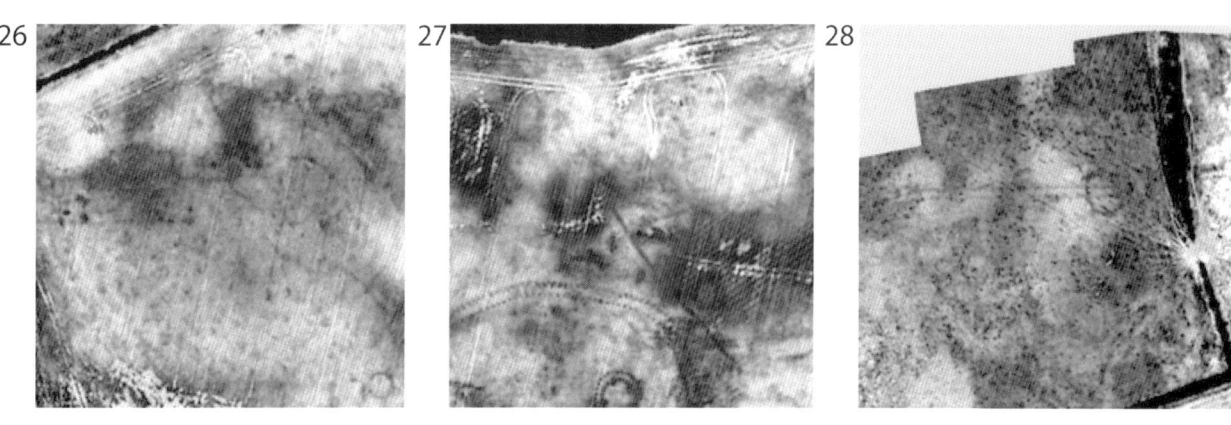

Figure 2.8 26: BNH43; 27: BNH17; 28: BNH15; 29: BNH16. Variable scales, see Fig. 4.1 for location (image © David Craig/ HeritageNI.com).

ditches terminating to the east in a large subcircular enclosure with two concentric ditches a similar distance apart.

This is not a final list, and further scanning and interpretation of the aerial images will, no doubt, reveal more sites.

2.2 Geophysical surveys

Although seemingly of great potential, geophysical survey has been particularly unrewarding in Ballynahatty. The inclusion of basaltic rock and the variable nature of the subsoil has effectively masked any meaningful response from magnetic prospection. The resistivity survey should have been more successful but produced indifferent results on an area scale, being able to detect a disused water pipe which crossed the excavation surface but not the 2 m-deep Neolithic postholes.

A large-scale survey in 2017 of the fields bordering the south side of Ballynahatty Road and the interior of the Giant's Ring was undertaken by the Romano-Germanic Commission, Frankfurt and UCD School of Archaeology (Davis 2017). This used a Sensys 16-sensor magnetic gradiometer rig producing high-resolution data acquisition (25 cm point spacing) and GPS capture. The report highlighted the problems of interpretation caused by volcanic rocks in the subsoil; however, it did indicate several potential anomalies. Comparison with the 1984 and 2018 aerial photographs has aided interpretation in three main areas, including the identification of a processional avenue leading towards the R. Lagan, indicating a third major focus of activity on the Ballynahatty plateau, which had not previously been identified (Figs 2.9–2.11).

The subcircular BNH14 (Fig. 2.6), to the west of the Ballynahatty Bog (BNH4) (Plate 10), has produced a distinctive magnetic response (Fig. 2.9) but does not detect the linear trackway, BNH15 (Fig. 2.8), which crosses it until east of the field boundary whereas it is clearly shown in the orthophoto to the west. The two images in Fig. 2.10 are of adjacent fields on the west side of Ballynahatty above Edenderry. The right image shows a parallel-sided avenue (BNH43) terminating in a double enclosure. The

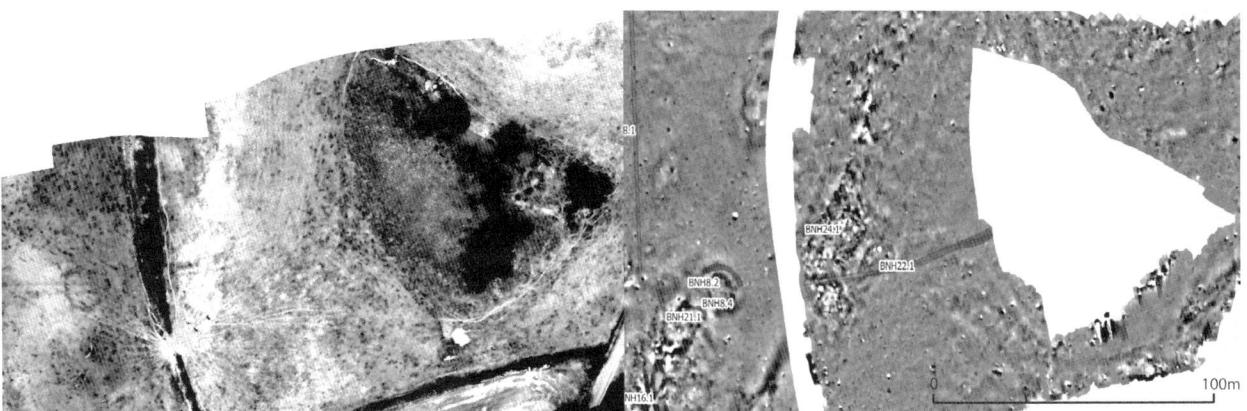

Figure 2.9 Relative visibility of BNH15 in orthophoto (left) and gradiometer survey (right).

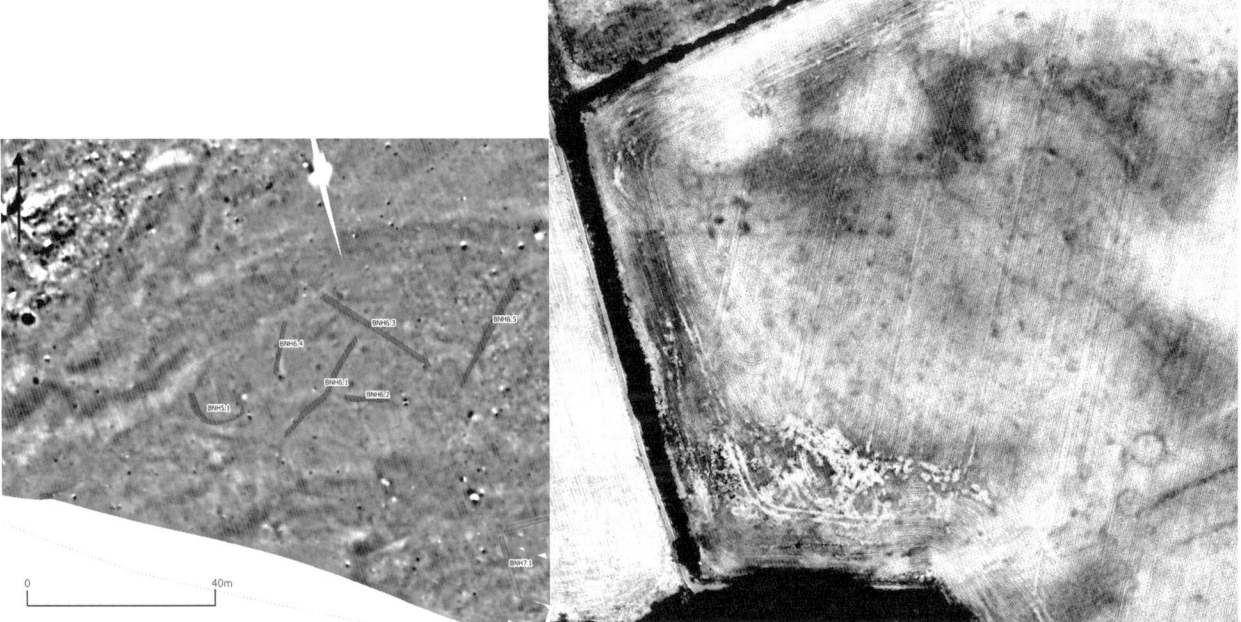

Figure 2.10 Extension of avenue BNH43 (orthophoto, right) into adjacent field (gradiometer survey, left).

gradiometer image on the left, which covers the field to the west, appears to show the extension of it curving southwest to meet the break in slope at an approximate right angle. Downslope is the R. Lagan. In the Giant's Ring (Fig 2.11), the central circular feature identified in the gradiometer image may reflect the inner ring identified in the 1984 aerial photo. However, the outer concentric rings do not feature. The gradiometer image also shows a number of lines running from the bank in the general direction of the interior. These may be attributable to the bank material which had been pulled down in the 19th-century land reclamation.

A resistivity survey in 1995 was undertaken inside the Ring along seven parallel transects, 1 m apart with a 0.5 m data interval (Fig. 2.12). Transects extended from the northern edge of the passage tomb east to the base of the bank. This survey showed, with reasonable consistency, two distinct, lower resistance zones approximating to the inner two rings in the aerial photograph, the trace of an outer ring and a wider low resistance area running up to the bank, which equates with the quarry ditch. The stacked profiles show an appropriate degree of curvature of the features demonstrating conformity with the bank, whilst the east end of the lower profiles imply increasing resistivity of the bank base. On the accumulated evidence, it is reasonable to assume that there are three concentric ditches – or possibly rings of conjoined postholes at the centre of the Giant's Ring. These approximate to the centre as defined by the outer embankment rather than by the passage tomb. A possible building sequence would then be:

1. passage tomb
2. embankment centred on passage tomb
3. concentric rings centred on the embankment.

The intriguing possibility is that these interior features may also relate to the use of the Giant's Ring at the time of the BNH5/6 enclosures.

Figure 2.11 Left: 1984 aerial photo of the Giant's Ring, showing internal concentric rings in the grass; right: the central circular feature identified in the gradiometer image may reflect the inner ring and shows lines running between the interior and the bank which may be attributed to 19th-century activity (left: Barrie Hartwell; right: Stephen Davis and Romano-Germanic Commission, Frankfurt).

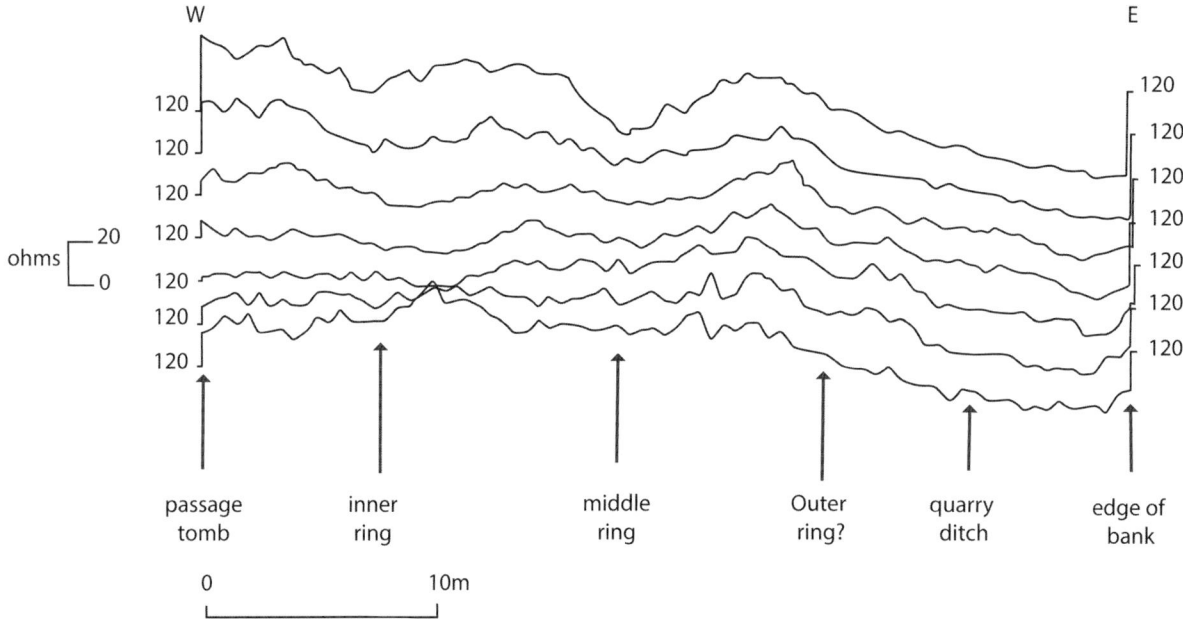

Figure 2.12 Resistivity profiles east across the Giant's Ring from the passage tomb to the embankment.

2.3 Stray finds

The timber circle complex at Ballynahatty, as has been detailed above, was only one of a large number of sites situated within the townland in antiquity. Artefacts have been recovered, both as stray finds and during earlier archaeological investigations, which enable the most recent excavated material to be placed in the context of continuous occupation through antiquity.

Given the extent of the archaeological remains here, it is surprising how few finds are provenanced to either Ballynahatty or the neighbouring townlands in the acquisition and recorded specimen files of both the Ulster

Museum and the National Museum of Ireland. The excavations previously carried out in the townland have also produced relatively few finds. The published investigation, which was undertaken in Bodel's field, Ballynahatty, by MacAdam and Getty, has been detailed above. This recorded the presence of three pottery vessels (and the possible existence of one other) accompanying substantial quantities of burnt and unburnt human and animal remains. The report only briefly describes the vessels, and no illustrations are included. MacAdam also recorded Bodel's description of the numerous sites found on his land (see Section 3.5.1.2) and some of the artefacts. There were a number of 'earthen urns' associated with bones or cremations, in cists, one of which had two arrowheads; stones axes; a 'yellow stone tapering at each end' with a central hole; 'a black stone, 6" long, knobbed at each end and hollowed between with a small hole passing through the centre of the intermediate stem'; and four stone rings of a 'black light substance, like jet ... fitting exactly one within the other', 4" in diameter and together having 'the appearance of a circular grooved disc' (MacAdam and Getty 1855, 364–365).

Alison Sheridan (2004, 27) refers to the finding of an oblique arrowhead (or petit-tranchet derivative) at the Giant's Ring, and quotes the information from Woodman that these tool types 'are almost certainly confined to monuments related to henges, *e.g.* ... the Giant's Ring' (Woodman 1994, 213). It may be that Woodman is referring to the petit-tranchet derivative arrowheads, which were recovered from the timber circle BNH5, although, as he refers to the Giant's Ring, it gives the impression that the artefact was recovered from the henge. This is an important distinction as no typically Late Neolithic artefacts have hitherto been found at the Giant's Ring.

2.3.1 The Ulster Museum

The Ulster Museum holds both acquisition registers and recorded specimen files, where finds are noted but retained by the finder. The computerised topographical records list three artefacts from Ballynahatty. They are two flint end scrapers (A16621/ A2.1:15.36 and A17538 A2.1:15.29) and a prehistoric blade (A16964/ A2.1:15.26), which is recorded as coming from the field between Shaw's Bridge and the Giant's Ring. The only recorded find in neighbouring townlands is a Late Mesolithic edge-retouched flint blade from Ballylesson (A751.1936/C4:106.29). Three further artefacts were listed in the recorded specimen files. A possible porcellanite polished stone axe (RS1999.13) from Ballynahatty townland has an unclear provenance – the finder states that it came from within the Giant's Ring, although the depositor states that she heard it was near the Ring and presumed it came from the field of the excavation (*i.e.* BNH5). The second recorded specimen, the worked flint artefact (RS2000.27), was found at Eden House (J334673) in Ballylessan townland while digging a flower bed. A flake with secondary working was also recovered from Ballycowan townland (RS2003.20).

A number of artefacts provenanced to Ballynahatty and held in the Ulster Museum include the Coarse Ware sherds discussed above (47.31 marked on the box). Two flint artefacts (A5079 and A5080) were found within 10 m of each other on the north slope of the east–west ridge in Ballynahatty during fieldwalking by C. Foley in 1989. The stone balls shown in Plate 13 are also thought to have come from Ballynahatty. The record notes, 'Mrs Dunlop of Edenderry House, Belfast remembers the three grooved stones as having been in the house for many years and believes they were found at the Giant's Ring'; the find was deposited in 1927. The museum also has a flint artefact from the ploughed field to the north of the Giant's Ring and another from the field adjoining the Giant's Ring to the northwest, close to the footpath to Edenderry.

2.3.2 The National Museum of Ireland

Apart from the remains of the two vessels detailed above, the only file relating to Ballynahatty is a flint end and side scraper with its edge produced by secondary pressure flaking on a bulbar flake and trimmed on one face (Reg. No. 1911:7, Store D1:16). It was 'found lying in a hollow in the Giant's Ring ... very near the dolmen'. A pointed stone (Reg. No. 1929:164, Store C2:13) and two stone axeheads (Reg. No. 1929:165, Store C2:13; Reg. No. 1934:76, Store C9:3) were recovered from Ballylessan townland and three flint implements (Reg. No. 1936:2585-7, Store C27:17) from Ballynavally.

2.3.3 The excavations

The four recorded excavations in Ballynahatty townland, and the long history of cultivation of the soils, had resulted in the discovery of relatively few artefacts. The excavations in 1855 recovered Coarse Ware and possibly also a Collared Urn. The excavations in the Giant's Ring recovered few artefacts, although these have an Early/Middle Neolithic character (Nelis 2003, 683). The stray finds recovered do little more than confirm the presence of prehistoric activity in the vicinity, evidenced by flint artefacts. The grooved stone 'balls' are interesting, although the context of their recovery is uncertain (Plate 13).

It is evident, however, that the Coarse Ware recovered during the recent excavation at the timber circle site BNH5 is only a small part of the Middle Neolithic ceramic repertoire likely to be distributed across the townland. The possible recovery of the Cordoned Urn also hints at the length of time that Ballynahatty was the focus for this type of activity.

2.3.4 Field collection

Since the 1980s, and especially during the BNH3/5/6 excavations, there have been more regular fieldwalking exercises with the resulting artefacts recorded and mostly lodged in the Department of Archaeology at Queen's University or held privately. The provenances vary from 8-figure OS grid locations to the more general 'Ballynahatty td'. This collection of over 250 artefacts goes some way to address the apparent dearth of recorded material. Over time much material has probably been found and removed but not recorded. William Gray (1899, 74), mentioned that the Giant's Ring has been a 'happy hunting ground' for antiquarians since the finding of the 1855 tomb.

For the purposes of this discussion, the finds are grouped, and the locations are distilled into three areas corresponding to the three fields north of the Giant's Ring in existence during the 1990s excavation. Field 1, on the west side, encompasses the Bodel farmstead site; Field 2 is central and includes the BNH5/6 site; and Field 3 is east and extends to the Giant's Ring car park. All field boundaries were removed in 2001. The principal collectors were the BNH5/6 excavation crew; Mike Baillie and Jim Mallory, Ken Pullin, John Davison and Gail Ritchie (Plate 14).

Field 1:
1 small multiplatform core; small prismatic core (for microliths); 69 debitage flakes including 3 primary and 22 secondary flakes and some angular debitage; 1 burnt flake; 2 hollow scrapers; 1 end scraper; 1 pointed flake (awl?); 1 subangular blade; 3 small broken blades; 1 squared flake with secondary working on dorsal side and edges.

Field 2:
1 pebble core; 58 debitage flakes including 4 primary flakes and 16 secondary flakes; 13 burnt flakes; 3 hollow scrapers; 1 side scraper; 1 end/side scraper; 1 Quartz end scraper; 2 blades with serrated edges; 1 small blade; 1 broken blade of an axe roughout; 1 butt of a broken ground stone axe with heavy end damage (possible wedge).

Field 3:
1 single platform core; 1 multi-platform core; 1 small prismatic core; 19 debitage flakes including 1 primary and 9 secondary flakes; 1 burnt flake; 1 end scraper; 2 side scrapers; 1 blade with single edge retouched.

In addition, in 1984 a 60 × 50 m gridded total collection was undertaken in Field 2 across the northern side of BNH3 and part of the field beyond. This gives a more accurate view of the productive capabilities of the area and highlights specialised tool fabrication:

14 debitage flakes including 7 secondary flakes; 4 end scrapers; 1 hollow scraper; 1 five-sided flake with combined hollow scraper, double concave scraper, serrated edge and point; 1 six-sided flake incorporating four worked concave edges, a small end scraper and a point; 1 six-sided flake with end scraper and shallow hollow scraper; 1 flake with worked shallow concave side; 1 cortical flake with shallow retouch and serrated edge on opposing side; 1 small hollow scraper with opposed narrow chisel end; 2 blades with serrated edges; 1 small bladelet; 1 microlith. Areas of burnt bone showed that ploughing had disturbed two cremations.

Gail Ritchie, a local field walker, has been collecting extensively from Edenderry and the Lagan to south Ballynahatty, and her finds of worked flint include a hollow based and two petit-tranchet arrowheads, scrapers and a Mesolithic axe (Plate 14, a–p). She has also found cord-impressed pottery. There is a cortex surviving on the proximal end of the Early Mesolithic axe (a), so it was probably made from a nodule found close to or in the limestone underneath the basalt exposed on the north side of the Lagan Valley. It illustrates the penetration of early Mesolithic settlers around the Irish coast and up the river valleys. Two other Mesolithic contenders would be a long narrow blade (b) with a thick keel and a core which shows the removal of small narrow blades using a punch (c).

The remainder of the material is Neolithic, 88 pieces of flake debitage including ten blades. The cortical flakes show, with one or two exceptions, that the cortex has been eroded, indicating that the nodules have been moved in the soil from their source under glacial action. This flint was obtained opportunistically, probably during cultivation of the fields or in exposures along the riverbank. The tools all show secondary working and include end scrapers (h–n) with three showing side as well as end retouch (l–n) and another with a thick keel which could be a modified core (h). There are three possible hollow scrapers (e–g), one small (g), one broad (f) and one which is also an end scraper (e) and so a dual tool. (d) appears to be a double hollow scraper on a cortical flake with a thick keel. Three thin flakes have been modified to produce a serrated edge – with two blunted on the opposite edge to accommodate the index finger. The butt ends of the two petit tranchet derivative arrowheads have been finely worked and rounded for attachment, although the points are broken (o–p).

Finds from the three excavations within the Giant's Ring earthwork have also been discussed previously. None has satisfactorily answered the questions posed by the excavators, partly due to the lack of recovered artefacts. This is perhaps not surprising as it has been suggested that monuments such as these were often purposely kept 'clean' of artefact remains (Thomas 1999, 74). At Balfarg henge, for example, no artefacts were found in the primary or secondary silts of the ditch (Mercer 1981, 66). Similar findings were also made at Avebury, where it was proposed that the paucity of finds from the interior of the enclosure and the primary ditch silts indicated that the enclosure was kept deliberately clean (Smith and Keiller 1965).

Field collection outside of the Giant's Ring illustrates a different picture. Although much of the flint used was of poor quality and probably collected opportunistically, there is ample evidence of reduction flaking – so stone tool fabrication was taking place across the plateau in Ballynahatty. The concave edges of the hollow scrapers are typically shallow, which suggests they are being made for a specialised task. These features are incorporated in the three hybrid, multipurpose tools found in the gridded search.

3

Environmental history of the Ballynahatty area

Gill Plunkett

3.1 Introduction

An infilled kettle hole lake (Ballynahatty Bog) to the north of Ballynahatty offered an opportunity to investigate the environmental history of the local area (Fig. 3.1; see Fig. 4.1a). The basin measures approximately 0.2 ha and today supports a fen vegetation, with a fringe of sycamore, birch and willow along its southern margin. The wet environment is favourable to the preservation of organic remains, including pollen produced by plants growing on and around the basin. The deposit was cored in 2006 to a depth of 3.3 m, enabling reconstruction of a vegetation history extending back to the early post-glacial period (Plunkett *et al.* 2008). The chronology of the upper part of the core was poorly resolved, however, and it was unclear if a sedimentary hiatus had occurred during the Neolithic. A parallel core was subsequently investigated further as part of a Masters dissertation, including geochemical analysis of a tephra (volcanic ash) layer and radiocarbon dating of additional levels (Trainor 2011). The tephra was identified as Lairg A, an ash erupted by

Figure 3.1 View north across the kettle lake basin from which the Ballynahatty core was taken.

Mount Hekla in southern Iceland, and dated to 4997–4902 BC (Pilcher *et al.* 1996).

Here, a new age-model for the Ballynahatty sequence has been created, incorporating the age of the Lairg A tephra and the radiocarbon dates obtained by Trainor (2011). Additionally, the model includes a modern (2006) age for the surface of the deposit but treats one radiocarbon date (UB-7119) as an outlier, as it requires a radical change in sedimentation to fit it within the model, a feature that is not evident within the stratigraphy of the sequence or the pollen record. The resultant age-model demonstrates fairly steady sediment accumulation, but with large age uncertainty above the radiocarbon-dated horizon (UB-7120) at 128–129 cm (Fig. 3.2). The pollen data reported by Plunkett *et al.* (2008) have been replotted against the mean modelled ages (Plate 15).

3.2 The Ballynahatty vegetation sequence

3.2.1 *The early post-glacial period*

The Ballynahatty pollen record begins at in the early post-glacial period, very soon after the end of the last ice age. The surrounding area was initially open and wet, characterised by grasses, sedges and bulrushes or bur-reeds but with some birch and willow reflecting wet woodland. Compared with other regional pollen diagrams from this time period (*e.g.*, Sluggan Bog, Co. Antrim; Smith and Goddard 1991), birch pollen representation is relatively low, suggesting that dense birch woodland may not have developed in this locality. Aquatic pollen (white waterlily, pondweed and water-milfoils) show that the Ballynahatty basin featured a shallow lake at this time. Sometime after 9000 BC, large increases in tree pollen indicate the establishment of broadleaf woodland, mainly of hazel but with poplars abundant for a short time. Oak and elm are represented to a lesser extent, indicating that they were minor components of the woodland, perhaps some distance from the site. The basin seems to have become in-filled with vegetation at this time, with aquatic and wetland taxa greatly reduced.

3.2.2 *Mesolithic*

By the time of the first known settlement in Ireland, the Ballynahatty area remained dominated by hazel scrub. After around 7000 BC, oak woodland expands slightly and soon after pine becomes more evident. Pine produces copious quantities of pollen, and the low values recorded at Ballynahatty imply that it was not locally abundant. In the mid- to late 7th millennium BC, however, an environmental shift to drier and colder conditions appears to have favoured the spread of pine onto bog surfaces in the north of Ireland (Torbenson *et al.* 2015). The Ballynahatty record may therefore be detecting vegetation changes within the wider region. The drier conditions may also explain a peak in microscopic charcoal that likely reflects wildfire.

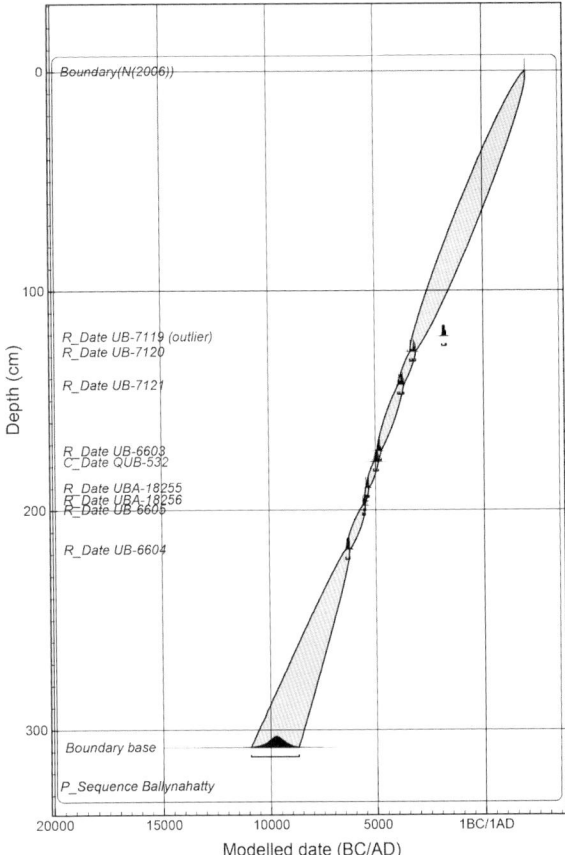

Figure 3.2 Depositional age-depth model for the Ballynahatty core produced in OxCal (Bronk Ramsey 2008; 2009) using the IntCal 20 calibration dataset (Reimer et al. *2020) and based on radiocarbon samples and tephra reported in Plunkett* et al. *(2008) and Trainer (2011).*

Apart from fluctuations in birch, possibly signalling changes in water-tables and its effects on wet woodland around Ballynahatty, the local environment changes little during the Mesolithic until alder becomes established sometime after 5500 BC. The significance of the spread of alder at this time has been debated but, at Ballynahatty, it possibly represents the further drying of the basin as soon afterwards ling, a heathland plant, expands.

Around 4500 BC, ribwort plantain makes a consistence appearance and charcoal values increase. Requiring light to thrive, ribwort plantain signals openings in the woodland and is a valuable indicator of woodland disturbance. It can also point to the presence of pasture or fallow land (O'Connell and Molloy 2019). A slight decrease in hazel abundance is concomitant with increases in oak and alder pollen values implying changes in the structure of the surrounding forest. Whether this was caused by natural environmental changes or by human activity is unclear. The re-appearance of aquatic pollen hints that water levels had again risen in the basin, possibly signalling a shift to wetter conditions. Pine pollen becomes sporadic from this time, suggesting that pine was by now locally absent.

3.2.3 Neolithic

As we move into the timeframe of the Neolithic period, evidence for woodland disturbance becomes more striking. Ribwort plantain values increase, along with grasses and a range of other light-demanding taxa that indicate openings in the forest. Oak, elm and hazel decline. The Elm Decline is a widespread phenomenon throughout Ireland and Britain around the time of the Mesolithic–Neolithic transition that saw a major fall in elm pollen representation, possibly triggered by a pathogen whose spread may have been facilitated by early farming populations. Because elm seems to have been a lesser component of the woodland around Ballynahatty than at other locations in the north of Ireland, the manifestation of the Elm Decline is subtle, but values fall by almost 50% around 3600 BC. At Ballynahatty it follows the start of farming, evidenced unambiguously by the presence of cereal pollen that testifies to arable activity in the area. Cultivation is continually attested for about 150 years but thereafter became more sporadic. Grassland remains important, however, suggesting continued occupation and farming in the area during the Middle Neolithic. By the Late Neolithic, the intensity of land-use seems to have declined: ribwort plantain values are reduced along with other open indicators and charcoal. Oak, in particular, appears to have benefitted from this diminished pressure but elm suffers a second decline, a feature also seen in other pollen records from the north of Ireland (Hirons and Edwards 1986; Smith and Goddard 1991).

3.2.4 Bronze Age

The top of the Ballynahatty pollen record appears to encompass the Early Bronze Age through to the start of the Middle Bronze Age. While open areas persist around the basin and cultivation continues to be recorded, some woodland regeneration is suggested by elevated levels of ash pollen. Ribwort plantain increases again in the early 2nd millennium BC, coinciding with increased mineral input into the basin that signifies soil erosion in the surrounding area (Plunkett *et al.* 2008). These changes point to a local increase in land-use. Sediment above this level was not analysed as the peat was poorly consolidated.

3.3 Discussion

The pollen record from Ballynahatty gives us insight into the changing nature of the locality since the end of the last ice age to the beginning of the Middle Bronze Age. The landscape featured deciduous woodland with a significant hazel component. While there is possible evidence for Mesolithic interference with the woodland, the main human impact on the environment is observed from the start of the Neolithic. From this time, areas of woodland were cleared and there is evidence for grassland and crop cultivation. Although farming seems to have declined in intensity during the Late Neolithic and start of the Early Bronze Age, woodland did not recover fully and the immediate landscape retained a largely open character. Farming increased again in the latter part of the Early Bronze Age.

4

Cumulative interpretation landscape map

4.1 The wider Ballynahatty landscape

In the historical descriptions of the Ballynahatty plateau, the focus has always been on the two visible sites of the Giant's Ring and the passage tomb. The exception is David Bodel's invaluable recollection of sites found in the 18th and 19th centuries. Although these can be placed within one or two fields, the specific locations are lost. By bringing together aerial and ground surveys, a remarkably detailed picture emerges of a mortuary landscape which had developed over 2000 years.

Table 4.1 List of BNH sites cross-referenced with NISMR numbers. The BNH sites are numbered on Fig. 4.1.

BNH No	Figs	NISMR No	Identification
–	–	DOW009	Multiple: ritual landscape (scheduled)
BNH1	–	DOW009:036	Giant's Ring (state care)
BNH2	–	DOW009:036	Passage tomb (state care)
BNH3	2.5.1	DOW009:079	Enclosure
BNH4	2.7.24	–	Ballynahatty Bog (kettle lake)
BNH5/6	2.7.25	DOW009:062	Timber enclosures, cist, cremations
BNH8	2.5.3	DOW009:080 DOW009:012	Flat cemetery, standing stone
BNH9–12	2.6.9	DOW009:071	Linear ring ditch cemetery
BNH13	2.7.23	DOW009:037	'1855' tomb (site of?)
BNH14	2.6.14	–	Ring ditch
BNH15	2.8.28	–	Trackway
BNH16	2.8.29	DOW009:081	Ancient routeway/ cursus
BNH17	2.8.27	–	Ditched field boundary

BNH No	Figs	NISMR No	Identification
BNH18	–	DOW099:078	Mound, multiple cist cairn
BNH19	–	DOW009:050	Megalithic tomb (site of)
BNH20	–	DOW009:051	Portal tomb (?) (site of)
BNH21	–	DOW009:053	Multiple cist burials (not located)
BNH22	2.5.6	–	Enclosure
BNH23	2.6.7	–	Pit/posthole cluster
BNH24	2.5.2	–	Pond
BNH25	2.6.8	DOW009:076	Ring ditch and post setting
BNH26	2.6.10	–	Ring ditch
BNH27	2.6.11	–	Ring ditch
BNH28	2.6.12	–	Ring ditch
BNH29	2.6.13	DOW009:077	Two enclosures
BNH30	2.7.15	–	Ring ditch
BNH31	2.7.16	–	Ring ditch
BNH32, 33	2.7.18	DOW009:074	Ring ditches
BNH34	2.7.21	–	Arc of enclosure
BNH36	2.7.17	DOW009:072	Timber circle (21 m)
BNH37	2.7.19	–	Cluster of sub-rectangular features
BNH38, 39	2.7.20	–	Small ring ditches
BNH40	2.7.22	DOW009:082	Sub-rectangular features
BNH41	2.5.5	–	Enclosure
BNH42	2.5.4	–	Enclosure
BNH43	2.8.26	DOW009:075	Enclosure
BNH44	–	–	Pit/post alignment
–	–	DOW009:035	Motte
–	–	DOW009:073	Routeway or cursus

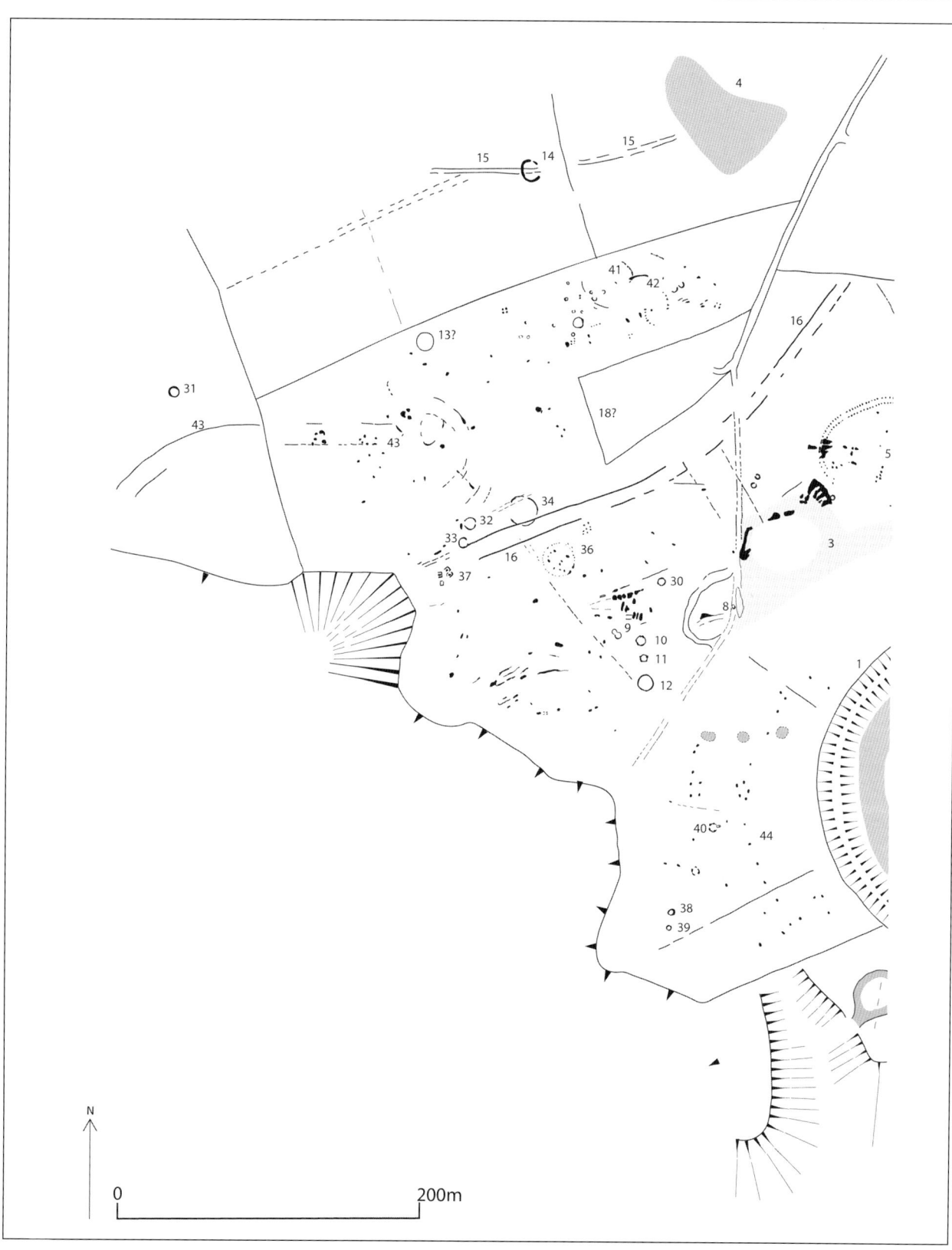

All the aerial photography and geophysical survey interpretations of known locations are summarised in Figure 4.1. It is not exhaustive, and only the principal sites in Chapter 2, Section 2.1.2, Figs 2.5–2.8 have been specifically identified. Each 'dot' represents a penetration of the subsoil and, therefore, a potential cist or posthole. In one case, these apparently isolated features can be interpreted as postholes forming a curving line (BNH44) which may have delineated the bank of the Ring on its west side. To the south, this aligns with a possible natural entry ramp

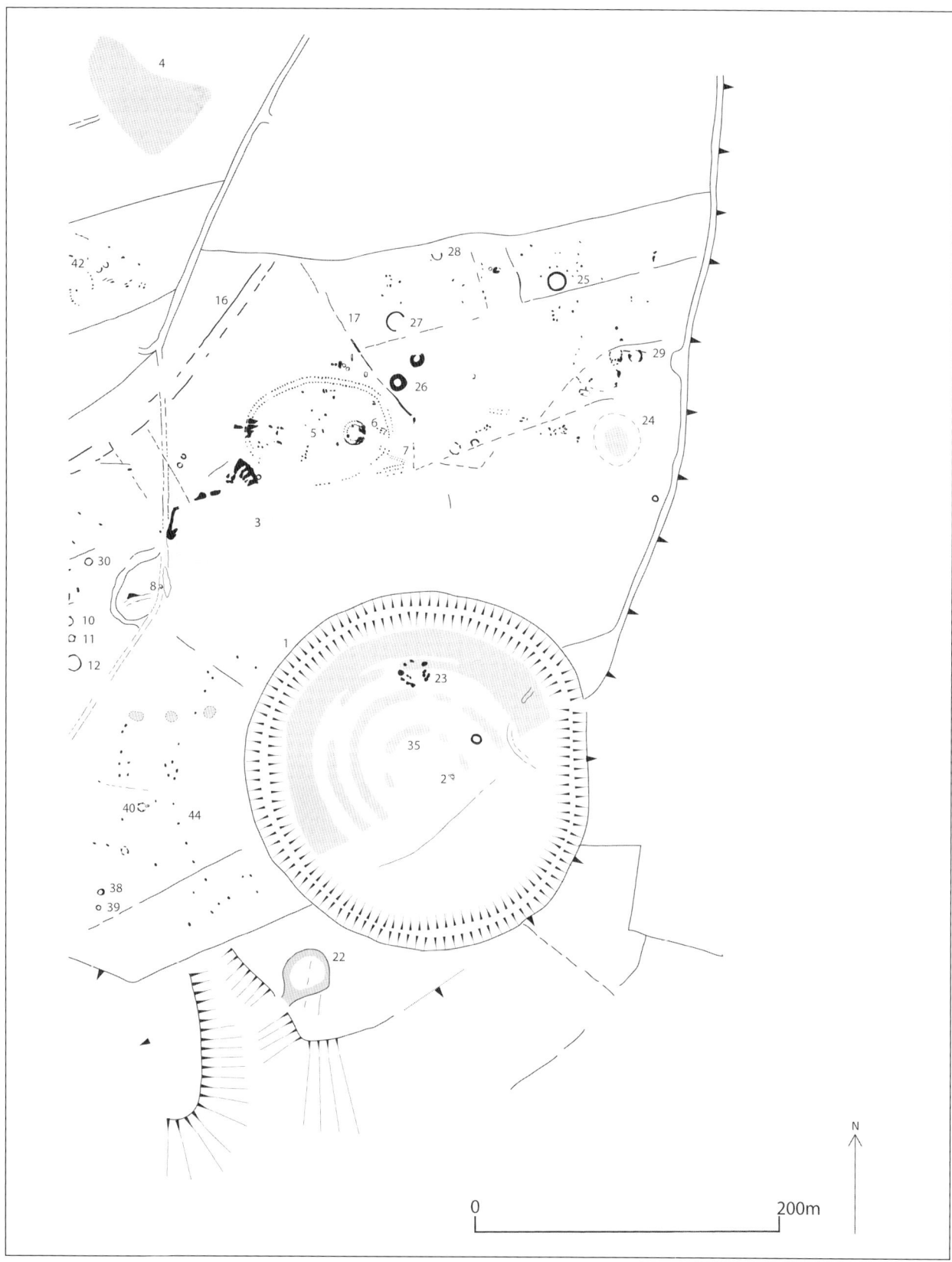

Figure 4.1 (above and opposite) Cumulative map of sites identified through aerial photography and geophysical survey. The plateau is defined on its east, south and west sides by steep slopes, although only two natural access points are shown by hachures. Another access route at top of the slope coming down the east side of the plateau from the north could have connected the Purdy's Burn with the entrances of the Giant's Ring and BNH5/6. The western edge of the BNH5/6 – Giant's Ring complex is defined by the post row (BNH44). BNH site numbers are shown.

onto the plateau. To the north, it stops at the ridge and could also have acted as a series of markers to move people through the landscape to the eastern approaches of BNH1 and BNH5. Although a 'busy' landscape, the corridor between BNH5/6 and the northern edge of the Ring remains remarkably clear of sites, emphasising the link between them. BNH5/6 and the henge may be contemporary entities in the ceremonial landscape.

Part 2

The excavations 1990–2000

5

Ballynahatty 5 and 6: excavating the enclosures

5.1 Overview

Part 2 brings together the results of the excavations. Chapter 5 describes the excavated features of the Late Neolithic timber enclosures, BNH5 and BNH6. Earlier funerary material from the Middle Neolithic and some unexpected medieval remains are also covered. At the end of the chapter the archaeobotanical remains from the excavations are presented and discussed. There are specialist chapters on the pottery, lithics, human remains and an analysis of the dating. Interpretation of the results is dealt with in Part 3.

The timber enclosures were first seen as cropmarks during an aerial survey in the dry summer of 1989 and were excavated over ten seasons from 1990 to 2000. The 1991 excavation examined another cropmark site at the western end of the ridge previously identified by aerial survey in 1984 (Fig. 2.1). The excavations (licence B106/99) were supported throughout by the Historic Monuments Division of NIEA (now Historic Environment Division of the Department for Communities) and the Department of Archaeology and Palaeoecology, Queen's University Belfast (now part of the School of Natural and Built Environment). The NIEA provided a regular grant and a site office and tool sheds. Queen's provided the transport, staff, equipment and laboratory facilities. For most seasons, the excavation formed part of the Archaeology teaching programme and was, therefore, largely staffed by students of Queen's with the essential addition of volunteers from the Ulster Archaeological Society, other universities and local 6th formers. Because of the large number of inexperienced diggers employed, for initial convenience, the excavation 1-metre grid was bypassed in favour of a more readily identifiable temporary 4-metre grid, which was given area numbers from 1–15. As the seasons progressed, the area numbers were retained and increased to a final tally of 100 (Fig. 5.1).

Most plans were drawn relative to these areas, each one being supported by a daybook which acted as a diary of work done. Over 750 plans and sections were produced covering 1700 contexts based on a single context recording system. All excavated soil was dry sieved on site through a 1 cm mesh, and 5.6 tonnes of soil samples were wet sieved in the Archaeology Laboratory at Queen's University under the supervision of John Davison and subsamples checked by microscope. In the early years, Cia McConway did post-excavation work, and the records were brought together as a final Data Structure Report for the NIEA by Sarah Gormley. Sarah was awarded an MPhil degree for her thesis *The Dating and Phasing of the Timber Circle Complex at Ballynahatty, Co Down* (Queen's University Belfast) in 2004. At various stages, a number of specialist reports have been produced, and these are acknowledged in the text. In retrospect, the weather during the summer seasons was relatively benign, with very few days lost through rain, which rarely puddled in the fast-draining soil. The field was usually cropped with winter barley or rape, and our access was always facilitated, for which we are indebted to the late James Thompson and family for the permission, unrestricted access and tolerance of our excavations for a decade. We would also like to thank the Dunlop family, whose house, replacing the Bodel's, sits in the centre of this great prehistoric landscape and who provided an unending source of water.

Although the aerial photos provided a near blueprint of the large enclosure (BNH5) and the smaller one within it (BNH6) (Plate 1), the easterly structures were partly obscured, and their relationship to the rest of the complex was uncertain. It was therefore termed the 'Annexe' and assigned a separate identifier – BNH7. When it became clear that it was part of the entrance structure of BNH5 the separate number was dropped, although the term 'Annexe' has persisted. This also meant that the duration

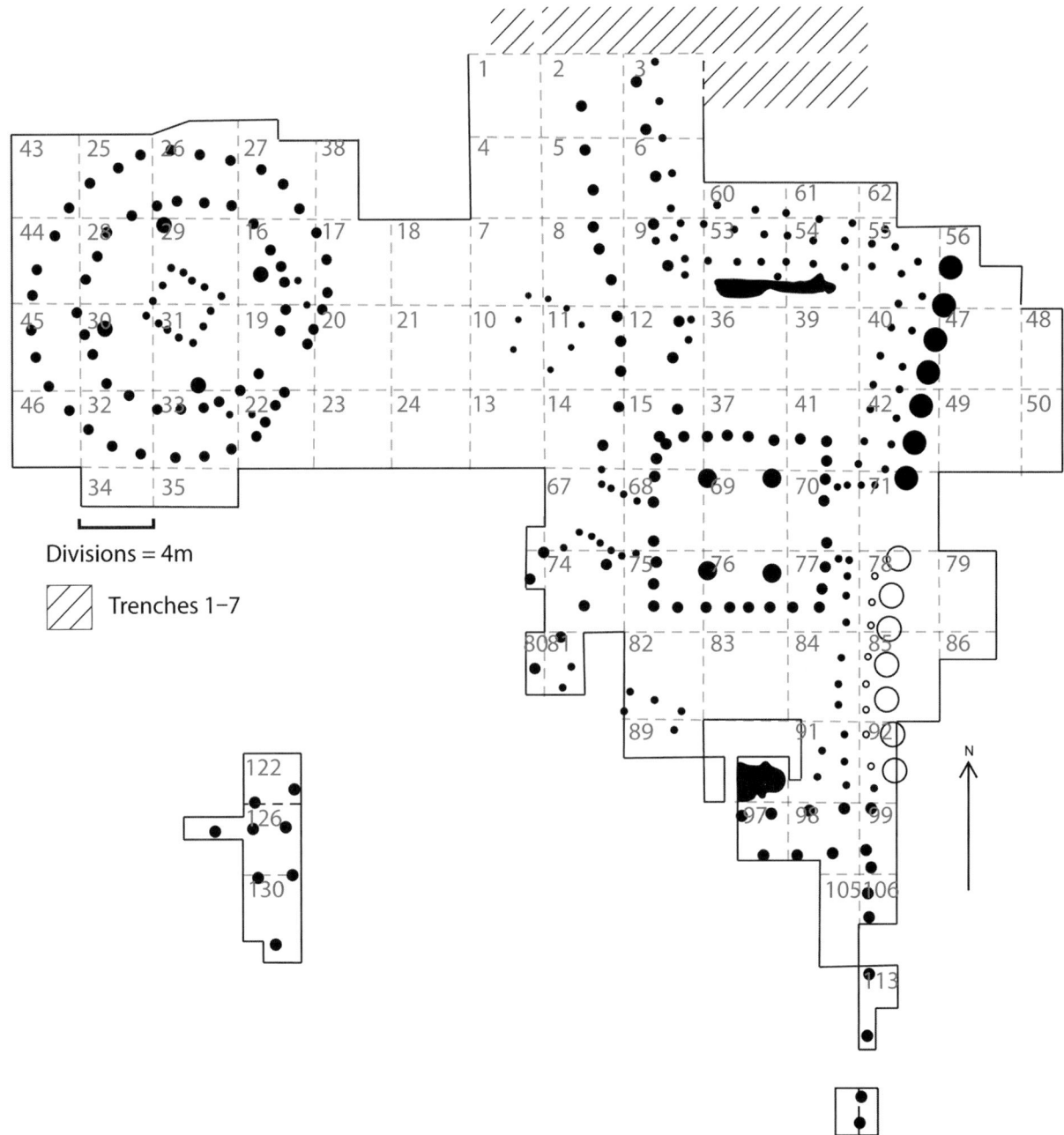

Figure 5.1 Area numbers allocated during the excavation based on a 4-m grid.

of the excavations, and the extent of the remains which were eventually uncovered, were not foreseen at the outset. The excavation initially focused on BNH6, the east side of the BNH5 and subsequently on the Annexe. The interior of BNH5 remains to be explored.

Figure 5.2 shows the excavated structures. As well as the outer enclosure (BNH5) and the inner timber circle (BNH6), a wedge-shaped structure (the 'Eastern Setting') was uncovered outside the entrance to BNH6, and complex sets of post rows were exposed in the Annexe area at the entrance to BNH5. To aid discussion of the numerous features excavated, the remains have been assembled by the 'feature group' names adopted as the excavation progressed and now retained to provide consistency with previous publications (Hartwell 2002, 527, fig. 1) although, with the benefit of hindsight, sometimes more appropriate terms could have been used. The postholes were not excavated in a regular order and, with the single context recording employed, locating any one of the hundreds of individual postholes is time consuming. To make navigation easier, all the postholes have been renumbered in a logical sequence and prefixed with identifying letters appropriate to their group. For example, 'EC1' refers to an Entrance Chamber posthole, 'OF3' is the Outer Façade and IR6 is the Inner Ring (when seen in the context of either BNH5 or BNH6).

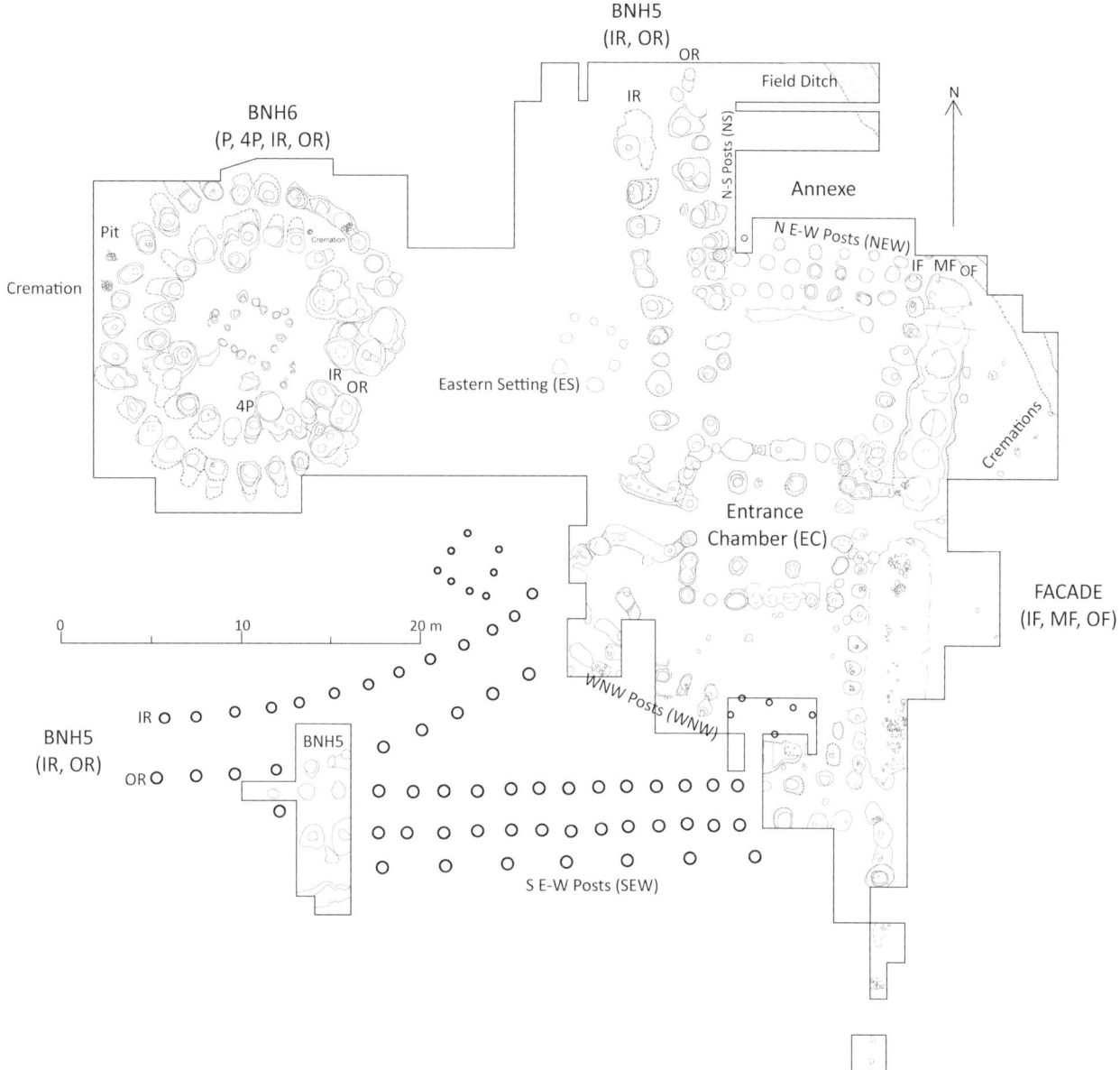

Figure 5.2 Site plan of excavation showing main feature groups. P: Platform; 4P: Four Poster; BNH6 IR: Inner Ring; BNH6 OR: Outer Ring; BNH5 IR: Inner Ring; BNH5 OR: Outer Ring; NS: North–South posts; NEW: Northern East–West posts; WNW: West–North–West posts; SEW: Southern East–West posts; IF: Inner Façade; MF: Middle Façade; OF: Outer Façade.

All these groups had been cut into the mixed sand, gravel and boulder clay subsoil on the top and northern slopes of the smooth east–west ridge; a natural topographical feature of a similar height to the artificially created bank of the Giant's Ring. An unquantified depth of the upper levels of all the features had been damaged by 18th- and 19th-century spade cultivation (Hartwell 1991, 14), and regular ploughing brings fresh subsoil to the surface on the top of the ridge to the present day. The damage caused by ploughing was found to be most severe on top of the ridge, adversely affecting the southern part of the site (Hartwell 2002, 526). The result of successive years of cultivation has not only been to remove or 'scramble' the upper levels of the features (an important factor when considering the feature depth measurements provided) but also to remove or obscure any potential stratigraphic links. Stratigraphy on the site is very restricted, and in most cases, there is no linking stratigraphy, with features being isolated; unfortunately, a common reality at timber circle sites (Gibson 2005, 132). In the following account, the depth measurements are best understood as a minimum depth, with an unknown amount truncated.

On excavation, the removal of topsoil revealed an interface layer (C2) made up of topsoil and subsoil disturbed by cultivation. Soil from prehistoric features was, therefore, incorporated into this interface layer to a degree. Several features of modern date were also found to be cut into this layer, including spade and plough marks, a field ditch (Plate 16) and a redundant waterpipe feeding a trough (Fig. 5.20). All the prehistoric features were found to be cut into the subsoil on the removal of C2, and these are presented under the relevant feature group headings. Burnt human and animal bone was widespread across the site and recovered from 59 contexts in varying quantities but, apart from the cremation deposits, most were unidentifiable.

Because the excavation was spread over a decade, the total excavated surface was never open at one time and was progressively backfilled to release the land back for farming. This is occasionally visible in the feature images. The large number of deep postholes presented real difficulties in excavation and recording. Sections could usually only be made and recorded in the upper 0.60 m because of the increasing difficulty of access, but this was often a zone of poor soil definition. Excavating beyond this (the maximum depth was 2.20 m) meant getting in the hole, and in the restricted space, the section could not be continued and the excavated surface was constantly being trampled. Sometimes it was possible to create a series of plans at increasing depth and reconstruct these into a section. As a result, the postholes are presented as complete sections, partial sections or simply profiles.

5.2 Passage Tomb related chamber and deposits of cremated remains

In 1994, a stone chamber, C591 was found near the eastern end of the prominent natural ridge when excavating posthole IF8 (C895) – one of a line of nine postholes running south to the entrance of the Annexe (Figs 5.3 and 5.29).

Figure 5.3 Chamber with line of Inner Façade to right and northern side of Entrance Chamber to left. Broken stones of the chamber roof lie on the baulk.

It was sub-rectangular in shape with interior dimensions of 1.10 m long by *c.* 0.40–0.50 m wide, 0.30–0.45 m high and aligned to the northwest. IF8 cuts through its northeast edge, removing much of this side (Plate 17). The two capstones and two floor stones survived, though badly shattered. The supporting stones of the southwest wall (C953), the rear wall, and the entrance structure also survived intact. The chamber contained two broken Coarse Ware pots with cremation deposits of a young child and an older adolescent (see Section 9.1.1.1; Plate 18) dated to 3350–2935 cal BC (95.4%, GrA14812 4460±40 BP).

Despite the chamber being built into a pit (Plate 19) and the capstone being just below the present cultivation level, it had a well-defined entrance. This was approached from the outside by a short, paved passage of two flat stones, 0.40 m long, terminating in a narrow threshold stone laid on end as a step from which there was a drop of a few centimetres to the floor of the chamber (Fig. 5.4). Immediately in front of this step, the entrance was further defined by two pillar-like miniature orthostats, joined by a narrow lintel on either side of a flat, irregularly shaped floor stone. There was no trace of supporting stones for a roofed passageway, although this area had been disturbed by a modern feature, C900.

The stones are of three distinct forms, though geologically all derived from the underlying greywacke of the Gilnahirk Group. The walls of the chamber are made up of weathered, rounded field stones (gritstone), which are typically found throughout the surrounding subsoil. The floor and capstones are large slabs of angular, horizontally split and quarried stone, 5–8 cm thick (silt/shale). This stone is not available on the Ballynahatty plateau and, although local, it may have been brought some distance from a suitable quarry. Every occurrence of flat or split stone in the excavated area is in a built context.

The entrance stones had been selected for their unusual pillar shape or colour. One is clearly a field stone; the other (NW) is more angular and darker (blue) in colour (Plate 18). Numerous small stones were packed between the walls of the chamber and the foundation pit in which it was constructed and in the entrance passage. This would have formed the lower levels of a marking cairn. These stones could have been easily removed from the entrance to facilitate access to the chamber for the interment of the pots and cremated remains – a similar arrangement to the 1855 tomb. The two Coarse Ware pots need not, therefore, have been placed in the chamber at the same time. These pots, each surviving in broken sections and containing cremated bone, were found resting on the floor slabs against the southwest (Vessel **1**) and rear (Vessel **2**) walls. Sherds of Coarse Ware pottery and quantities of burnt bone were found throughout the remaining chamber, in the fill of posthole IF8 and part of the chamber cut by that posthole. The west end of the chamber was filled by C956, a loose reddish-orange fill, which contained stones and burnt bone, but no potsherds, and was cut by a modern pit (C900).

The Coarse Ware pots were in a fragile condition, having been damaged by the construction of the later post-hole but also, in part, by the very poor quality of the fabric. The cremation deposits were removed during excavation and the surviving pottery fragments were supported on a plaster framework. However, much of the shattered bases disintegrated into dust, crumbs and smaller sherds as they were removed from the chamber, leaving only the rims and upper body (see Plates 48 and 49). Approximately one-third of each pot and its cremated contents survived intact, suggesting that they were initially positioned more centrally in the chamber, in the destruction zone, and had subsequently been moved in their damaged state against the southwest wall. The stone against which Vessel **2** was resting was loose and, on being removed, revealed an area of disturbed soil and a quantity of burnt bone from the cremation deposit, some pieces of which were quite large. This confirms that Vessel **2** was not in its original position; otherwise, bone could not have been deposited under the stone. It would seem likely that, when the chamber was disturbed, the stone fell from the wall and the cremated bone from the smashed pot was incorporated under the stone as it was replaced; the pot carefully pushed back against it.

Once the posthole had penetrated the floor of the chamber it was widened to facilitate access and excavated to the full depth. The post was then inserted, its position shown by the decayed mould, and the primary fill (C896) was packed around it, incorporating cremated bone and sherds of Coarse Ware pottery from the broken pots. Fieldstone from the dismantled northeast side of the chamber and the shattered remnants of the floor and capstone were also put into the fill. At the subfloor level, the position of these packing stones defines the diameter of the original post (*c.* 0.40 m). C896 also fills the interior of the chamber. Only a few small cairn stones were found in C896, suggesting that the marker cairn had been removed prior to digging the posthole. The post builders, therefore, probably knew of the location of the chamber and, although moving the posthole 0.30 m to the north would have saved much effort, rigid adherence to a fixed plan demanded the chamber's partial removal. However, repositioning the surviving pots with their cremated remains to the opposite side of the chamber before infilling does show some respect for their predecessors.

When post IF8 was eventually burnt, the fill C896 and the chamber became covered over with C779, which contained some fragments of the burnt bone of an adult. Much later, the west end of the chamber was filled by C956, containing only a little burnt bone, and this is cut through by C900, containing modern glazed pottery and some burnt bone. All the bone presumably originated from the damaged chamber.

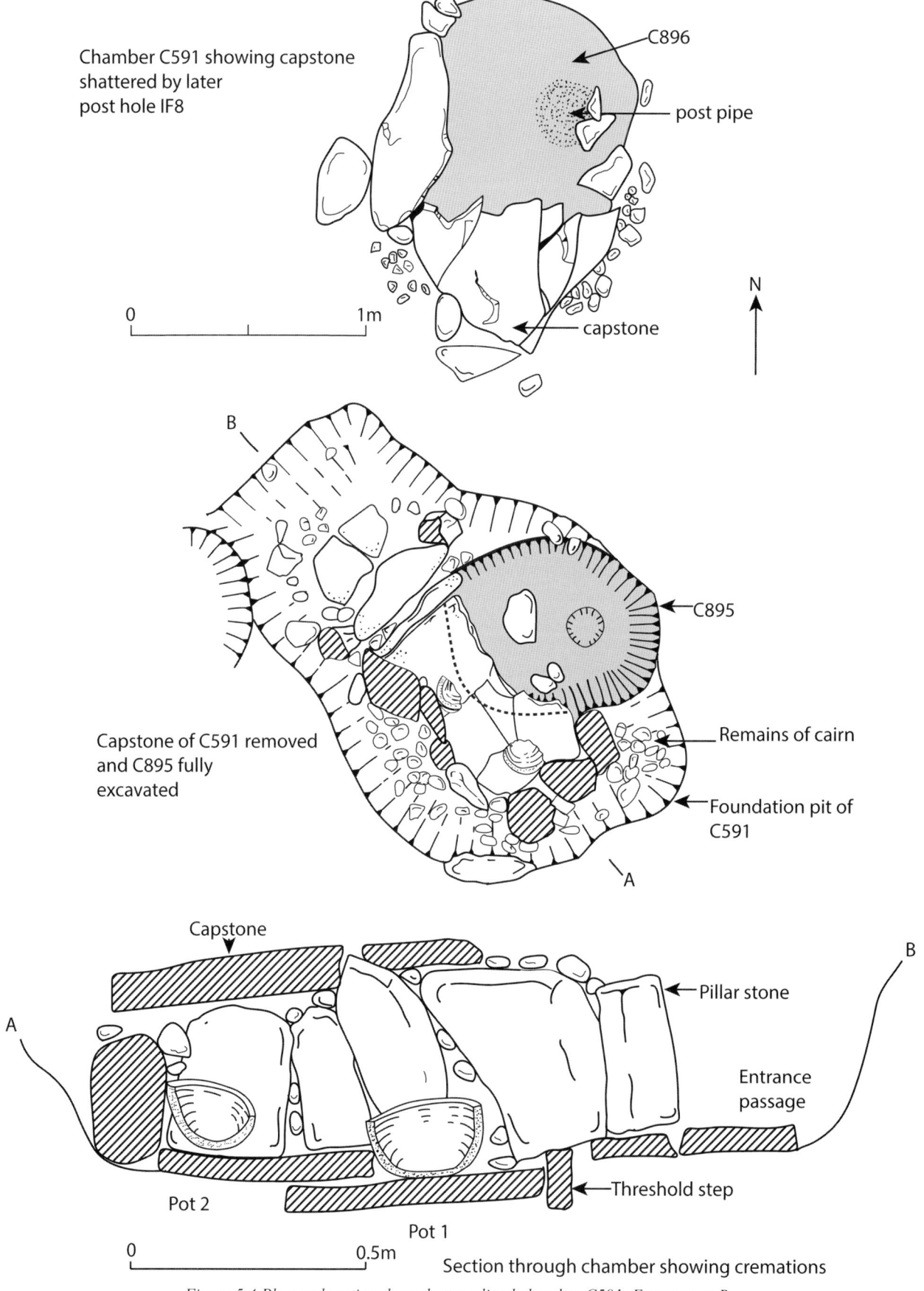

Figure 5.4 Plan and section through stone lined chamber C591. Entrance at B.

Just 7 m away to the east was a line of four small, shallow, equally spaced cremation deposits, representing at least two adults, one mature and a juvenile. Each deposit had suffered varying degrees of plough damage (Plate 20). All were revealed initially by flecks of bone in an orange sandy fill. Moving east and set 1 m apart, C1012 had no surviving structural remains, whereas the lower part of C1014, the best surviving example, was intact (Fig. 11.1, 1). Two flat stones, placed longitudinally side by side with a narrow transverse stone on the northern side, in total only 0.40 m long, were placed on the floor of a pit of about the same depth. An edging of split stones, no more than 0.10 m high, formed the symbolic walls of the 'chamber', which was just wide enough to hold a Coarse Ware bowl containing burnt bone. The third pit, C1016, consisted of two fieldstones and a scatter of burnt bone and the fourth, C1018, contained a single stone and burnt bone. The line may have continued under the unexcavated baulk and field fence, though evidence of this is now lost (see section 13.3.5). All the pits were ill-defined, but C1014 and C1016 had clusters of small stones, suggesting that miniature cairns may have originally marked them.

The 'chamber' of C1014 was clearly not meant to be accessed at a later stage. Indeed, the relationship of these four cremation deposits is atypical in that they are equally spaced 1 m apart in a straight line, which has a northeast orientation in contrast to the other Middle Neolithic sites in Ballynahatty, which are found in an apparently random scatter across the southern part of the townland with east or west orientations. Although all the 'chambers' were damaged, C1014 suggests that none of these structures would have had a capstone, though the contents of the pots may have been protected by some form of lid. The remains in C1016 were radiocarbon dated to 3360–3100 cal BC (95.4%, UBA-42747; 4513±28 BP).

5.3 Cremation deposits and pit

Two cremation deposits (C588 & C943) and a pit (C998) were uncovered during the excavation of BNH6 (Fig. 5.5). C588 lay between the inner and outer rings of BNH6. The ramp cuts of postholes IR19 and IR20 in the inner ring splay to avoid it, showing that it was in place before the construction of BNH6 and that the builders were aware of its existence (Hartwell 1998, 39). The cremation deposit

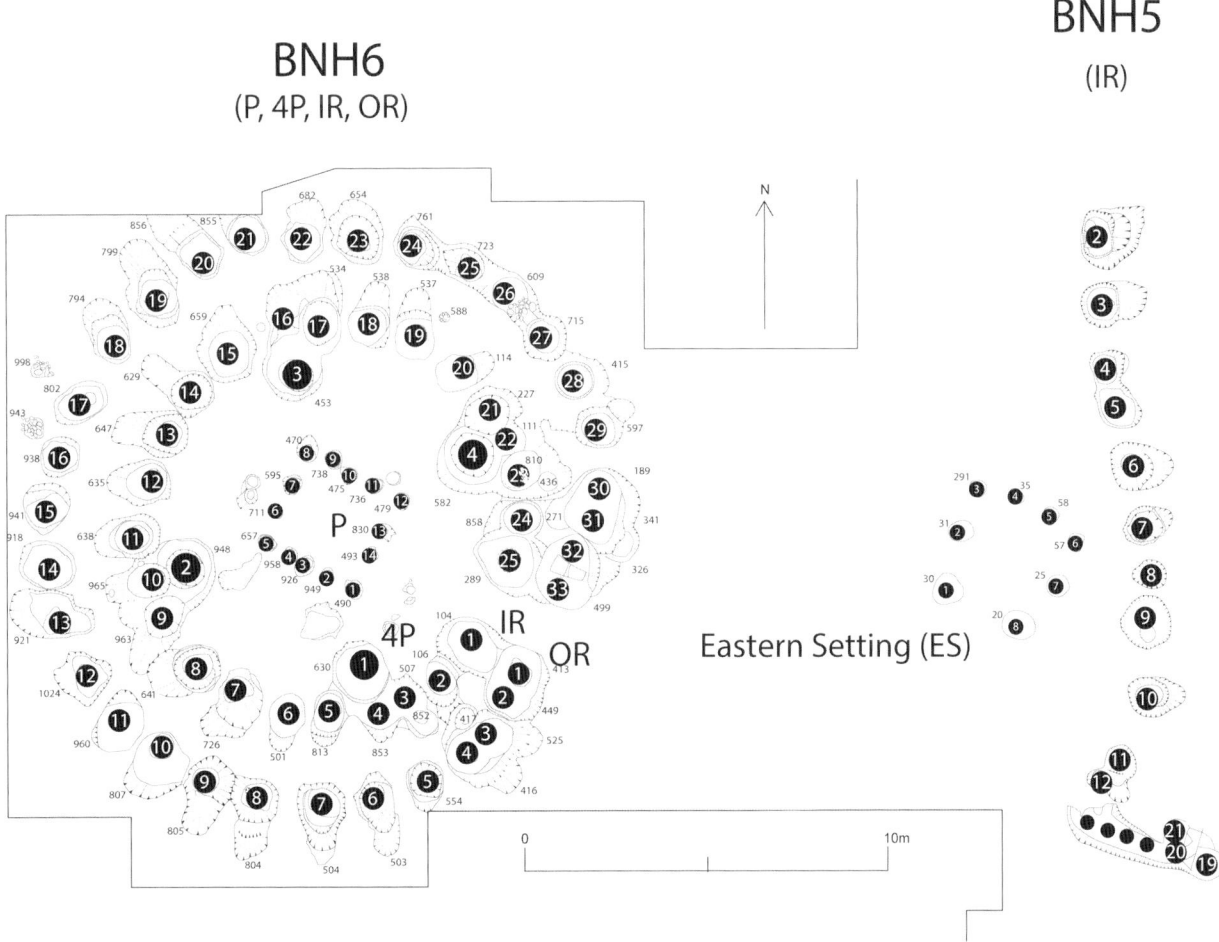

Figure 5.5 Plan of BNH6 and ancillary structures, showing excavated features and original context numbers and renumbered postholes. Cremation deposits C588 and C943 and stone pit C998 are also shown.

was placed into a shallow scoop, approximately 0.18 m deep and 0.55 m at its widest point, and the cut was lined and covered with split stones (Plates 21 and 22). There was no pot, but the compact shape of the deposit suggests that it may have been originally contained in a bag. Sufficient bone survived for identification as a female aged 30–40 years (see Section 9.1.1.7) and radiocarbon dated to 2875–2630 cal BC (95.4% UBA-42749; 4400±31 BP).

The cremated remains, C943, were uncovered immediately outside the outer ring of BNH6 (Fig. 5.5). The shallow cut, 0.16 m deep and 0.45 m wide, was lined with square-cut stone on which the cremation deposit was placed. The main deposit was that of a 4–5 year-old child, but there were also several fragments of an adult (see Section 9.1.1.8). The deposit was dated to 3340–3025 cal BC (UBA-42748; 4467±29 BP).

A pit (C998) containing a 'hoard' of split stones but no cremated human remains was recovered at the rear of BNH6 immediately opposite the entrance, between postholes OR17 and OR18 and close to cremation deposit C953 (Plate 23). The ramps of these two postholes diverge to avoid the pit, which occupies one of the three nodal positions from which the BNH6 complex was laid out (Hartwell 1998, 39).

5.4 The timber circle (BNH6)

The aerial photograph taken in 1989 (Plate 1) clearly shows the inner cropmark, named Ballynahatty 6 (BNH6), as a circular enclosure comprising a double line of pits and four central posts with an overall diameter of 16.00 × 14.50 m. The cropmark lay approximately 100 m north of the Giant's Ring, on a level area on the north slope of a low ridge which runs east–west (Hartwell 1998, 39). It was fully excavated and consisted of a series of large posthole cuts (86 in total) and a number of other features. The excavation showed that the cropmark was the remains of an elaborate timber circle with a double concentric ring of postholes (Figs 5.2 and 5.5). The 'Inner Ring' (IR) consisted of 25 posts and the 'Outer Ring' (OR) of 33 posts. Six smaller posts were found between the inner and outer ring in the entrance area, and in the interior of the circle were four larger posts (the 'Four-Posters', 4P1–4) and a central square setting of 14 smaller posts (Platform, P1–14). Beyond the timber circle entrance was an arrangement of eight posts in a wedge shape, the 'Eastern Setting' (ES). The entrance of the timber circle faced southeast and the opening of the Eastern Setting faced southwest onto this axis. The entrance through the outer enclosure, BNH5, was also part of this unitary design. The two cremation deposits (C588 and C943) and the stone-filled pit (C998), previously described, were also discovered in this area. A number of other scoops and features were discerned lying to the west of the BNH6 structure.

All the postholes in BNH6 have been renumbered in a logical sequence, clockwise from the entrance as 4P1–4 (Four-Posters), IR1–25 (Inner Ring) and OR1–33 (Outer Ring). The central platform has been numbered P1–14 from the south corner (Fig. 5.5).

5.4.1 Four-Poster setting

Four large postholes were uncovered in the interior of the BNH6, set at equal intervals, creating a square setting immediately within the inner circle (4P1–4). These postholes measure an average of 1.00 m in diameter and 2.40 m in depth. The large 'U' shaped holes contained multiple fills and, like the postholes of the inner and outer circle, had ramp cuts and a 'primary' but also a 'secondary' fill evident. The primary fills were mixed orange-brown sand and gravel soils with occasional stones. The secondary fills contained a considerable concentration of charcoal and much stone rubble and were defined by large funnel-shaped holes created during the removal of the posts. All the postholes have substantial ramp cuts, which must have been necessary to control the raising of such large posts. It is estimated from the depression at the bottom of the holes that they could have held posts measuring at least 0.50 m in diameter at the base and 7.20 m above ground (Fig. 5.6). Each of the Four-Posters are cut by the Inner Ring of postholes, providing one of the few opportunities to assess a stratigraphic relationship between two feature groups. A good understanding of the morphology of these postholes can be seen in 4P2.

5.4.1.1 Posthole 4P2

Initial removal of the topsoil and C2 revealed a layer of dark soil, patches of charcoal, stones – some as large as c. 0.30 m – and a concentration of darker material on the edge (Fig. 5.7a). Thirty centimetres below this (Fig. 5.7b), the dark soil had resolved itself into discrete areas of loose, charcoal-rich soil and stones, the largest of which corresponded to 4P2. This is the secondary fill – a mix of charcoal, charcoal-stained soil, stones, voids and clean orange subsoil – the latter of which was indistinguishable from the primary fill. By this stage, over 150 stones had been removed and at a depth of 1.00 m, concentrations of large stones also marked the position of IR9 and 10 (Fig. 5.7c).

Large stones (c. 0.35 m) naturally occurred in the surrounding subsoil and some were incorporated into the primary fill but in much lower concentrations. The variability in the fills at least partly reflects that of the subsoil, which varies dramatically in depth and area. This does not explain the very high stone concentration in some secondary fills where additional stone must have been introduced from the surrounding area, perhaps by surface collection. Another naturally occurring feature is iron pan and its associated pale grey, green, or yellow leached layers, which contrast with the orange, iron-rich natural subsoil. This is found wherever the pit conditions favour water movement up and down the soil profile. In

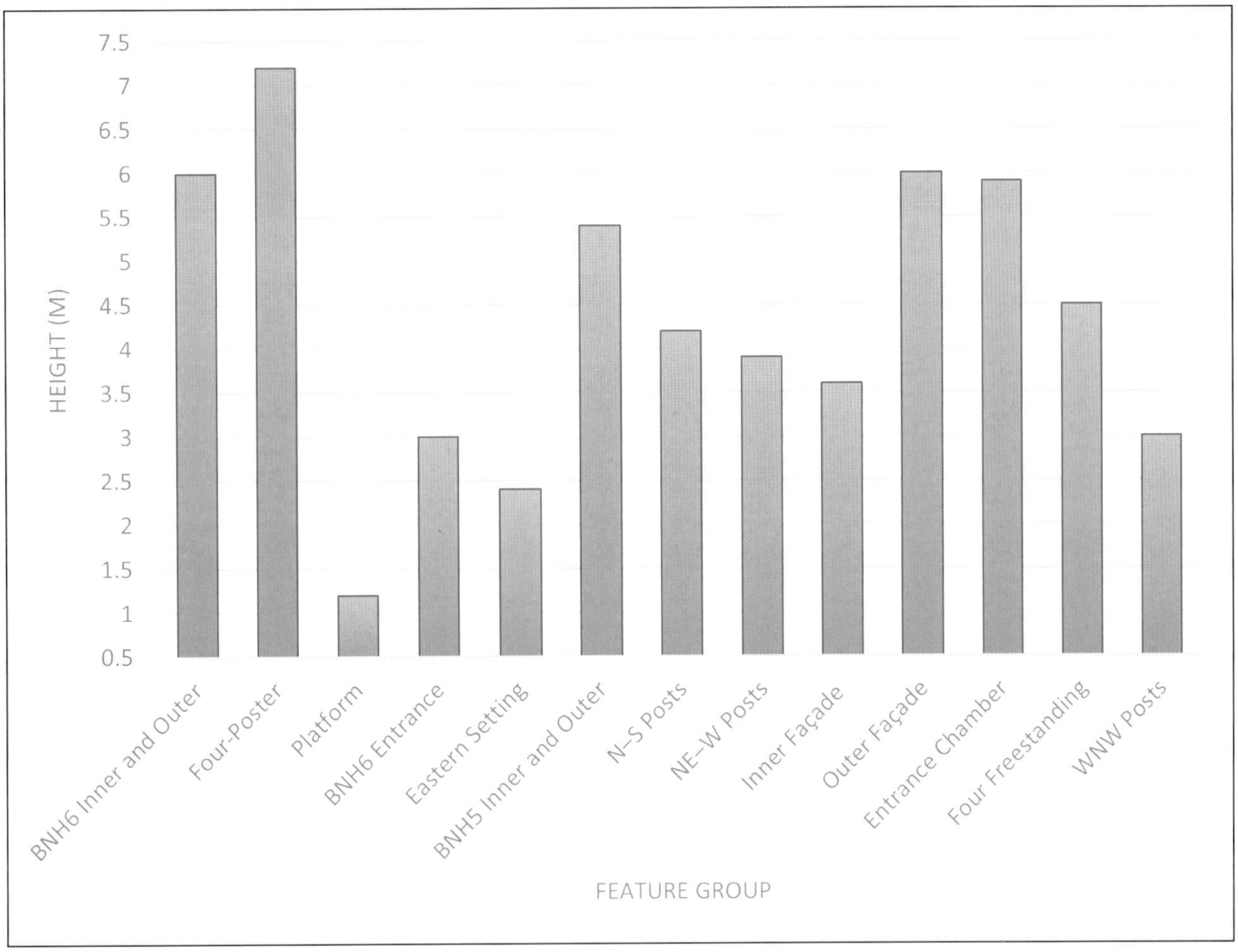

5. Ballynahatty 5 and 6: excavating the enclosures

Figure 5.6 Estimated height of posts above ground based on average excavated posthole depth and height of 3 × depth. No account has been made for differential loss of upper layers of features due to ploughing, therefore, these should be seen as minimum heights.

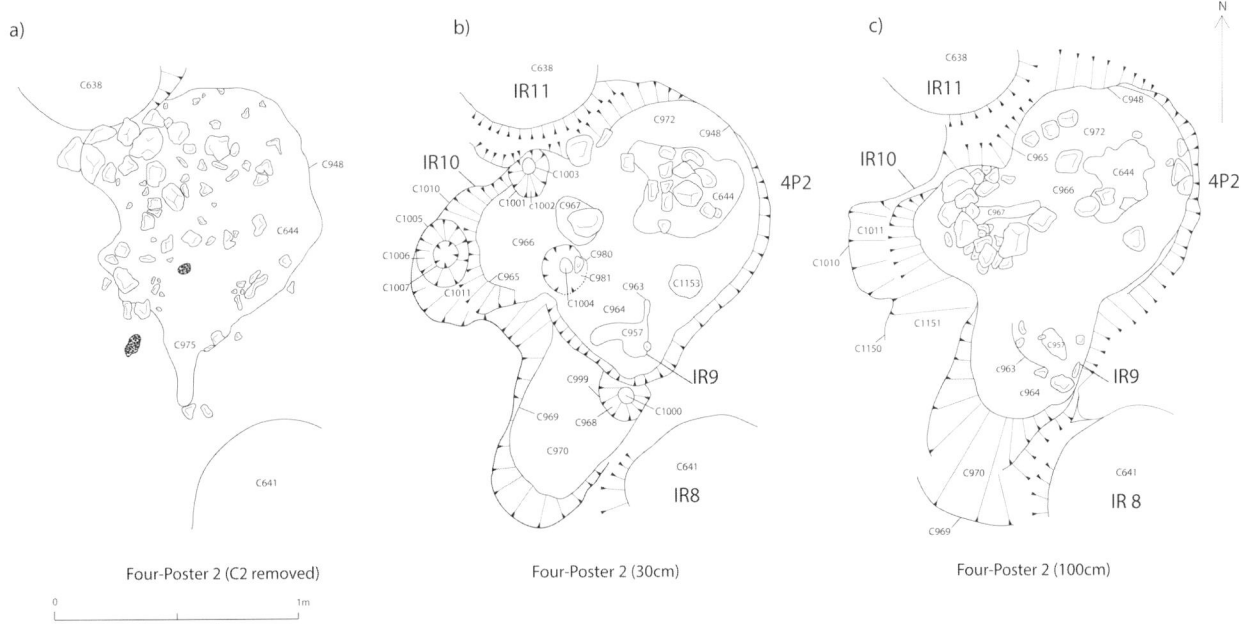

Figure 5.7 Excavation plans of 4P2 showing levels: a) below C2; b) at 0.30 m; c) at 1.00 m depth.

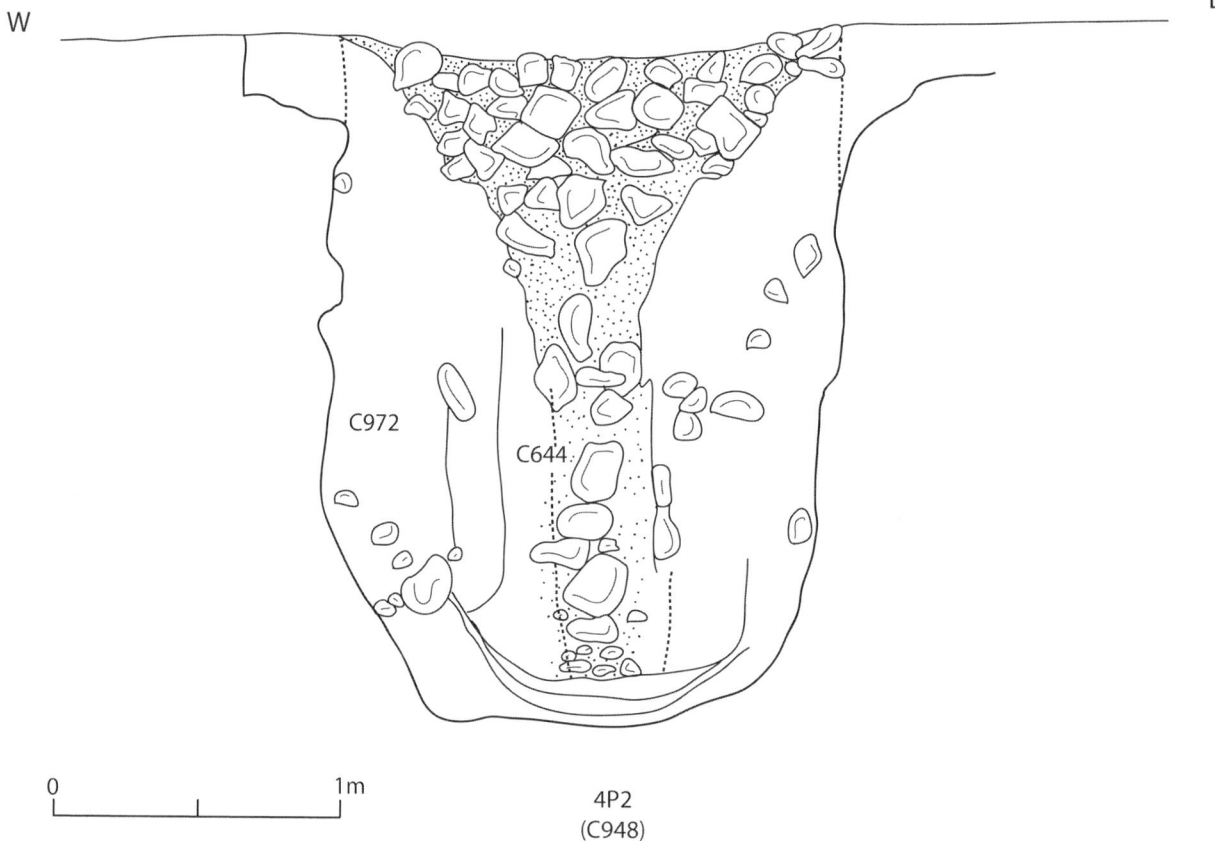

Figure 5.8 East–west section through 4P2.

this case, the secondary fill acts as a wick, with lenses of iron pan forming typically at the lower end of the postpipe. Sometimes this defines the edge of the postpipe but can also, confusingly, run into the primary fill.

Iron pan can be seen at the bottom of the section through 4P2 (Fig. 5.8), which clearly shows the funnel-shaped cut, with concentrations of stone and charcoal in the secondary fill, running down to the bottom of the postpipe. These stones do not represent the packing of the post but were the subsequent fill after its removal. The north–south section through 4P2 (Fig. 5.9) shows the relationship to the Inner Ring of BNH6 posts (IR10), which, although with less stone and more slumping, has a high charcoal component in the secondary fill to the bottom of the post pipe, showing that this post was also removed.

From the sharp edges and barriers at the intersections, it is clear that no two of these postholes were open at the same time. Figure 5.9 shows how IR10 and its secondary fill (C967) cut into the primary fill (C972) of 4P2 at the bottom of the hole. The pit wall of IR10 has been considerably flattened adjacent to 4P2 to avoid cutting too deeply into C972. IR10 was therefore erected after 4P2. That IR9 was also erected after 4P2 is not clear on the plan. The postholes barely intersect, and the line between the two fills could not be seen. Ramp C969 of IR9 appears to be cut into the fill of a larger ramp, C1150 of 4P2. In the upper layers, such intersections were particularly difficult to see, but this is reasonably probable. Postholes of the Inner Ring do not intersect one another. IR9 appears fully circular, and although IR10 is distorted, this is in response to 4P2, the width of c. 1.00 m is similar to the diameter of IR9. However, the thin wall between them collapsed during the original excavation leaving only a small section at the bottom. The top of this slopes down towards IR9, indicating that this was open when IR10 had already been refilled, although this is not as definite as the precedence of 4P2. IR10, therefore, precedes IR9.

5.4.2 Central structure

Fourteen smaller postholes were uncovered at the centre of the timber circle, arranged as a square measuring 2.00 × 2.00 m (Fig. 5.10). The postholes measured an average of 0.40 m wide and were uniformly 0.40 m deep. Rather than being 'U' shaped, these postholes were found to have tapered towards the base. There was some evidence that posts were removed from the postholes and the voids were replaced, in some cases, with a deposit of stone cobbles and soil. There was a noticeable lack of charcoal in the backfill of these postholes (Fig. 5.11). Nominally, the post heights would be 1.20 m, but allowing for depth of topsoil would bring the height to at least 2.00 m. Taking the number of postholes per side, the structure is five holes

5. Ballynahatty 5 and 6: excavating the enclosures

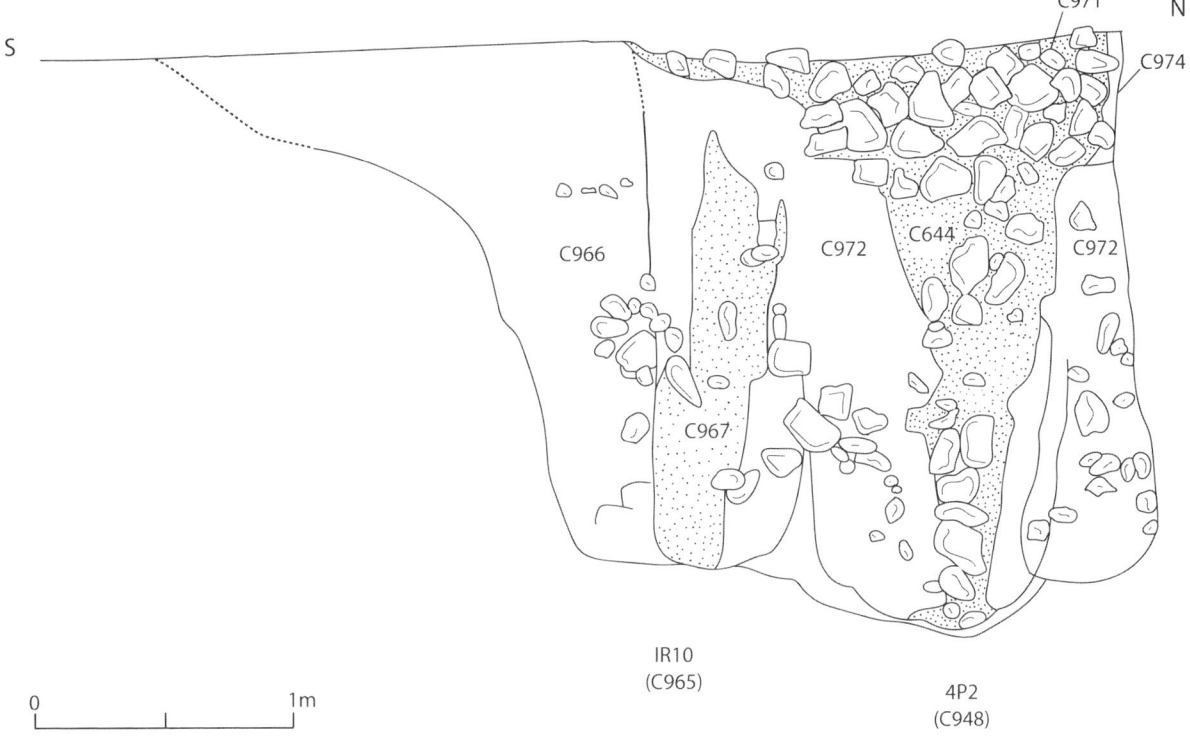

Figure 5.9 North–south section through IR10 and 4P2.

by four, although the dimensions are similar. The reasons for this are discussed in Chapter 11.

5.4.3 BNH6 Inner and Outer Ring

The remains of the inner and outer rings of the timber circle were revealed on the removal of C2. The postpits were initially identified as roughly circular patches of sandy, charcoal-flecked soil, surrounded by a ring of orange-brown gravelly sand, which, as redeposited natural, was often difficult to distinguish from the surrounding natural subsoil (Hartwell 1994, 11). The postholes were found to follow a standard pattern, averaging 1.20 m in width and 2.10 m deep in both the Outer (Fig. 5.12) and Inner Rings (Fig. 5.15). They were typically oval in plan and 'U' shaped in section, with steep, near vertical sides. A depression in the floor of the posthole marked the actual position of the post, probably caused by a combination of its weight and the disturbance caused by manually adjusting the position. Care was clearly being taken to align each post accurately and the width of the hole facilitated this fine adjustment. The holes were then filled with a 'primary fill', which was intended as packing around the post. This soil was composed of the sands and gravels of the subsoil and was, therefore, indistinguishable from it, with the exception that there were occasionally stratigraphic layers in the natural. In contrast, the fill was mixed – any layers here being erratic, diffused and confusing. On the surface, these features initially became visible as brown patches with a dark, charcoal-rich border caused by the overlying soil being drawn down into the centre of the posthole as the contents compacted (Plate 24).

If the post rotted in the hole, the postpipe was filled with a grey, looser soil; occasionally with voids, charcoal-flecked especially in the upper levels and with a charcoal-rich plug from the burnt post at the top. Over time, the bottom of the postpipe and the surrounding fill were leached to a green-yellow layer with lenses of iron pan and this was another clear indication of the position of the post in the hole.

Shallow cuts were revealed at the edges of most of the postholes and are interpreted as ramps created to facilitate the raising of the large posts. In some cases an indentation was evident on the lower wall of the posthole opposite the ramp where the post had slid down the ramp and slammed into the wall, gouging the natural as it was pulled upright. This was particularly evident at the south of the enclosure, where the subsoil was fine sand (Hartwell 1994, 13). Indeed, it was even possible to detect the striations made by the original picks as, here, the fill peeled easily away from the walls without the need to trowel the sides aggressively. OR7 also had a deep gouge on an adjacent side of the hole, which bears witness to the difficulty of controlling the erection of these heavy posts (Fig. 5.12).

These holes held posts measuring an estimated 0.50 m in diameter and 6.00 m in height. This measurement is based on the 1:3 ratio employed elsewhere (Gibson 2005, 135) and may be an under-estimation as a portion of the top of each posthole is missing. The postholes of

Figure 5.10 Excavated postholes of the central platform and Inner Ring of BNH6, entrance at the far side. The pit containing split stones is situated on the central axis at centre bottom. The stone 'cairn' at right of the picture is the unexcavated stone plug in the secondary fill of 4P2. The soil level has been lowered to define the edge of the posthole but this shows how the posthole would have been marked after the final removal of the posts.

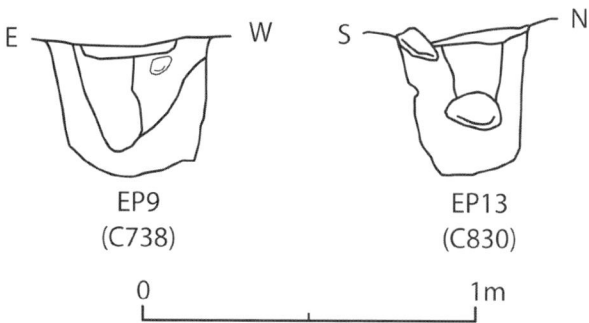

Figure 5.11 Postholes EP9 (C738) and EP13 (C830) of the Central Platform.

both the inner and outer rings were found to have been regularly spaced around their circumference, at intervals of 1.60 m (Hartwell 1994, 13). There is some variation in depth, which may be accounted for as an adjustment to compensate for any variation in the length of the posts, thereby maintaining a constant height.

In many postholes there was evidence of a 'secondary fill', similar to that of the Four-Posters, which proved to be more complex than simply the remains of a post rotting *in situ*, as was initially thought (*ibid.*, 11). The dark, sandy, funnel-shaped fill was rich in charcoal with concentrations of cobbles and boulders, sometimes in layers. This seems to result from the post above ground having

Plate 1 Cropmarks photographed in 1989 reveal BNH5 with BNH6 visible within it on the northern slope of the ridge overlooking the Giant's Ring (QAD/17/7/89/I/11) (photo: Barrie Hartwell).

Plate 2 Multiperiod cropmarks in Ballycarn townland (QAD/ 17-7-89/ VI/12) taken in 1989 (photo: Barrie Hartwell).

Plate 3 Southern Ballynahatty and the Giant's Ring at the end of the 1996 season of excavation. BNH6 is on the right of the excavated surface, with part of the Annexe uncovered at the left. The undulating field wall gives a good impression of the ridge on which it sits (photo: Barrie Hartwell).

Plate 4 Orthorectified image of Ballynahatty td (Image © David Craig/HeritageNI.com).

Plate 5 View of Ballynahatty townland taken by a drone looking towards the northwest and demonstrating the setting of the Giant's Ring on the southeast corner of a plateau above the Lagan Valley, where it abuts the top edge of a steep, tree-covered slope. Beyond the Ring, south Belfast and the Antrim Hills are visible, where a number of flint quarries are recorded on the slopes. Divis Mountain is central, with Cave Hill to the right (DJI_0064) (image © David Craig/HeritageNI.com).

Plate 6 View to the west of the Ring showing the position of the pond at the east end of the ridge (DJI_0049) (image © David Craig/HeritageNI.com)

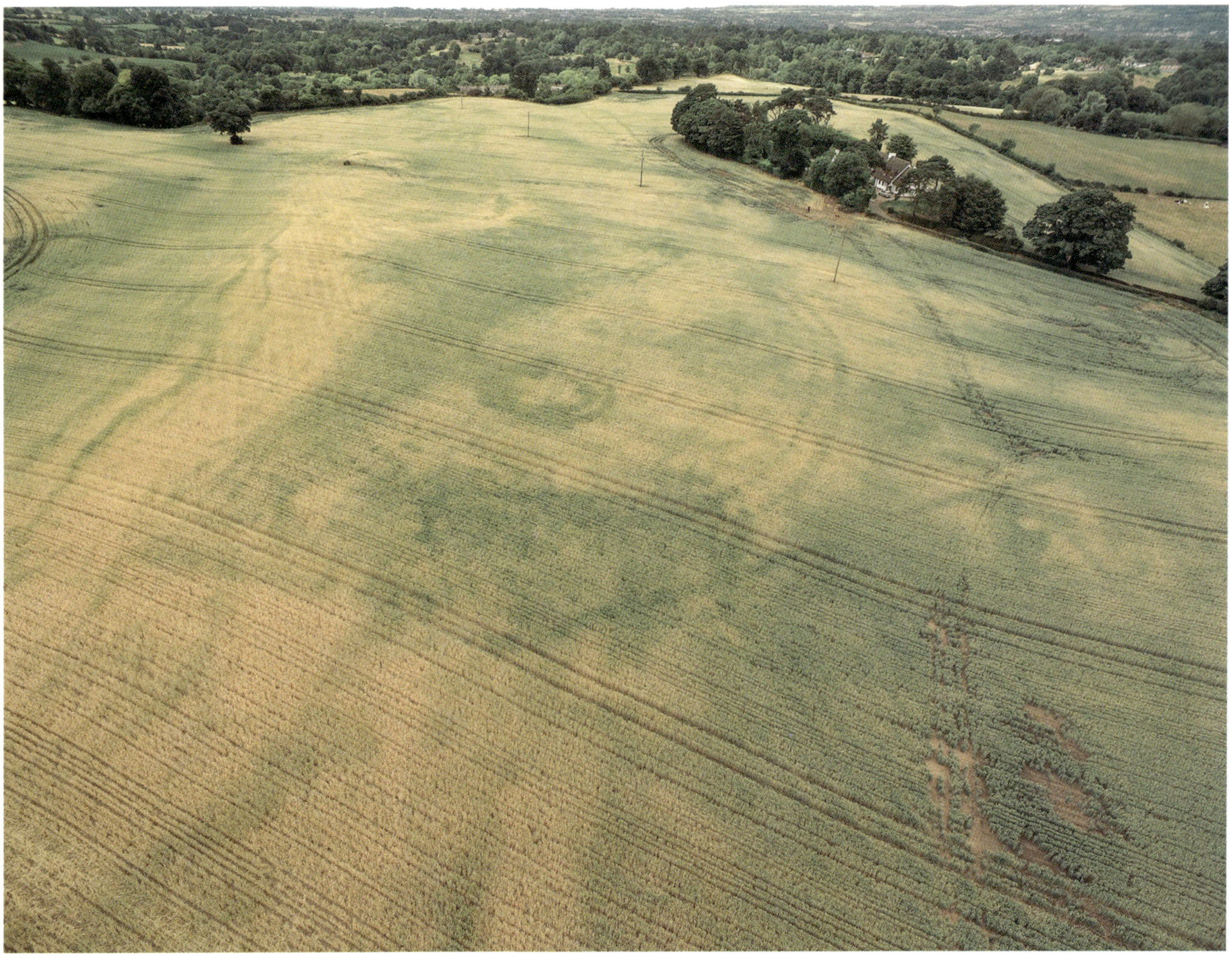

Plate 7 The BNH5 enclosure visible in 2018 (excavated in the 1990s) with BNH6 at the near end. The area of trees marks the position of the Bodel's farmhouse (DJI_0045) (image © David Craig/HeritageNI.com).

Plate 8 The BNH5/6 enclosures in 2018 from the west (DJI_0907) (image © David Craig/HeritageNI.com).

Plate 9 The area of Bodel's fields in 2018 where so many sites had been destroyed by 1855. Two parallel ditches (BNH16) of a possible cursus stop short of the plateau edge. Beyond, at the bottom of the slope, the mill village of Edenderry and the River Lagan is visible (DJI_0281) (image © David Craig/HeritageNI.com).

Plate 10 View northeast across Ballynahatty Bog (BNH4) in 2018 towards Belfast Lough and the Irish Sea (DJI_0284) (image © David Craig | HeritageNI.com).

Plate 11 The Giant's Ring in 2018, looking southeast towards the Co. Down hills. The standing stone is in the right foreground at the end of the ridge. The southern edge of BNH5 can just be seen at the left of the picture (Image © David Craig/HeritageNI.com).

Plate 12 Ring ditches BNH30, 31, 10, 11, 12 (image © David Craig/HeritageNI.com).

Plate 13 Grooved stone balls, possibly recovered from the Giant's Ring.

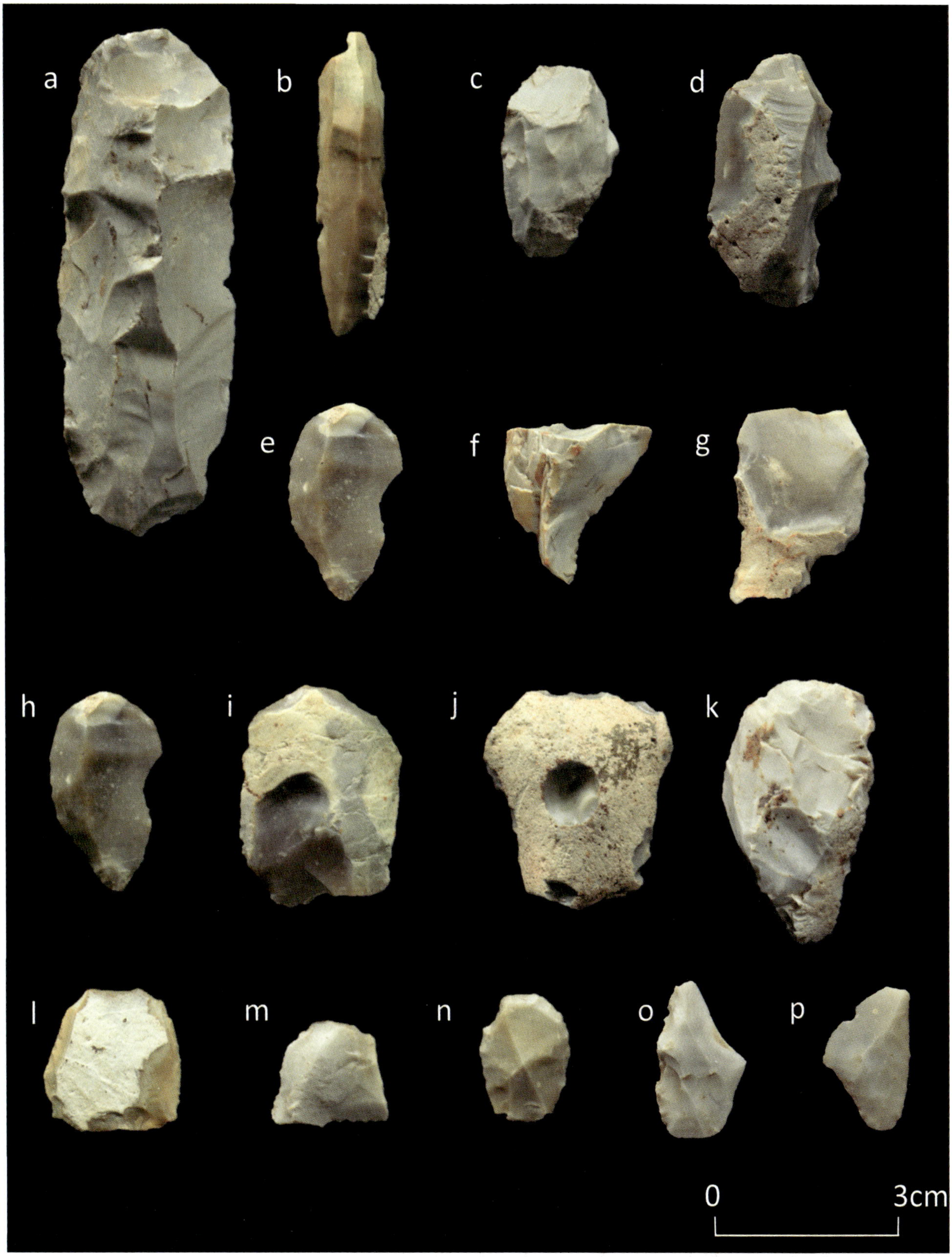

Plate 14 Selection of worked tools found by Gail Ritchie in Edenderry and Ballynahatty areas.

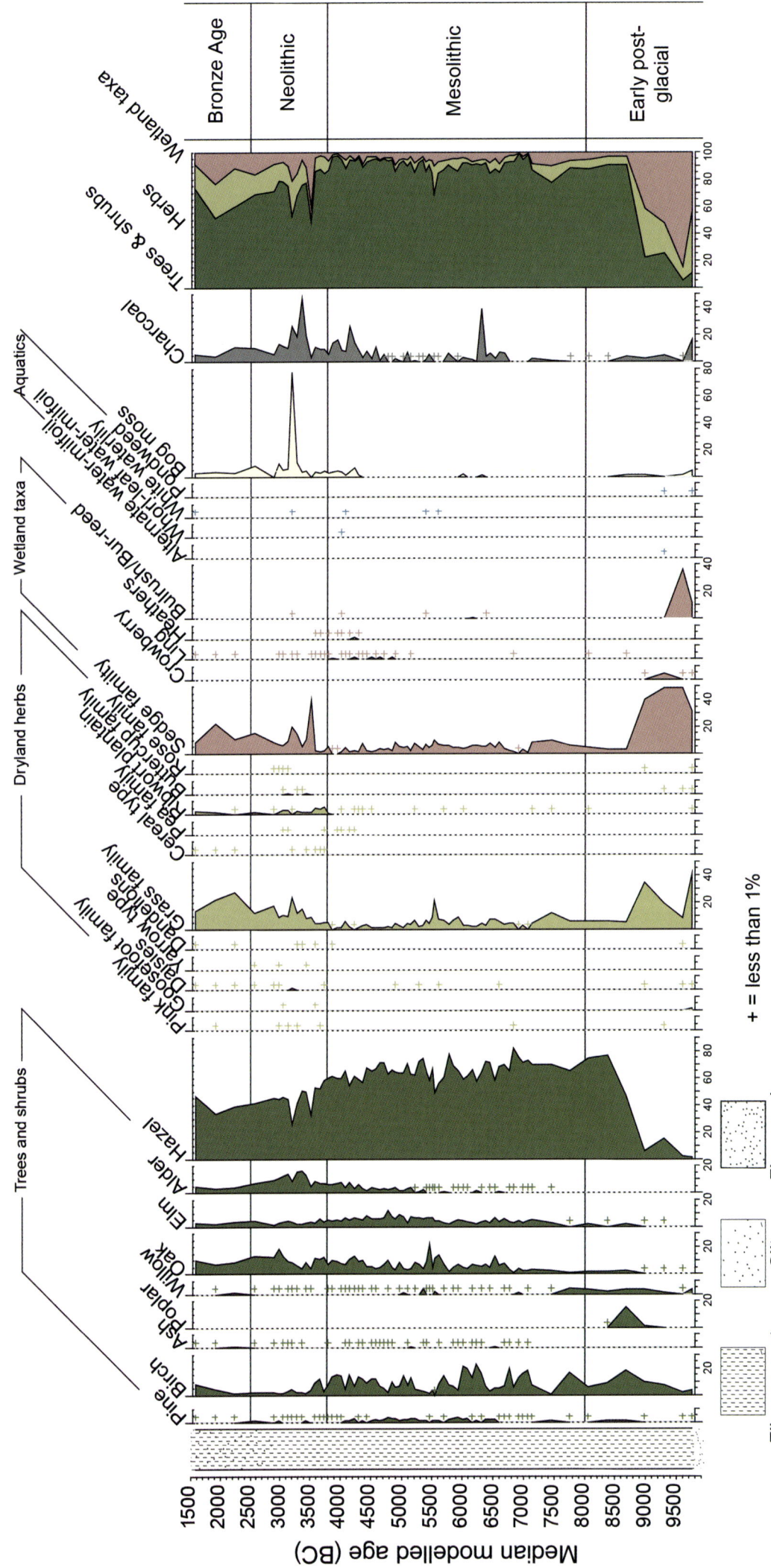

Plate 15 Selected pollen percentage data from Plunkett et al. (2008) replotted against the revised chronology presented in this volume. The main stratigraphic features of the core are also shown.

Plate 16 Early modern field ditch.

Plate 18 Chamber C591 during excavation with the entrance in the right foreground and one remaining Coarse Ware pot in situ.

Plate 17 Chamber C591, floor.

Plate 19 Chamber C591, pit with stones removed.

Plate 20 Cremation deposits C1014, C1016 and C1018.

Plate 22 C588 after removal of cremation showing the carefully arranged split stone setting.

Plate 21 Cremation deposit, C588.

Plate 23 Pit containing split stone and pebbles.

Plate 24 Surface identification of postholes.

Plate 26 OR25 during excavation, showing the stony, compacted natural on either side of the trench. Beyond it is an attempted section through the boulder-filled upper secondary deposit of OR24.

Plate 25 Stones supporting the fill of OR27.

Plate 27 Postholes on the north side of the entrance to BNH6. In the centre, at a higher level, is the base of an intermediate posthole.

Plate 28 Slot between OR1 and OR2, revealed after removal of C835.

Plate 29 Charred plank from the secondary fill of OR32–33.

Plate 30 Cobbled floor of northern Annexe.

Plate 31 Grooved Ware pot BNH4 from northern Annexe.

Plate 32 NEW postholes between the Inner Façade (foreground) and BNH5 OR/IR at rear. NEW4–6 are emerging under the baulk on the right.

Plate 33 NEW postholes during excavation showing the stone-filled secondary deposit.

Plate 34 Section through a NEW posthole showing the stone column created in the secondary fill.

Plate 35 The Annexe Façade from the entrance looking north. The stone filled spine marks the Outer Façade behind which is the excavated line of the Inner Façade. IF8 at the near end has broken through the Coarse Ware chamber. At the far end the Inner Façade articulates with the NEW posts.

Plate 36 Looking south along the gently curving Façade to the Giant's Ring. The two ranging rods at the far end are on the line of the Middle Façade, which subtly changes direction to point directly at the passage tomb, just visible to the left of a tree. The posts and tape to the right of the excavation represent the line and entrance to the BNH5 enclosure.

Plate 37 The southern end of the Façade, facing the Giant's Ring was cleared until the postholes were defined. The more extensive, darker area is the end of the Outer Façade. The bulge at the end, and the three conjoined postholes beyond it, is the line of the Middle Façade (MF6–9). The postholes on the right are the southern end of the Inner Facade (IF17 and 18), the start of the WNW (WNW9 and 17) and the SEW postholes (SEW7 and 12).

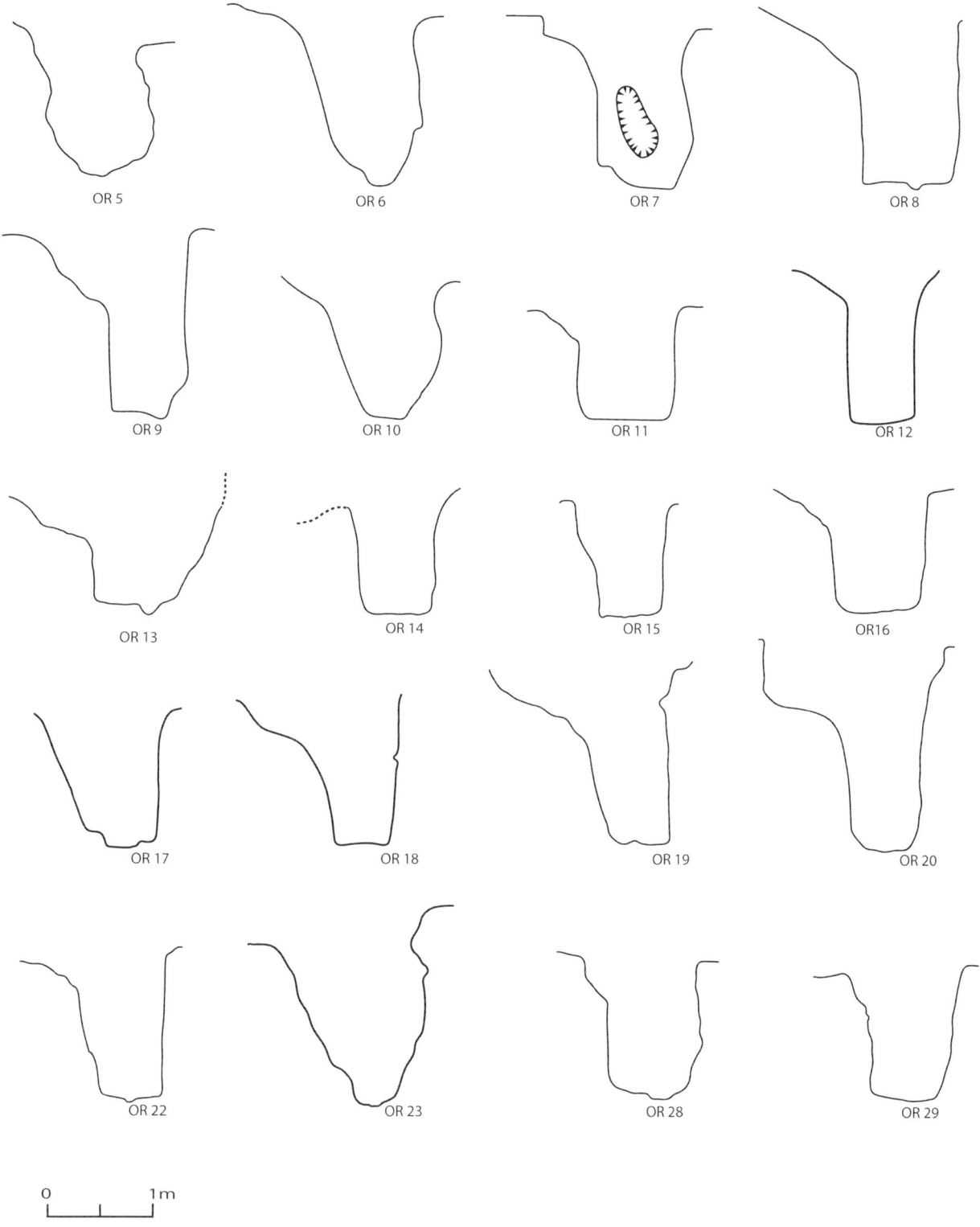

Figure 5.12 BNH6 Outer Ring: radial posthole profiles, right side towards the centre.

been burnt and the earth-fast stump removed from the posthole. The abundant charcoal, cobbles and boulders were then poured into the void until filled. Experimental work carried out during the excavation has shown that the removal of the stump can be accomplished with relative ease (see Section 11.7).

5.4.3.1 Outer Ring (OR1–33)

OR5–29 were laid out as individual postholes in a 15.20 m circle with an entrance to the southeast (Fig. 5.5). The subsoil on the Ballynahatty plateau is very variable, as can be expected of glacially derived material sorted and cut by runoff in post-glacial conditions. Nowhere is this more apparent than in the northeast sector of the Outer Ring (OR) of BNH6, where the line of postholes coincided with a particularly thick and indurate area of compacted cobbles, grit and sand (Figs 5.13 and 5.14). This proved resistant even to a modern metal pick and certainly caused the Neolithic workers to modify their strategy of digging individual postholes. Having smashed through this layer digging out OR24, it proved easier to work horizontally along the circumference for the following three postholes (OR25–27) by levering off chunks of concretion, thereby excavating a trench to a depth of about a metre. Individual holes were then dug into the more friable sand and clay mix beneath. On the one hand, this would have facilitated raising the posts, as the trench would have acted as a ramp, but against this, the absence of much of the fourth side of the hole would have presented a problem of stability until all the posts were in place. This was partly solved by closing each hole with a wall of cobbles to retain the post packing before the next post was raised. This is most clearly seen between OR26 and 27 (Fig. 5.14, I–J, Plates 25 and 26), where a pile of loosely packed stones continued down beyond the ledge. Chunks of the compacted layer were found throughout the fill. This indurate layer also inhibited the digging of posthole ramps on the eastern side of BNH6.

In common with other postholes in BNH6, the post positions within their holes could usually be seen after the removal of the ubiquitous sub-plough zone, C2, by a concentration of charcoal and stones. However, definition of the actual posthole cuts did not become apparent until an overall depth of c. 0.40 m had been reached. The position of the postpipe could be traced down the profile as a darker and softer soil, flecked with charcoal which lessened with depth. This contrasted with the harder, orange primary fill, which had been rammed in to support the posts. The difficulties presented by the archaeological excavation of deep postholes have already been discussed, but the presence of this trench did allow better accessibility and gave the opportunity to record full sections. This went some way to clarifying the nature of

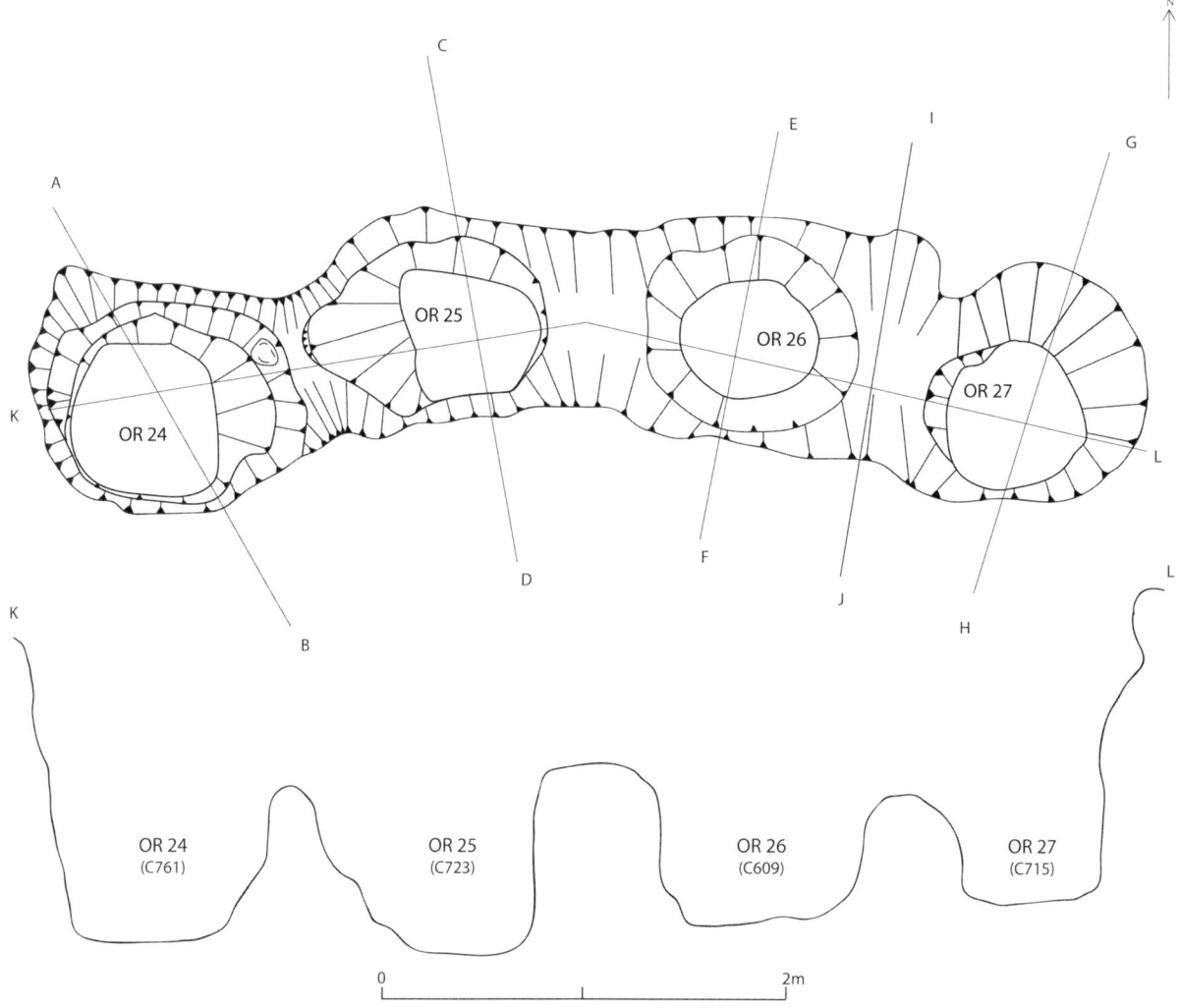

Figure 5.13 OR24–27: plan and longitudinal section.

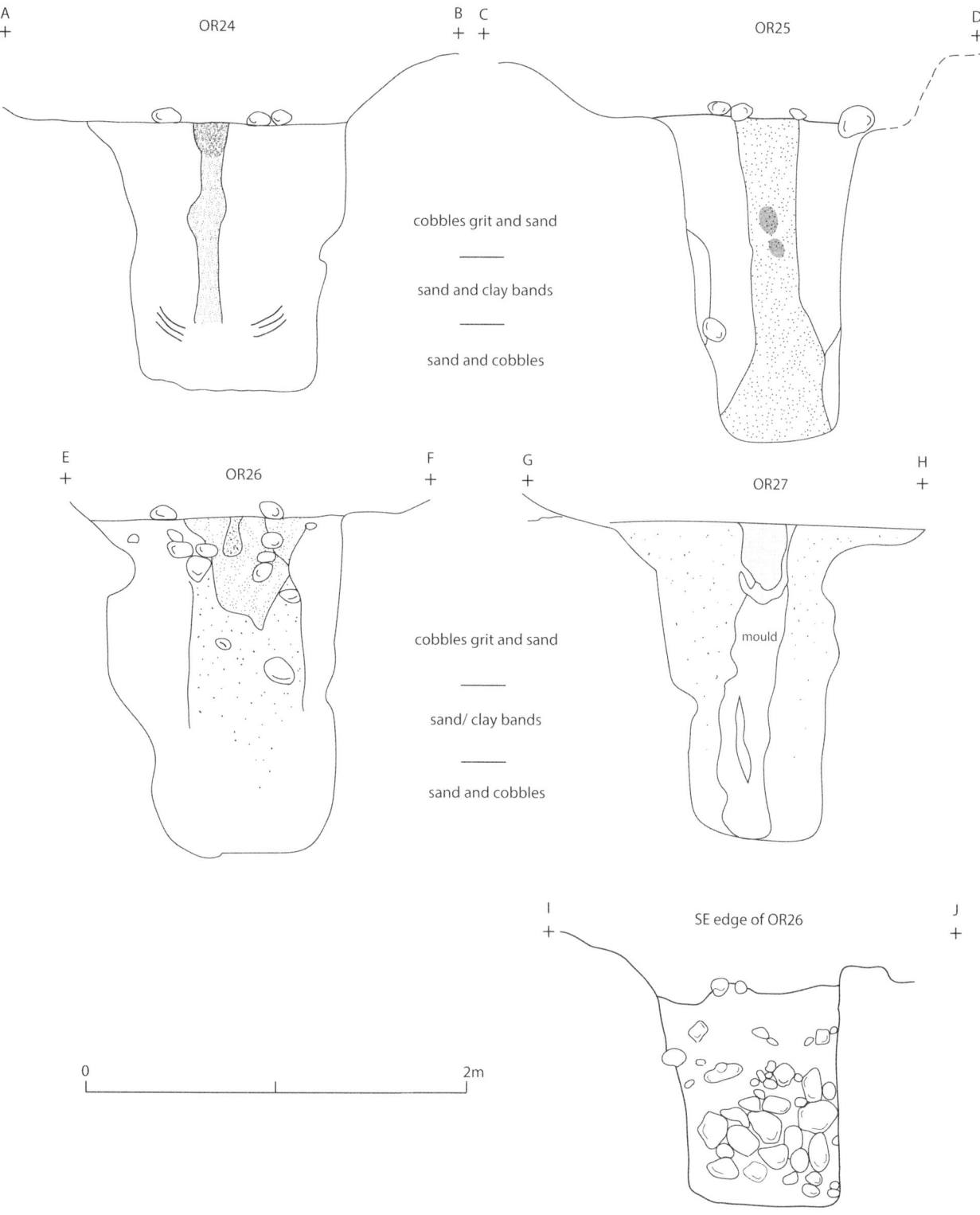

Figure 5.14 OR24–27: cross-sections.

these postholes. OR27 showed a post mould of softer, darker soil with an elongated vertical cavity indicating extensive settlement, which had pulled down the original burnt charcoal cap of the post into a cone infilled with a grey-tinged soil. The earth-fast sections of this outer post had rotted *in situ*, with the aerial parts being burnt to ground level. OR26 had a secondary fill incorporating stones, and a wider, disturbed postpipe, with the two fills merging towards the bottom. This strongly suggests that the post butt was removed. OR24 had the characteristics of OR27 but had charcoal running down the post pipe, suggesting that this was backfilled after the post had been removed. OR25 was ambiguous. The implication is that, whereas the intention may have been to remove all the earth-fast timbers, in practice, some may have been too rotten and possibly started to disintegrate in the process and so were left *in situ*.

5.4.3.2 Inner Ring (IR1–25)

The 25 postholes of the Inner Ring (Fig. 5.15), describe a circle of 10.8 m, are largely uniform in character and similar to those of the Outer Ring. They incorporate ramps which radiate outwards, although there are some exceptions. The postholes IR4, 10, 16 and 22 are noticeably slighter, probably because they are hard against the Four-Poster setting, and a more substantial cut would have affected their stability. As already discussed, the ramps of IR19 and 20 diverge to avoid cremated remains, and at the entrance, OR1–2 and OR24–25 are paired postholes reflecting the structures in the Outer Ring, except that there is no flattening of the circle. Although the distance between the Inner and Outer Ring is a consistent 2.20 m, it narrows to 1.60 m at the entrance. Burnt pig and cattle bones were found in four different secondary contexts in the IR postholes and burnt human bone in one primary context.

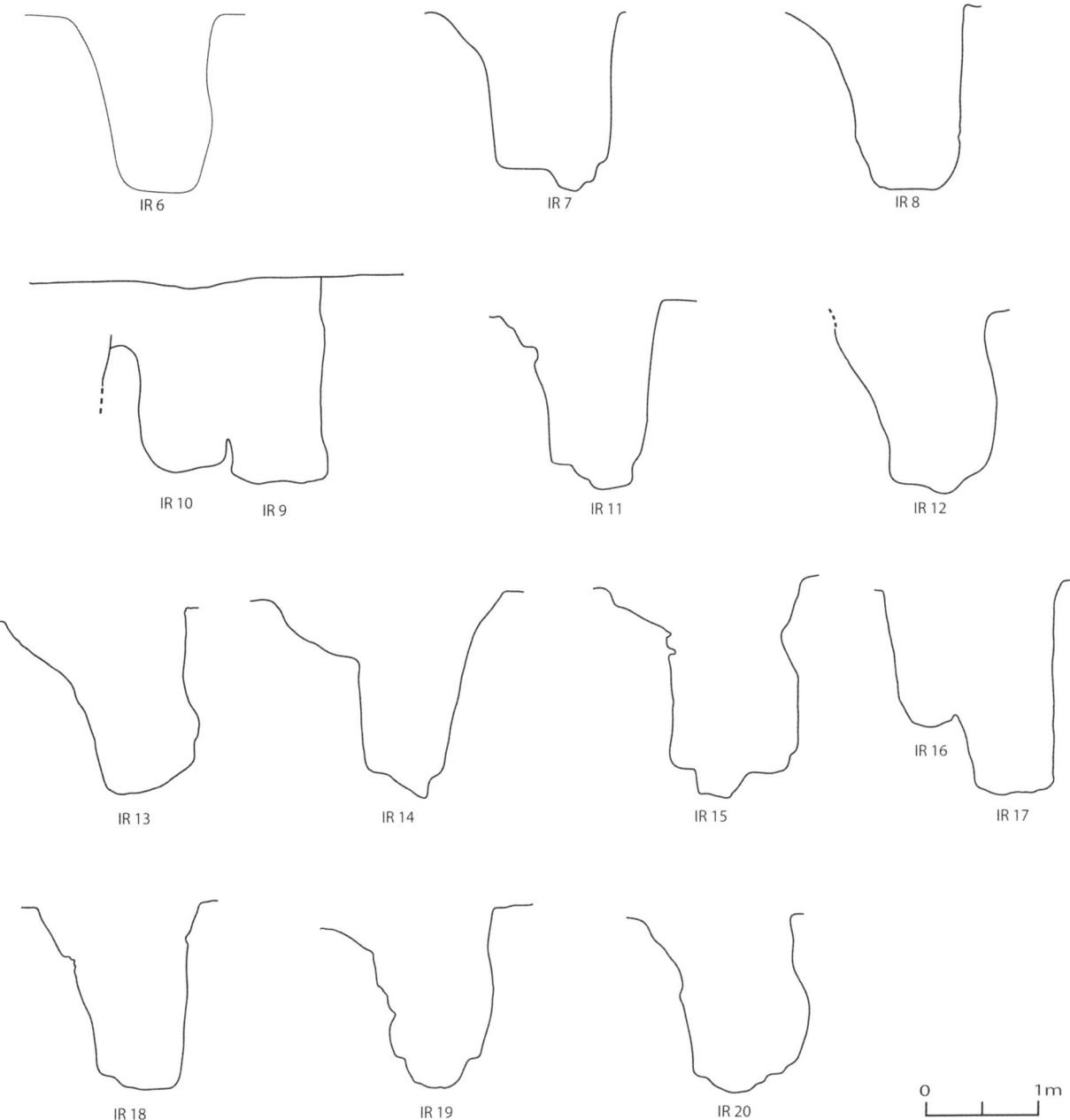

Figure 5.15 BNH6 Inner Ring: radial posthole profiles. Right side towards the centre.

5.4.4 The entrance to BNH6

The Inner and Outer Rings of postholes from BNH6 come together at the well-defined entrance. The posthole organisation on either side of the entrance was similar, so only the south terminal of the post rings needs to be described here as the subsoil produced a clearer footprint (Figs 5.16–5.18). The pattern of two concentric rings of individual postholes changed on either side of the entrance where OR1–2 and OR3–4 were dug in paired postholes. This was reflected in OR30–31 and OR32–33 on the north side of the entrance (Plate 27). The orientation of these two groups also differed as the outer ring was flattened to produce a distinct entrance façade. There were also slots connecting OR1–4 and OR30–33, and part of a charred plank was retrieved from the secondary fill of OR32–33 (Plates 28 and 29).

The slot between OR1 and OR2 was particularly clear and contained C835, a soft, black charcoal-rich soil with large chunks of charcoal and burnt flint. This is the only direct evidence of infill between the posts and relates to the entrance façade alone, which has been selected for special emphasis.

The initial interpretation of these two areas was difficult because of the conjoined nature of the pits, the mix of fills and disruption to the upper levels. After several

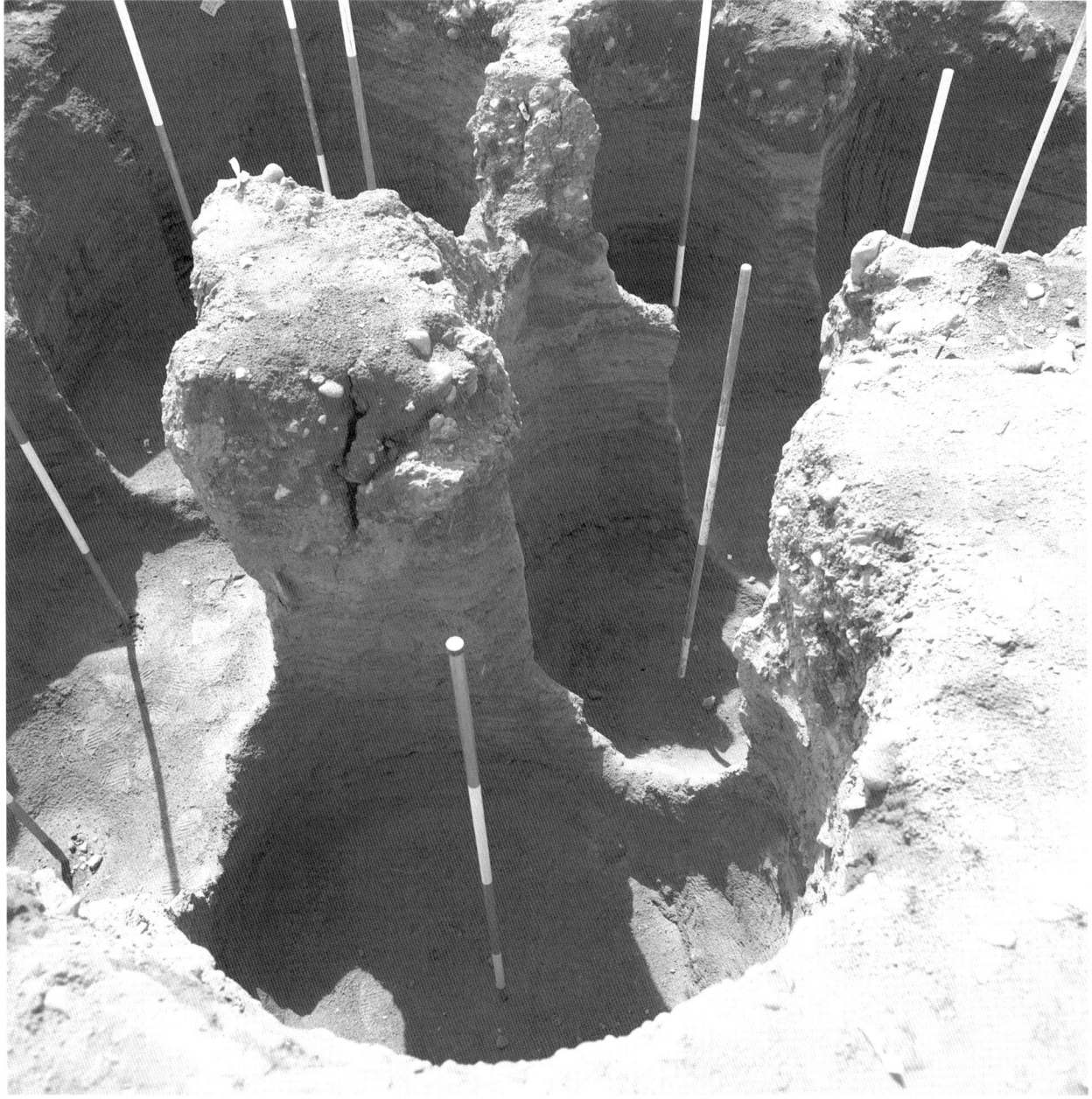

Figure 5.16 BNH6 south terminal postholes from north. Each 2 m ranging rod represents a 0.30–0.40 m diameter post and emphasises the technical achievement of digging the holes and raising a series of 8.00 m posts in such a confined area.

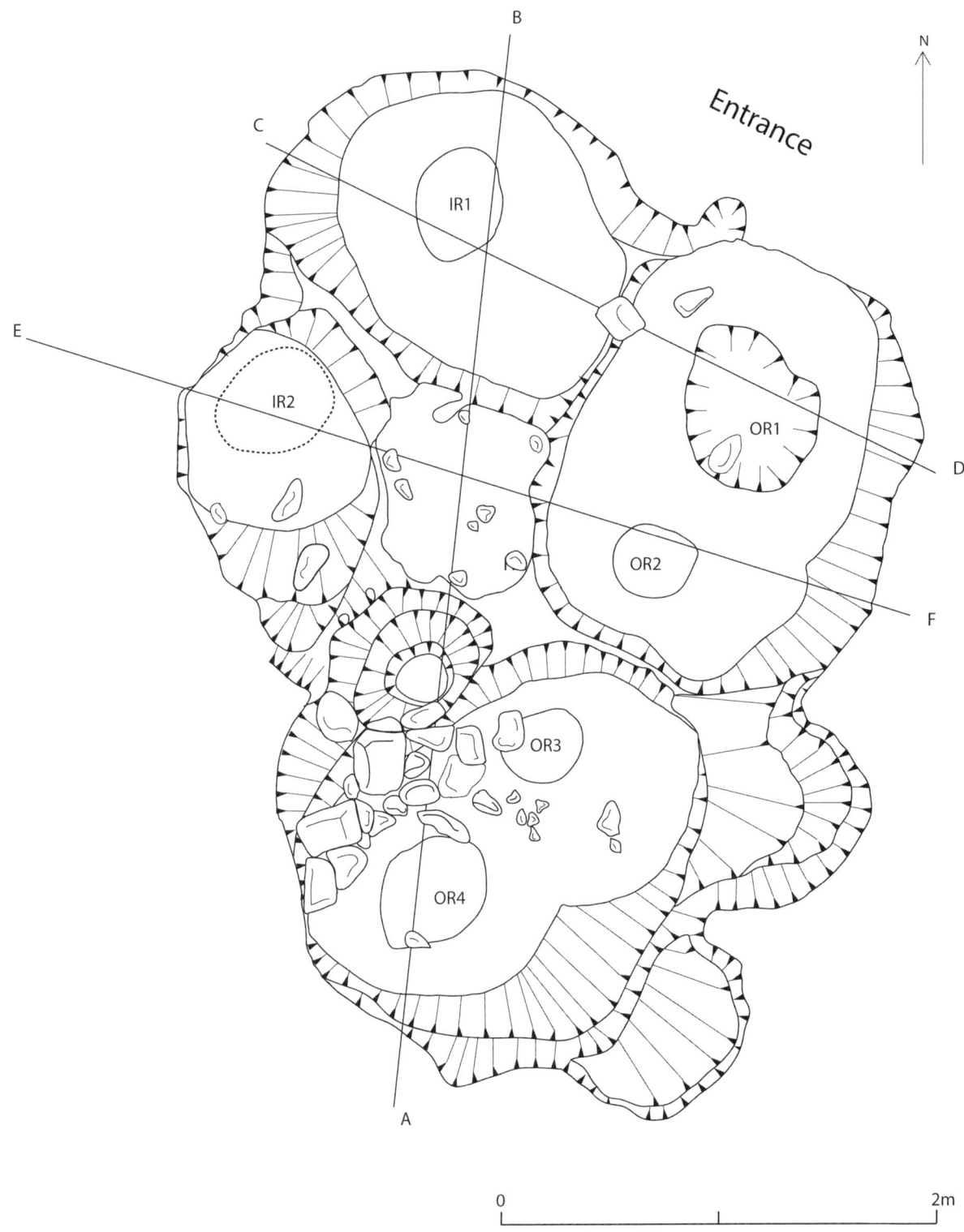

Figure 5.17 BNH6 entrance: south terminal (sections in Fig. 5.18).

attempts, the edge of the combined pits was established, defined by the indurate upper levels of the natural subsoil. There was mixed evidence for the removal of posts here because of considerable slumping and ill-defined interfaces. If the posts had rotted *in situ* OR 1 and 2, for example, would have lost nearly a third of the bulk of the combined posthole. This would account for the surface charcoal deposit being pulled down into the post mould (Plate 28) and the survival of the slots. There was also evidence in some cases for considerable mixing of the upper layers, which could be due to post removal or, at least, attempted post removal.

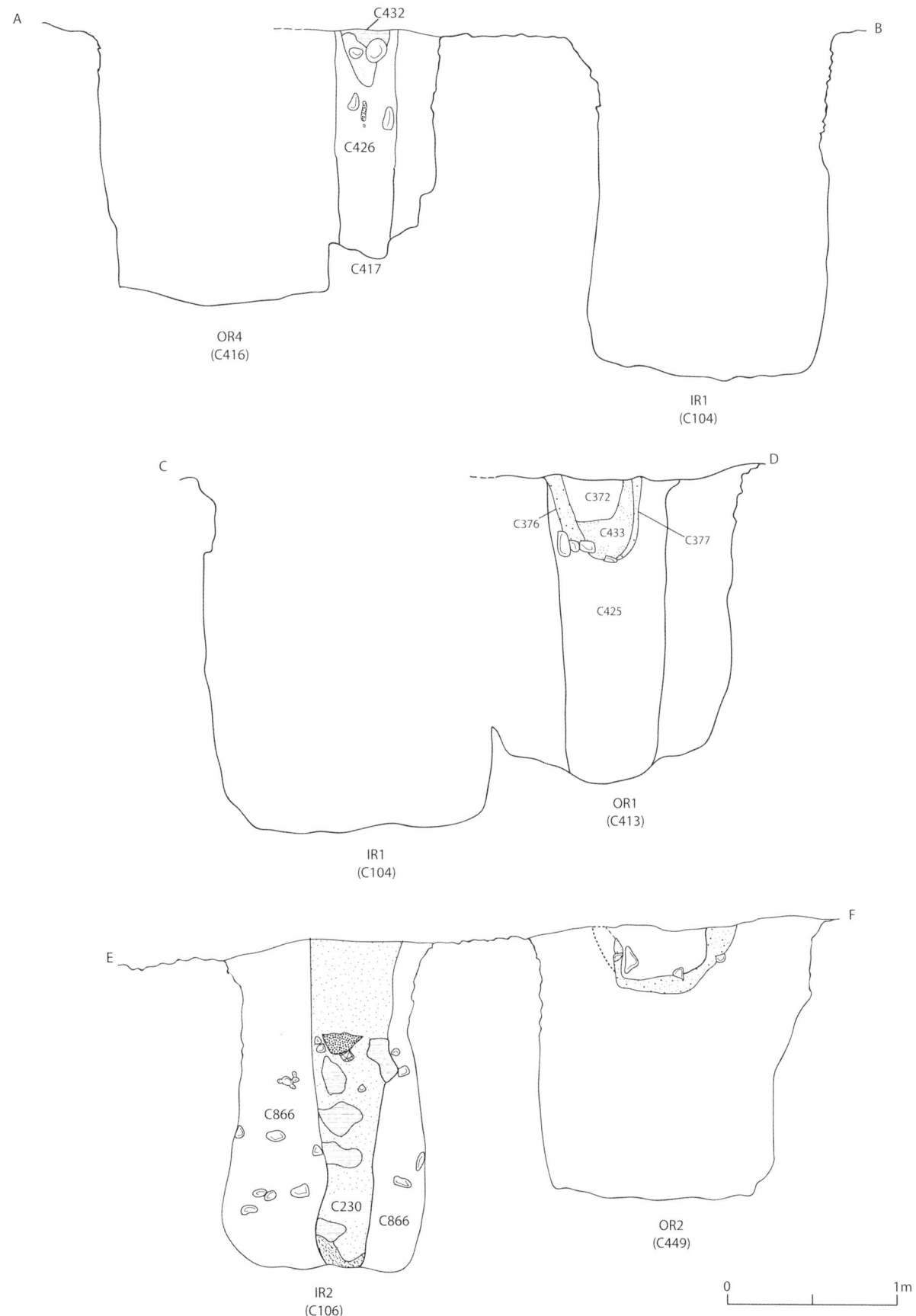

Figure 5.18 BNH6 entrance profiles: OR1, 2, 4 and IR1, 2 (see plan in Fig. 5.17).

5.4.4.1 Intermediate postholes

The remains of six posts (Fig 5.19) were uncovered between the Inner and Outer Ring of posts, three at either side of the entrance and running diagonally between, and cut by, IR4–OR3 and IR22–OR31 and are, therefore, earlier. Although these posts were shallower and tended to taper inwards towards the bottom, the fills followed the same pattern as those of the inner and outer circles. Indeed, the fills of IR3 and the intermediate post were so similar that their relationship was ambiguous. It seems unlikely that these posts would have been erected before the placement of the inner and outer rings of posts if they were part of the same building phase. It may be that these features represent the developed entrance of an earlier timber structure based on the same plan as the Inner Ring, the evidence of which has largely been obliterated by the later, deeper postholes (Hartwell 2002, 526). Just 200 m to the west, the 2018 drone survey discovered a precedence for this. A circular, pit-defined enclosure, approximately 20.00 m in diameter, appeared as a cropmark (see Fig. 2.7, 17). Although larger than BNH6, the single ring of about 40 postholes had a similar southeasterly entrance emphasised by conjoined postholes to form a short façade.

5.4.5 Eastern Setting (ES)

Eight postholes were uncovered outside the entrance to BNH6 (Figs 5.20 and 5.22) and just north of the axis between the entrances of BNH6 and BNH5 (ES1–8). As excavated, the holes had a 'U'-shaped profile measuring an average of 0.90 m deep and 0.50 m wide and may

Figure 5.19 Posthole IR3 at centre, which cuts two of the intermediate postholes which are not as deep.

Figure 5.20 The Eastern Setting, adjacent to the BNH5 enclosure. BNH6 is to the left (west). A modern farm water pipe crosses the structure.

have held posts standing up to 2.70 m above ground. In retrospect, these postholes were incompletely excavated, and although taken to their full depth, only the secondary fill of the postpipe was removed, leaving the primary fill *in situ*. This orange-brown sandy fill contained charcoal flecks, flint and occasional stones of varying sizes from gravel to cobbles. Most of the postholes were topped by a grey 'plug' 0.20–0.50 m deep, and in the case of ES5 and 8, these each contained charcoal, flint and a scraper. ES2 and ES7 had a horizontal band of charcoal at 0.60–0.70 m depth, and ES7 had a 0.10 m deep depression in the base containing concentrated charcoal, which could represent deliberate charring of the original post butt.

The four postholes defining the rear of the structure (ES3–6) formed an arc 3.00 m long. Utilising the two end postholes, the east and west sides were similarly 3.00 m long and consisted of three postholes terminating in ES1 and ES8. The sides converged towards the

Figure 5.21 The IR and OR postholes of the outer enclosure (BNH5) looking south. The Eastern Setting is on the right and the entrance (before excavation) is under the standing figure.

south as though set out as a segment of a circle. There were no intermediate postholes on the fourth side, so the structure was open-ended. Whether or not this structure was roofed is open to question. The restricted entrance would allow the complete wider rear of the interior to be seen from a very limited position. If this managed view was important, the structure is more likely to have been roofed. Bridging the 3 m gap between the walls with a series of lintels to produce a flat roof would have presented little difficulty and this, in turn, could have supported a pitched roof and more weatherproof result. An area of disturbance which mirrors the 'Eastern Setting' has been located by geophysical survey on the south side of the axis projected from the centre of BNH6 through the entrance. The position and orientation of these structures show that they were an integral part of the BNH6 approach.

5.5 Outer Enclosure (BNH5): feature analysis

5.5.1 Overall shape and extent

Ballynahatty 5 (BNH5) was first recognised as a cropmark in 1989. The feature appeared to be a large enclosure of paired pits, roughly oval and measuring 73.00 × 101.00 m in size on the top and northern slopes of the east–west ridge. Although partly obscured on its southern side, the regular oval shape is distorted in two areas where the circuit can be clearly seen. There is a distinct flattening in the section running south to the main eastern entrance and in the centre of the northern circuit there is a slight angularity and disruption in the smooth line of the outer row of postholes which could be a side entrance on the downslope and in the approximate direction of Ballynahatty Bog. In the early seasons, the excavation investigated a short portion of the enclosure's eastern end (Fig. 5.21) before it was realised that there was a substantial entrance structure, the 'Annexe', attached to the east side. The importance of the entrance to BNH5 then became apparent as it was laid out on the southeast axial line of the entrance to BNH6, yet it had to link with the Entrance Chamber in the Annexe, which had been reorientated to the east–west line of the ridge (Fig 5.24). All the major elements came together at this point. However, disruption in antiquity and modern cultivation made this a difficult area to excavate and interpret. First attempts were made in 1997, then again in the final 1999 season, but we were forced to return in February 2000. Despite this, lower levels of the north side of the entrance remain unexcavated, but a clearer interpretation is now possible. A short stretch of the enclosure was also investigated south of the entrance. Together this only represents a 31.00 m section of the complete circuit of 266 m, 23 postholes of over 280: just 8% (Fig. 5.22).

Figure 5.23 shows longitudinal sections through the inner and outer circuit of postholes running downslope from the entrance towards the north. The postholes of the BNH5 enclosure are an average of 1.20 m wide and 1.80 m deep and could have held posts an estimated 0.40 m in diameter, standing 5.40 m above ground (Fig. 5.6). There is some variation in the depth of the holes, which may simply reflect the variability in the length of timber posts used and a desire to standardise their height, so the tops appeared uniform. If so, the posts must have been on-site before the postholes were dug. Where accessibility allowed, the postholes were half-sectioned before being fully emptied. Like those of BNH6, the enclosure postholes were comprised of a primary fill packed around the post to hold it in place in the pit. As elsewhere on the site, the postpipe could take one of two forms. If the post rotted *in situ*, there would be a dark charcoal-rich plug, possibly with stones, below which was a column of friable brown soil with charcoal flecks, especially in the upper levels. Lower in the section, there were occasional voids formed as the post rotted and the pipe sides slumped to narrow the profile. If the post was removed, there was usually much more disturbance to the top of the section, sometimes evidence of a bell-shaped pit, a secondary fill of charcoal-rich subsoil in the postpipe and concentrations of stone in the upper levels, sometimes running down the postpipe. There may also be some voids formed as the contents settled (Fig. 5.25, G–H, K–L). Ramp cuts are not as obvious as those recovered during the excavation of BNH6, but IR2 (C441), IR7 (C664), IR10 (C279), OR1 (C213) and OR6 (C221) show that the posts were erected from the outside. Because of difficulties in locating the edges of the postholes in this area, the natural was lowered by 0.30+ m. When assessing the original depth of these postholes, approximately 0.40 m (including the turf layer) should be added. For examples of this, see the longitudinal profiles (Fig. 5.23) and IR14 (Fig. 5.25).

The upper levels of posthole OR11 (Figs 5.24 and 5.25, G–H) initially showed as a charcoal-rich layer with stones on the north side and a much heavier charcoal concentration to the west and narrowing down to C1428, where it could be recorded in the section. The section placement was off-centre and recorded at the top edge of the funnel containing charcoal-rich secondary fill and some settlement voids in the post pipe, indicating that this post had been removed. OR12 (Fig. 5.26) is characteristic of the Outer Ring in having a ramp leading into the posthole from the outside (here to the southeast). Opposite the ramp, there is a distinct depression where the sliding post has hit the soft natural and been pulled upright. Another depression in the base shows where the rounded butt of the post finally came to rest. The primary fill (C1598) is red-brown, redeposited natural around a dark, blackish-brown secondary core incorporating small stones (C1597) with a more concentrated charcoal plug in the top 0.60 m. The lower section of the primary fill has slumped into the soft core of the secondary fill to produce an artificially narrow profile.

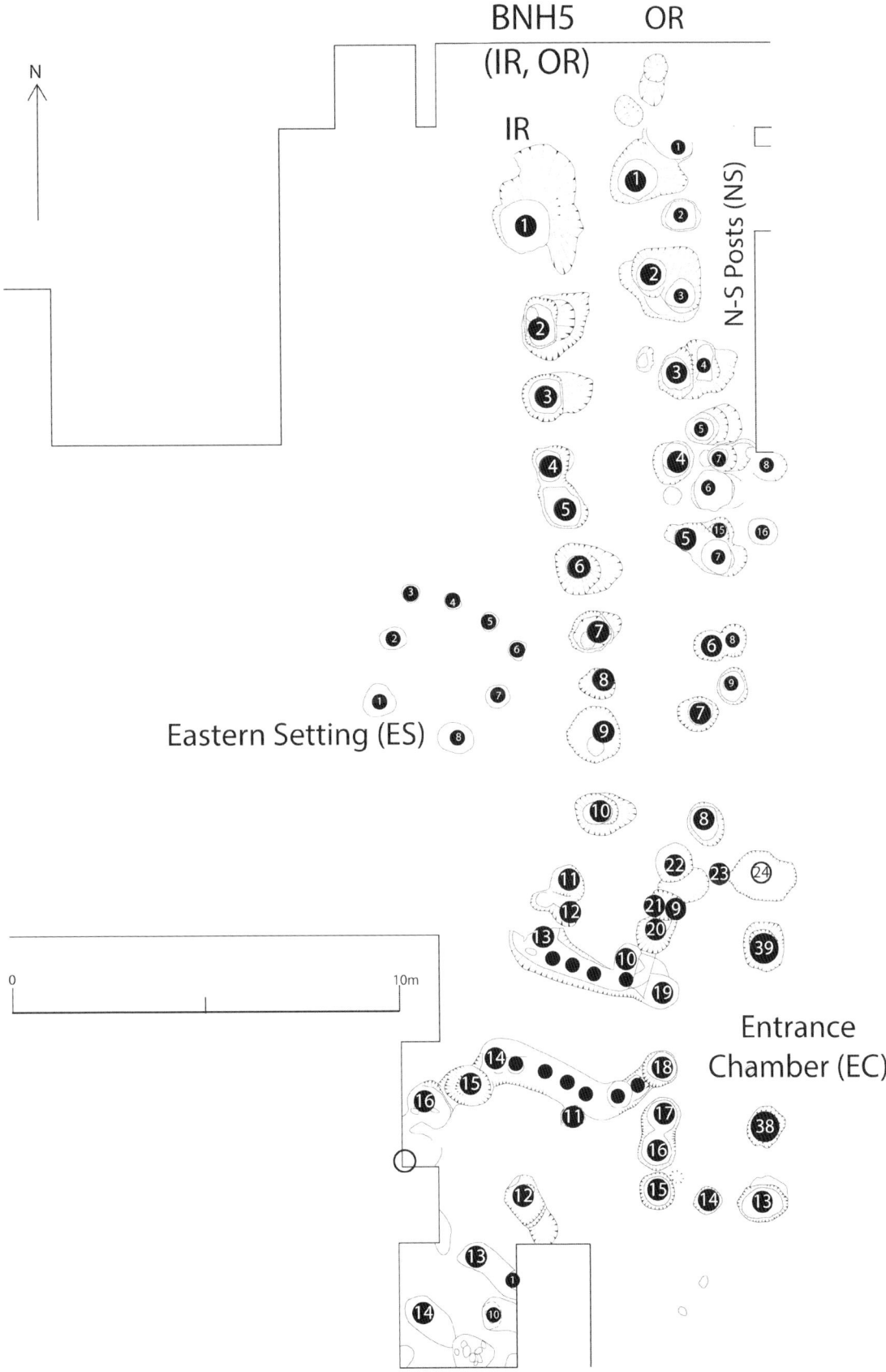

Figure 5.22 Plan of BNH5 excavated features.

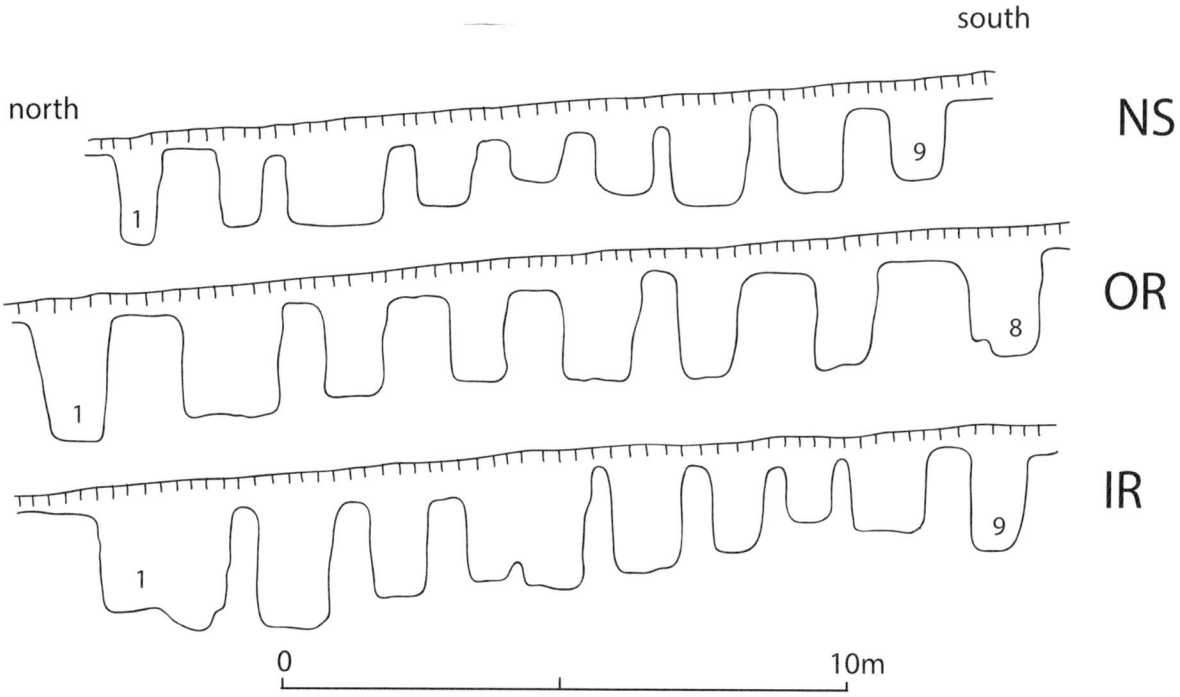

Figure 5.23 Longitudinal section through BNH5 IR and OR, and NS postholes.

The profile of IR14 (Fig. 5.25) has the characteristic concave depression in the lower side, showing that this post was erected from inside the enclosure. The secondary charcoal-rich fill (C1606) was a wide, shallow funnel narrowing with depth. IR15 (Fig. 5.25), though partially excavated, shows a cone of mixed secondary fill with irregular charcoal lenses characteristic of extraction. Figure 5.27 shows the relationships of the inner and outer ring of BNH5 and the north–south (NS) line of posts. Here, the centre points of posts IR1 and OR1 are 3 m apart and the NS1 less than 1.20 m from OR1, although this measurement diminishes to the south. The surviving depth of posthole IR1 is 2.00 m with a truncated ramp to the east and a soft secondary fill to the post mould containing variable amounts of charcoal to the bottom of the section. There were several dense concentrations of charcoal, including chunks, a cluster of stones (>0.10 m) and a lump of clean natural. A funnel, nearly a metre wide at the present surface, ran down to the post mould. Although there may have been some settlement, this funnel had been cut, as can be seen in the truncation of layers within the primary fill, to aid extraction of the post butt. This post had been removed relatively cleanly in contrast to OR1. This posthole showed a wide extraction funnel of 1.75 m at the surface, running two-thirds down the section, which had removed the southern lip of the hole on the eastern side, biting into the natural. The secondary fill was similar to the primary fill but softer and contained more flecks of charcoal with a concentration at 0.3 m depth. Stones (>0.15 m) were found throughout the secondary fill as well as clean lumps of natural, one of which was over the position of the post pipe, confirming that this post had not rotted *in situ* but had been removed, although not without considerable difficulty. OR1 was also 2.00 m deep, but on the east side had a more substantial ramp and the characteristic concave wall on the opposite side gouged by the sliding post. Later, a north–south (NS1) posthole was dug through this ramp to a surviving depth of 1.40 m. Cattle and pig bone were present in an OR posthole.

5.5.2 *The entrance structure to BNH5*

The entrance structure consisted of an Inner and Outer Ring of large postholes and two linear slots of smaller posts joining them to create the entrance through BNH5 (Figs 5.24 and 5.25). The importance of the linear axis is paramount and indicates that this was laid out before the Outer Enclosure was in place. The size and orientation of the slots mirror that of the Eastern Settings and were conceptually similar. The component parts differ in the number of posts the slots contain. The slot on the south side (C1608) was fully excavated, but the evidence for posts was more limited due to the level of disturbance in antiquity. The bases of two posts, C1736 and C1735, did survive at the western end, indicating a diameter of 0.30–0.40 m and set just 0.05 m apart hence the use of a slot rather than individual postholes. Lined up in the trench, they would probably have produced a solid wall. The northern slot contained large amounts of charcoal, and the partially excavated line of six postpipes could be seen. Although the dimensions seem to indicate that these were smaller posts of *c.* 0.15 m diameter, this was probably due to the settlement of the fill. The profile E–F

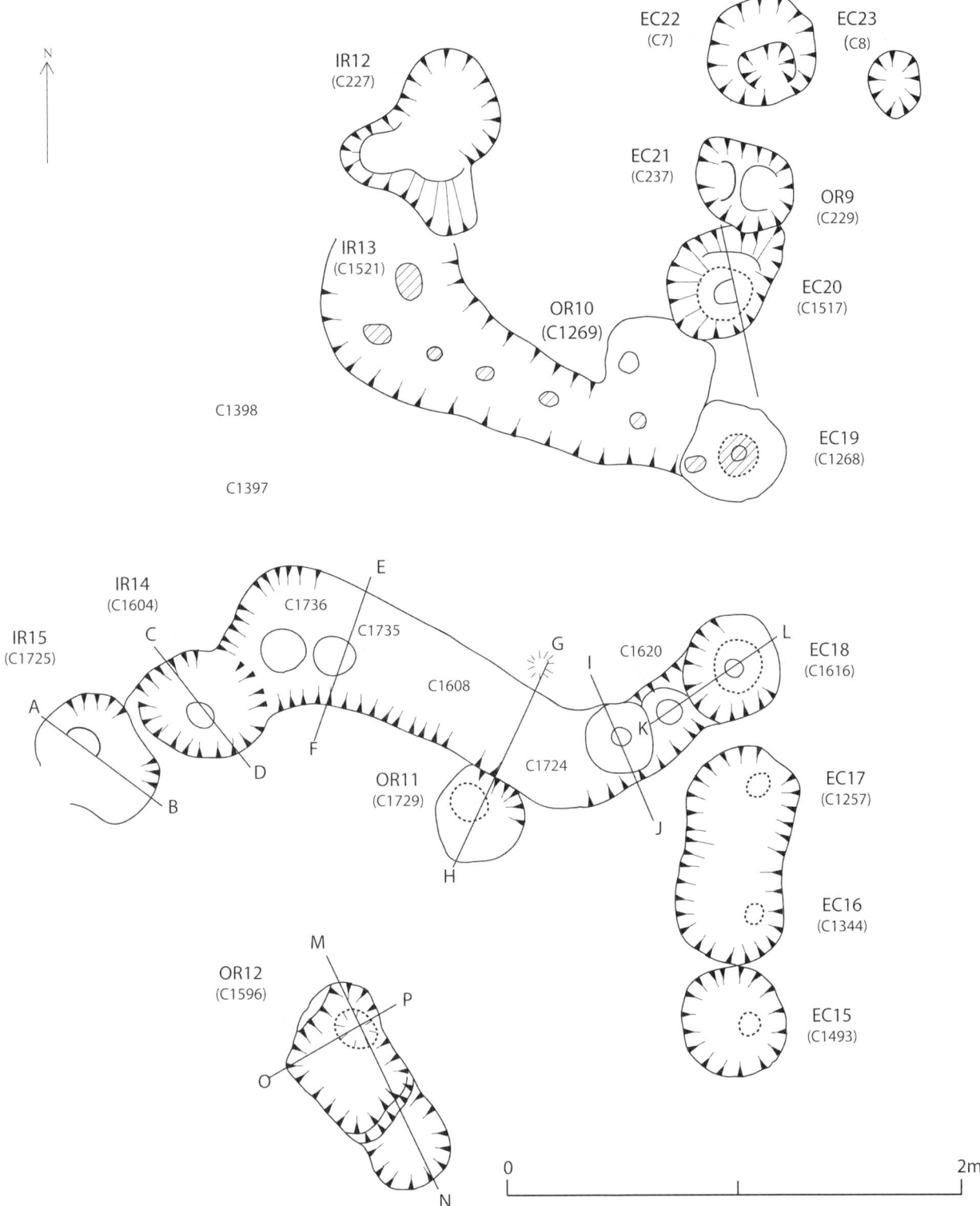

Figure 5.24 Plan of BNH5 entrance showing adjacent Inner (IR) and Outer (OR) Ring postholes, the western postholes of the Entrance Chamber (EC) and exit, and the intermediate posts which connect them.

(Figs 5.24 and 5.25), taken across the southern slot at C1735, shows the characteristic concave indentation of a heavier post having gouged the sides during erection from the south. Section G–H (Fig. 5.25) shows a similar profile across C1608 and the adjacent OR11 (C1428). If a post had been erected in OR11, it would have blocked the erection of a post in the slot, so it is most likely that the entrance structure was earlier than the BNH5 enclosure. The partially excavated northern terminal posts of BNH5, IR13 (C1521) and OR20 (C1269),

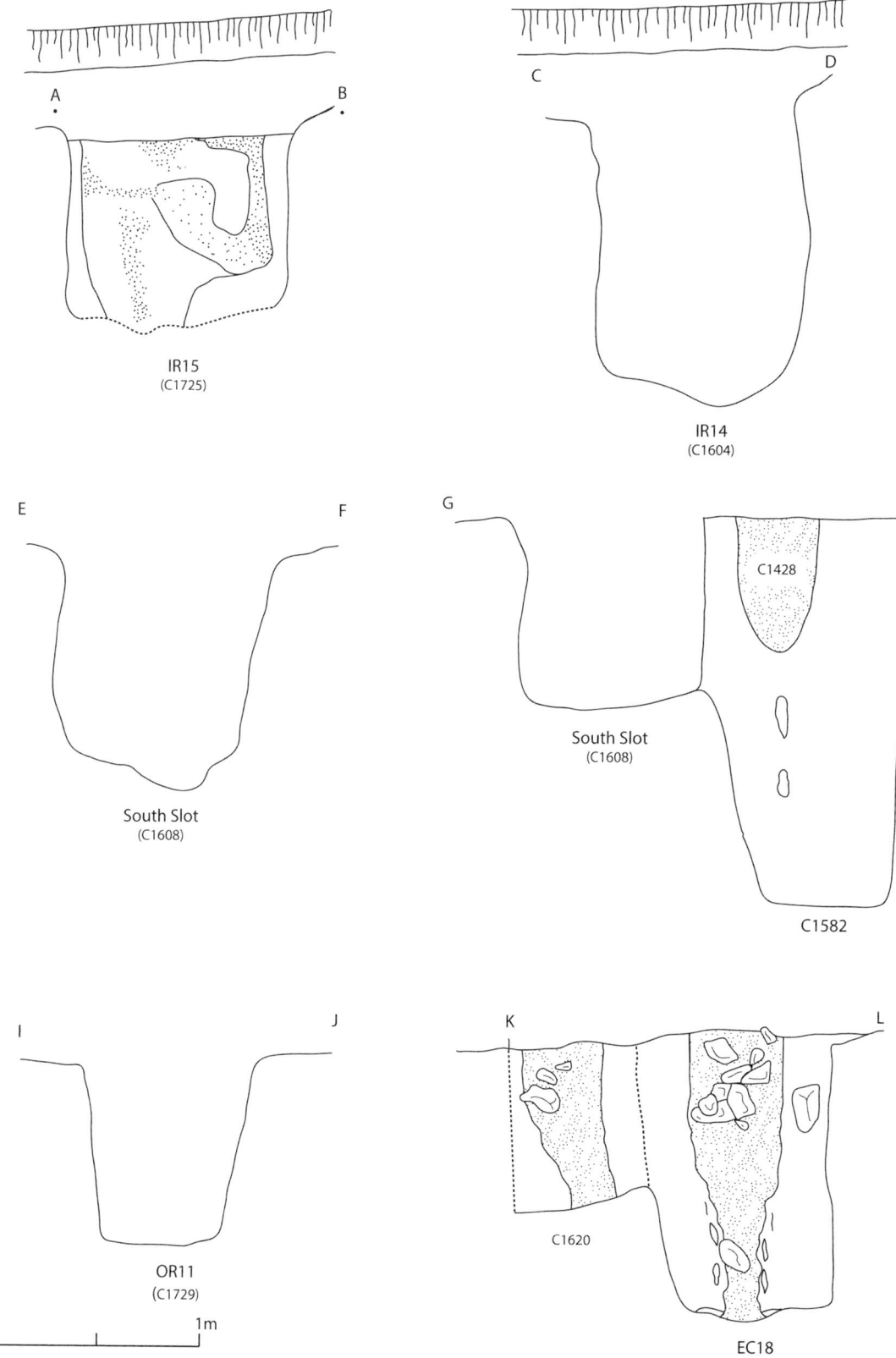

Figure 5.25 Section drawings of BNH5 postholes (see Fig. 5.24).

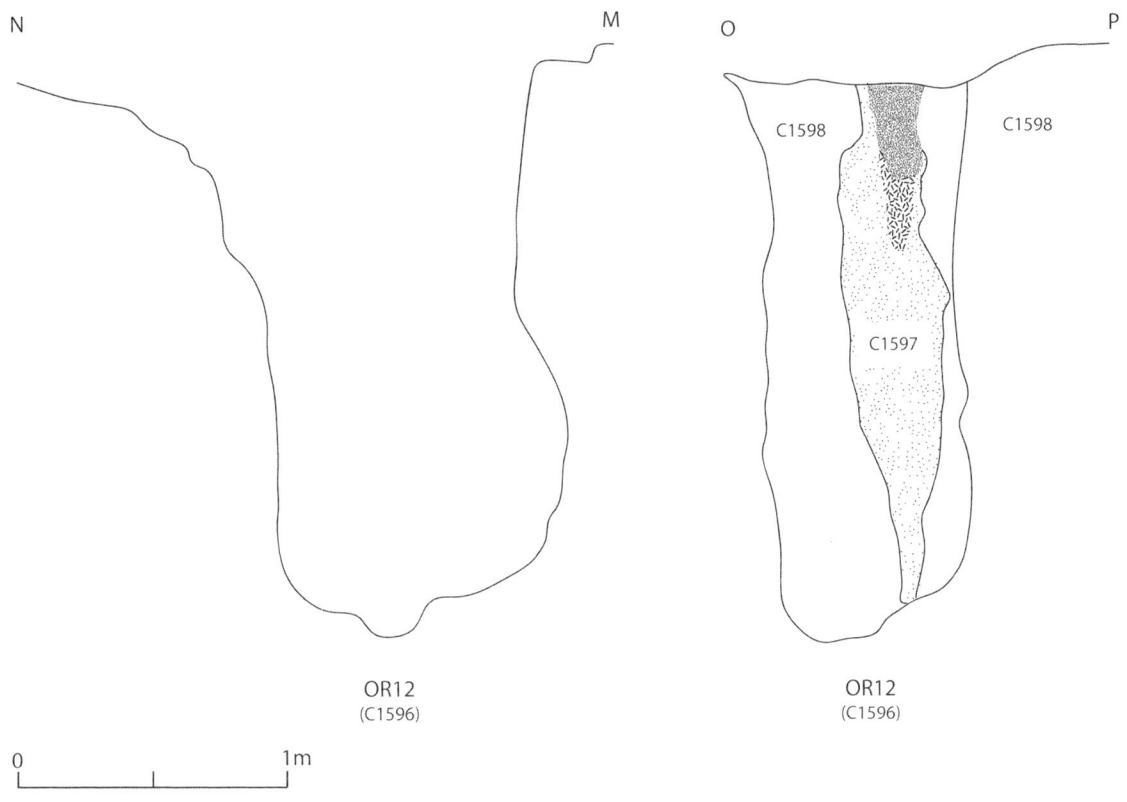

Figure 5.26 Profile and section through BNH5 OR12 (see Fig. 5.24).

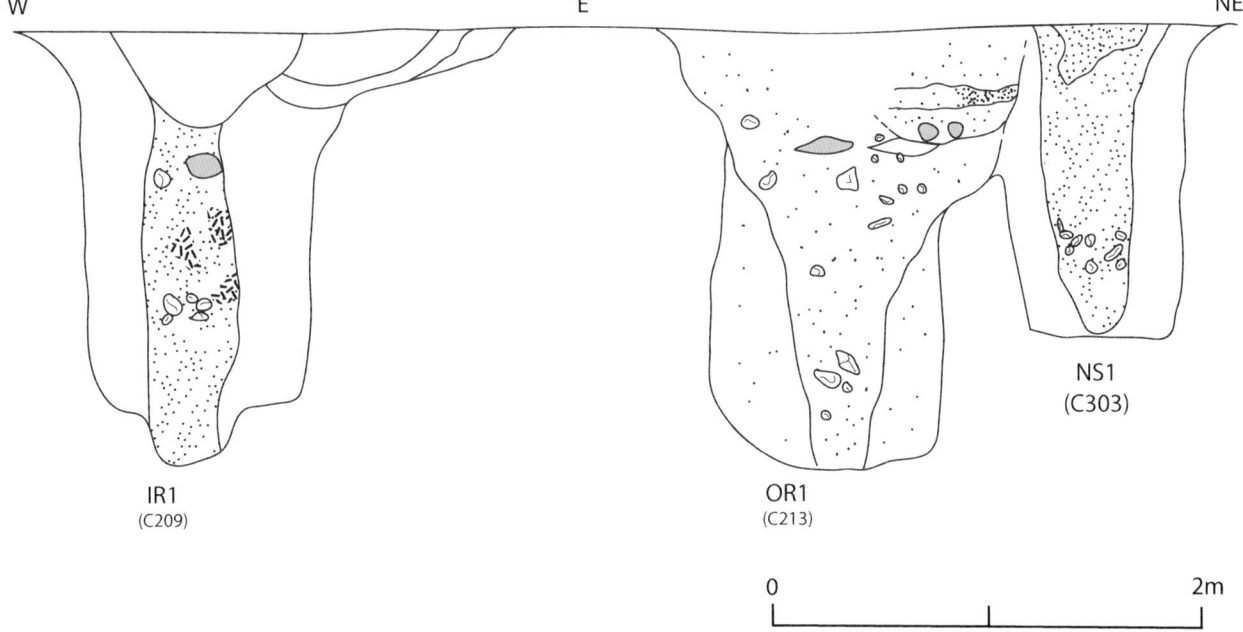

Figure 5.27 Section showing IR1 (C209) the relationship of NS1 (C303) to OR1 (C203). Shaded areas are lumps of natural.

mirror those of the southern terminal, IR14 (C1604) and OR11 (C1729).

For the Neolithic builders, the later development of the elaborate entrance structure – the Annexe, and particularly the square Entrance Chamber that sits within it – caused problems due to an awkward re-orientation from a southeasterly direction to an easterly one along the main axis of the ridge. If we think of the Entrance Chamber as having an east entrance and west exit, the exit had to articulate with the entrance to the Outer Ring of BNH5 (Fig. 5.22). This pivoted on the post at the north side of the exit (EC19), established as an eastern extension of the slot. This entailed a diagonal extension to the east end of the south slot with two intermediate posts (C1724 and C1620) to bridge the gap to EC18 on the south side of the exit (Fig. 5.25, I–J and K–L). C1620 in section K–L (Fig. 5.25) illustrates the charcoal-laden secondary fill running down the profile and incorporating stones to a depth of 0.40 m. On the north side, the westerly posts of the Entrance Chamber conflicted with the Outer Ring of BNH5, resulting in posthole EC21 (C237) being dug between the two rings. On excavation, it appeared that EC21 shared a posthole with OR9 of BNH5. Because of the similarity of their primary fills, it was impossible to determine their interface and, therefore, which was earlier. But, whereas OR9, on the Outer Ring of BNH5, contained the characteristics of a post which had rotted *in situ*, the post pipe of EC21 contained grey-black charcoal-rich soil with cobbles, pebbles, and some flint typical of an extracted post. The only way this could have functioned is if the OR9 post had been deliberately cut down to allow for the erection of EC21 of the entrance chamber, thereby leaving the base to rot. The Entrance Chamber, therefore, belongs to a later building phase and was not part of the original concept.

A feature of the south slot (C1608) was the placement of a large number of stones in the fill on one side, but it was unclear if this was deliberate packing for the posts or part of the secondary fill. One explanation would be that a trench was dug in the slot along the line of the posts and the post butts were pulled over into it and removed. This trench was then filled with stones giving the appearance, intentionally or accidentally, of deliberate placement. On the other side, there was considerable disturbance along the slot associated with the final removal of the posts, but the postpipes remained as distinct entities. When first exposed, these postpipes appeared as a continuous charcoal-rich groove that coincided with modern the plough line direction. Numerous east–west plough lines also ran across the site showing the level of modern disturbance. However, it remains a possibility that this groove could be evidence of planking at the entrance. The slots, therefore, provide contradictory evidence for the construction and removal of the BNH5 entrance, but it would fit the character of the other feature groups if this was a planked entrance which had subsequently been removed.

The western end of the northern slot was partly overlaid by a shallow, rectangular-shaped pit 1.15 × 0.83 m in extent and densely packed with charcoal. This had no clear relationship with the features below and was of indeterminate date. It was cut by a shallow trench 4.00 × 0.73 m and 0.03 m deep containing stones and some charcoal and interpreted as the remains of modern spade cultivation. Burnt human remains were recovered from C431 and C1418.

At the interior of the BNH5 entrance, patches of hard-packed sand and cobbles survived, probably relating to the original concentrated footfall, with areas of charcoal and some evidence of burning (C1397 and C1398) relating to its subsequent destruction.

5.5.3 North–South postholes (NS)

A tangential line of nine postholes was uncovered running approximately north–south immediately outside the outer line of the BNH5 enclosure (Fig. 5.22). The line is thought to continue north beyond the limit of the excavations but stopped within 8.00 m of the entrance to the south. In form, these North–South postholes resemble BNH5 with a distinct secondary fill. The postholes of this line were smaller than those of BNH5 and, measured an average of 1.00 m wide, 1.40 m deep and were oval in plan. NS1 cuts through the ramp of OR1 (Fig. 5.27) and is, therefore, of a later phase of construction. The secondary fill contained charcoal flecks throughout, with a cluster of small stones and charcoal pieces towards the bottom. NS4 also cuts a ramp, that of OR3, and showed the characteristic funnel with a mixed backfill, including stones (Fig. 5.28). Although morphologically and sequentially similar, functionally, it is difficult to see how the N–S group integrates with the rest of BNH5 and so remains enigmatic.

5.6 The Annexe area

The complexity of this area, revealed by the excavation, was not apparent from the aerial photographs. Initially termed the 'Annexe', this was an elaborate entrance that, in its final phase, involved a re-orientation and incorporated a massive façade and square 'Entrance Chamber' (Figs 5.29, 5.30 and 5.38). This area, like BNH5 and 6, is characterised by a lack of linking stratigraphy between features. The area is defined by a number of posthole groups: Northern East–West (NEW1–25), West–North–West (WNW1–17), Southern East–West (SEW1–13), Inner Façade (IF1–18), Middle Façade (MF1–12), Outer Façade (OF1–7), Entrance Chamber (EC1–40) and the enclosed areas of the northern and southern Annexe. The secondary fill of these postholes contained stones and cobbles and provides the most dramatic visual evidence for the removal of posts.

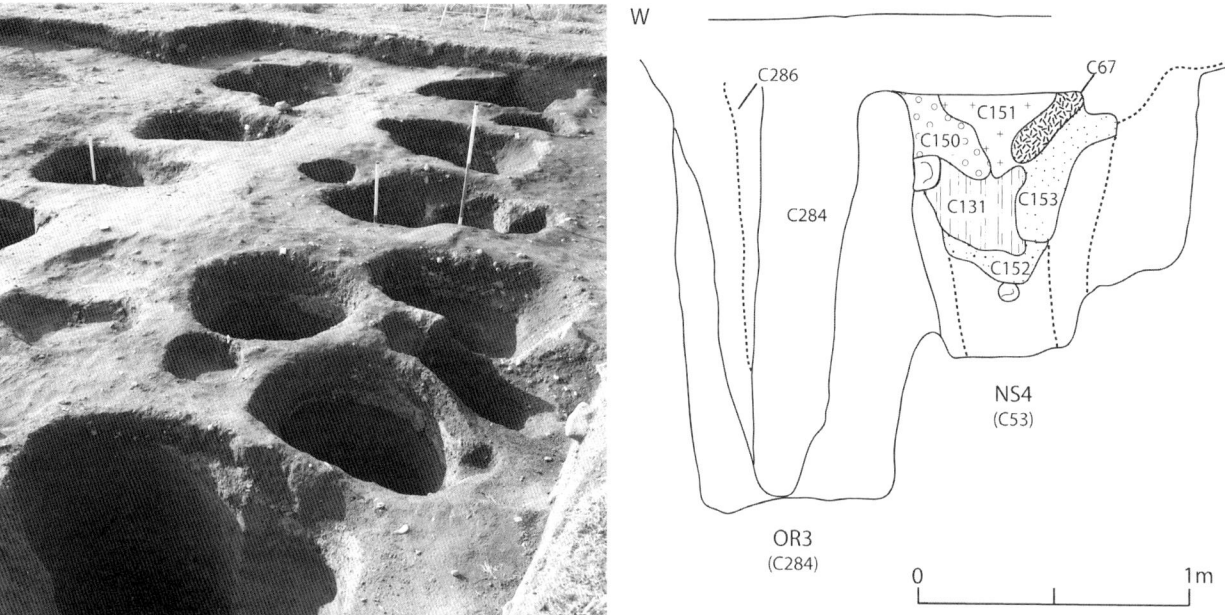

Figure 5.28 Left: BNH5 (IR and OR) and NS postholes from south. Ranging rods show IR3, OR3 and NS4. Right: NS4 cuts OR3.

Abundant signs of activity were uncovered in the northern part of the Annexe after the removal of C2. The features had been heavily truncated, however, making interpretation difficult. An extensive stony deposit (C500), disturbed by ploughing, had the appearance of a cobbled floor (Plate 30). On, between and under this were spreads of charcoal and charcoal-flecked soil (C555) and larger pieces of charcoal. Several charcoal-rich scoops, some with burnt bone, were also recorded in this area, including C522, C786 and C771. At least one pit (C777) was modern, containing glazed pottery. Immediately south of the NEW postholes, and parallel to them, was a long shallow spread, 6.00 × 0.50 m in extent but only 0.06–0.10 m deep. This was charcoal-rich but contained no finds. A great deal of Grooved Ware pottery was recovered in two clusters of 23 and 16 sherds in the south-central area, particularly from the spread below the floor. This included a complete base and flattened side wall of a small, finely made Grooved Ware pot (BNH4). Burnt flint and some scrapers were also found and discoloured subsoil shows the presence of scattered hearths (Plate 31).

The finds from the Annexe to the south of the Entrance Chamber, which is close to the eroded top of the ridge, consisted of glazed pottery and a shot gun cartridge lying on the subsoil showing that modern ploughing had removed all traces of surface prehistoric occupation. Truncated upper levels of the SEW postholes appeared immediately under the plough zone.

The area between the WNW posts and the SEW posts (Fig. 5.29) was not excavated but the probability is that any evidence would have been destroyed – although this is unlikely to have been an occupied area. The SEW posts do not fully integrate with the BNH5 main enclosure at the west end and are seen as a cosmetic operation to impress people passing below between BNH5 and the henge.

The WNW posts would have been retained as the functional southern edge of the Annexe.

5.6.1 Northern East–West postholes (NEW)

Three lines of posts defined the north of the Annexe area (NEW 1–22), running approximately from east to west (Fig. 5.29). The postholes were roughly oval in plan and, when sectioned, were 'U' shaped in profile (Fig. 5.31). The postholes average 0.60 m wide and 1.30 m in depth. These postholes were filled with a gritty orange, sandy primary fill and a charcoal-rich secondary fill deliberately layered with cobbles (Fig. 5.31, Plates 32–34). It is estimated that the postholes could have held posts which stood to a height of 4.00 m above ground level (Fig. 5.6).

The middle (NEW7–14) and south (NEW15–22) rows were paired and linked to posts OR4 and OR5 of the BNH5 Outer Ring and with a similar spacing. The North–South posts (NS5–7) interfere with this arrangement and would have caused confusion if they were contemporary – they do not seem to have any relationship to the NEW alignment, although they are tangential to BNH5. The paired NEW rows narrow towards the east, where NEW14 and 22 articulate with the main façade (Plate 32). The northern row, NEW 1–6, is parallel to the middle row but has a wider spacing, does not link to BNH5 and probably had a different function as a line of stand-alone posts – not a replacement, but an embellishment. The inclusion of substantial quantities of stone in the secondary fill of all three rows is a distinctive feature and was deliberately layered with charcoal down the vacant postpipes after the removal of the posts (Plates 33 and 34). All three lines of posts have been treated in the same way, suggesting that they were ultimately treated as a unified structure. It is likely that these features together represent the northern limit of the Annexe, although this cannot be stated conclusively as the post rows lay at the edge of the excavated area.

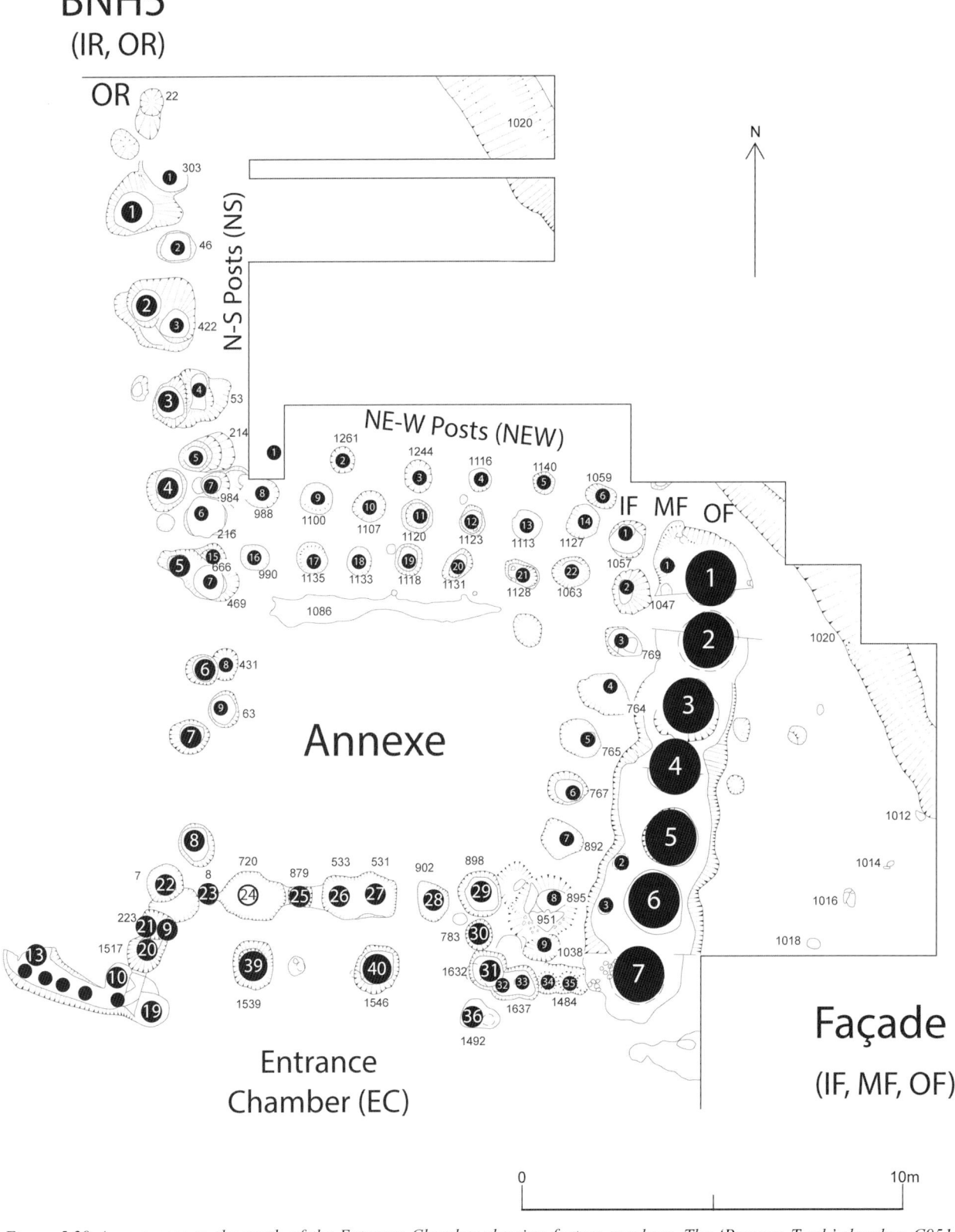

Figure 5.29 Annexe area to the north of the Entrance Chamber showing feature numbers. The 'Passage Tomb' chamber, C951, is between IF8 and IF9.

5.6.2 West–North–West postholes (WNW)

A double line of postholes running approximately west–north–west (WNW1–17) was uncovered in the south of the Annexe area (Fig. 5.29), and a representative selection was excavated (Fig. 5.32). The line of posts met the BNH5 enclosure at its northwest end, and WNW10 of the south

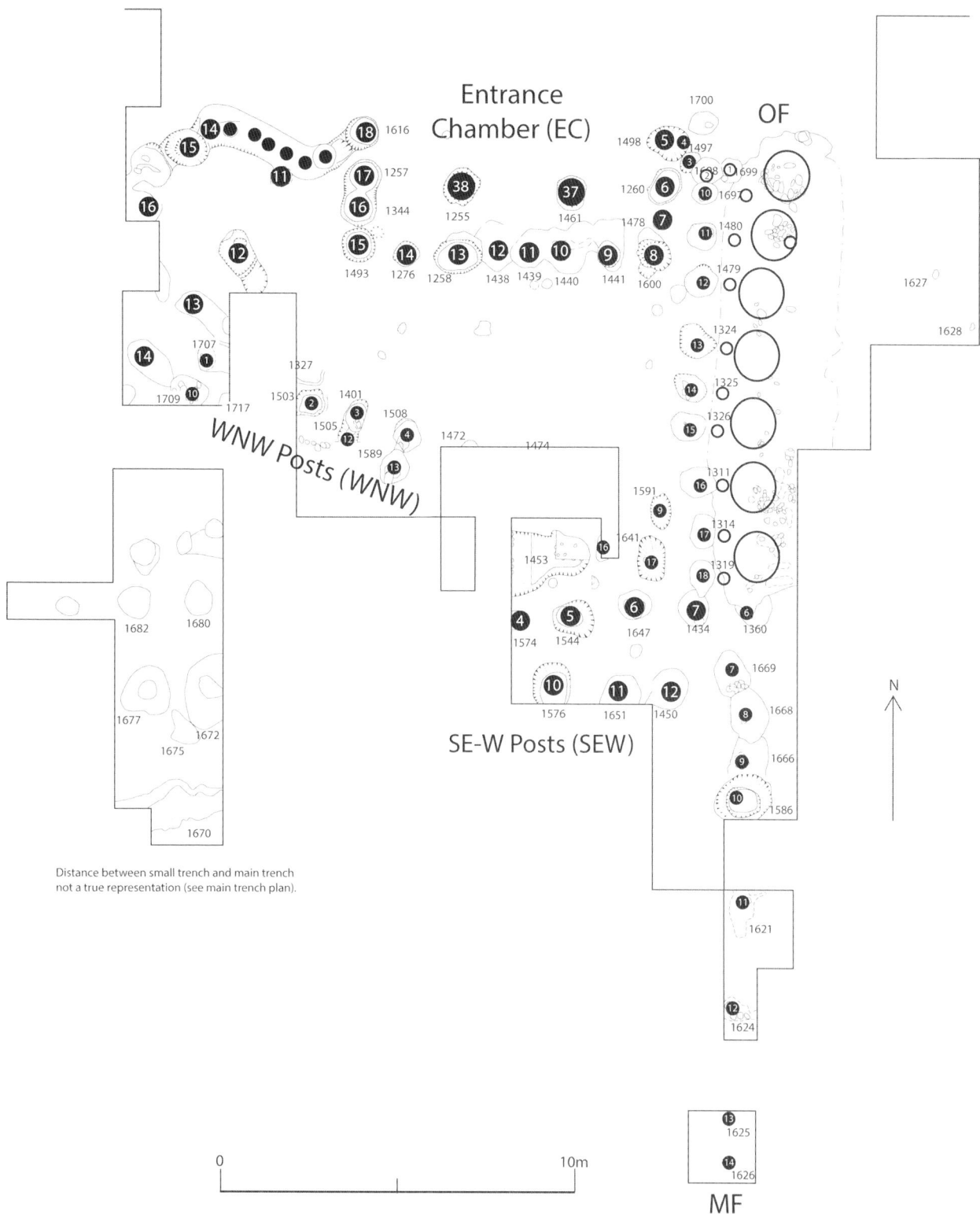

Figure 5.30 Southern area of the Entrance Chamber, the WNW posts and SEW posts showing post and feature numbers.

row aligns with the Outer Ring of BNH5 at OR14. Ten postholes of this feature group were revealed, and the outline of the primary fill and secondary fill was identified and recorded in plan. Two pairs of postholes and a single posthole were fully excavated: half-sectioned and recorded and then emptied (WNW4 and 13, WNW9 and 17, WNW2). The postholes, roughly oval in plan, with an irregular 'U' shape in profile, were an average of 0.80 m wide and 1.00–1.50 m deep, although they had been subject to substantial plough damage, being on the top of

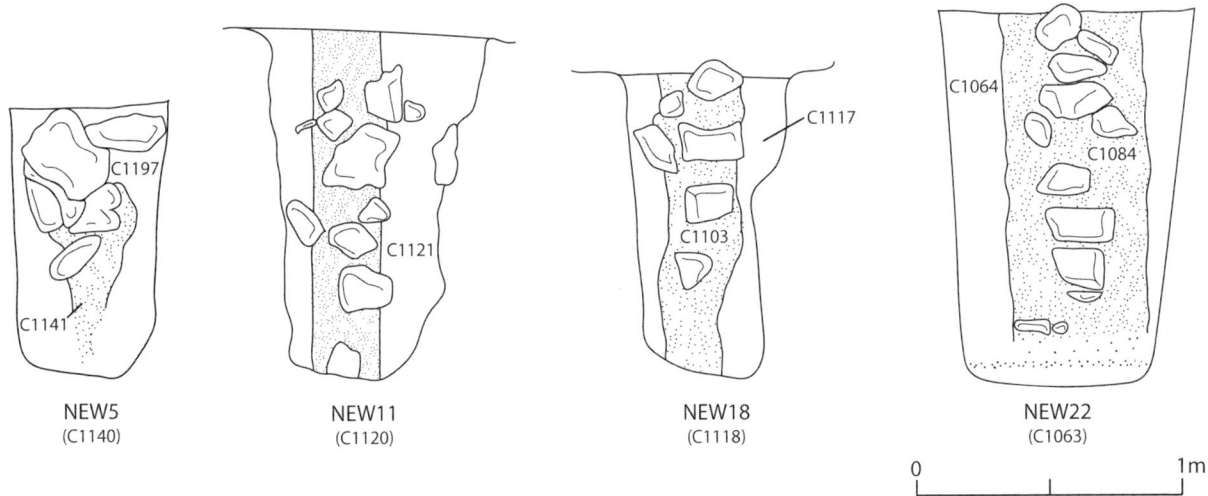

Figure 5.31 Selection of sections from the Northern East–West (NEW) posts which define the north of the Annexe.

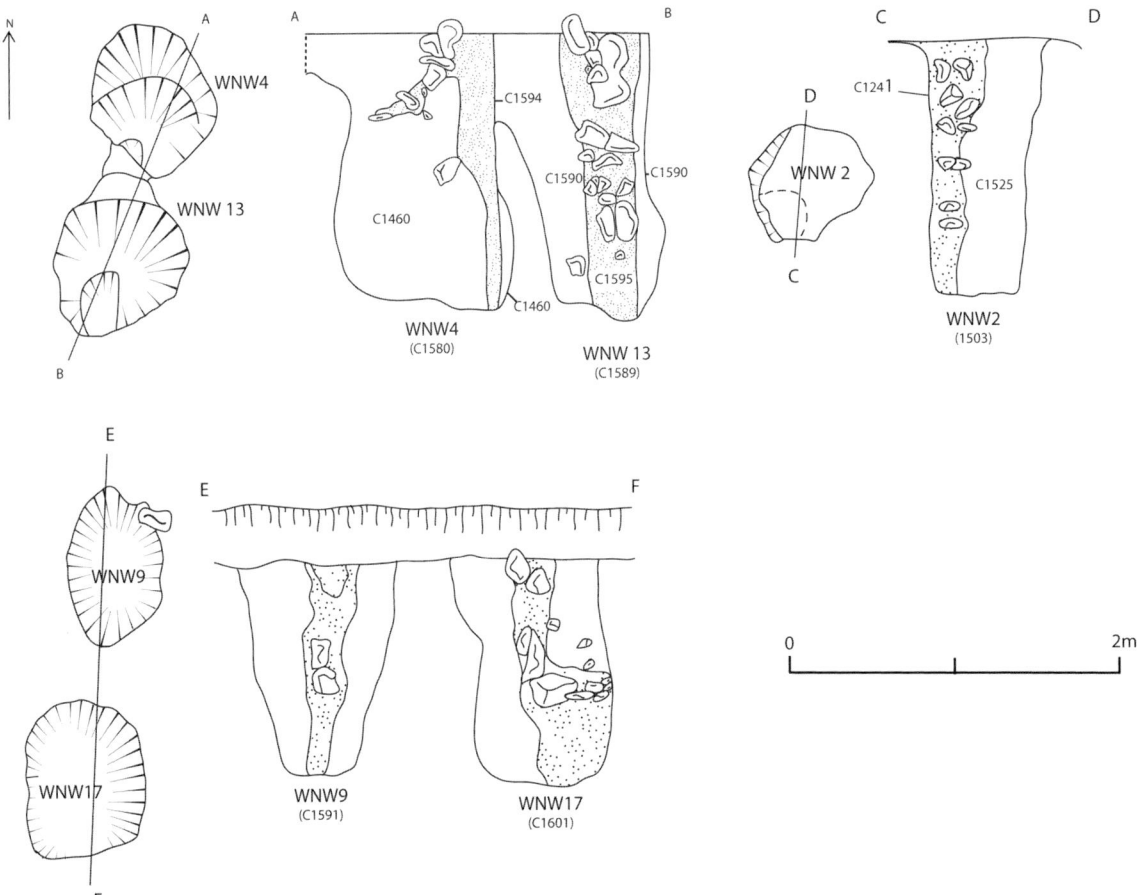

Figure 5.32 Plans and sections of WNW4 and 13, 2, 9 and 17.

the ridge. Again, the postholes were filled with a distinct secondary fill of charcoal and stones, in character like the NEW postholes. It is estimated that they could have held posts which stood at least 3.00 m above ground (Fig. 5.6). WNW4, 13 and 17 have suggestions of a ramp and show the characteristic gouge in the opposite lower wall caused by the impact of a post sliding into the posthole.

The WNW postholes were constructed later than the BNH5 Outer Enclosure, as WNW10 cuts BNH5 posthole OR14. Postholes WNW4 and 13, when fully excavated, were found to be conjoined. Posthole WNW4 cuts the ramp of posthole WNW13, showing that the south row of postholes (consisting of WNW10–17) was dug, and the posts erected and backfilled with primary fill prior to the excavation of the north row of postholes (consisting of WNW1–9). However, this was most probably part of the same construction event.

5.6.3 Outer Façade and entrance (IF, MF, OF)

A complex façade structure was uncovered at the eastern entrance to the Annexe, which would have both drawn attention to the entrance and screened all that was behind it from view. In this area were the two lines of an Inner (IF) and Middle (MF) Façade and a massive Outer Façade (OF).

The outline of 18 postholes of the north–south Inner Façade (IF1–18) was uncovered with the removal of C2 (Figs 5.29 and 5.30). The postholes were oval in plan and c. 1.10 m wide. Nine of these postholes (IF1–9) to the north of the entrance were fully excavated (Plate 35) with an average depth of 1.20 m they could have supported posts of an estimated 0.30 m in diameter and 3.60 m in height above ground (Fig. 5.6). The postholes were an irregular 'U' shape, mostly with vertical sides but in some cases sloping in towards the base. The holes were filled with a primary and secondary fill, the latter sometimes packed with stones in the upper levels (Fig. 5.32). In character, these were similar to NEW7–22 and WNW1–17. An entrance through the façade was created with the construction of a slot-defined passageway, lined by a row of smaller postholes on either side. There was a slight constriction of the entrance (EC4 and 32) at the inner end of the passage which was similar to the outer gap in the entrance into BNH5. The north–south positioning of the façade and its central entrance marked a major realignment of the axis of approach to conform to the east–west orientation of the ridge (Hartwell 2002, 549).

A second, parallel row – the Middle Façade (MF1–5), was detected on the north side of the entrance which had been almost completely obliterated by the removal of the Outer Façade posts (Fig. 5.29). As only the bottom of some postholes could be identified, there was not enough fill remaining to distinguish a secondary fill and, therefore, if these posts had been removed post-fire or during the construction of the Outer Façade. It is most probable that the Middle and Inner Façade are contemporary and fulfilled the same function as the Middle and Inner NEW posts and the Inner and Outer WNW posts as one cohesive structure. The Middle Façade continued south of the entrance and MF6–14 is a further extension of it, orientated towards the passage grave (BNH2) in the centre of the henge (Plate 36). The area further south was tested by excavation, but no more postholes were located. MF14 is therefore the southern terminal (Plate 37).

5.6.3.1 Outer Façade (OF)

The subsoil in this area was a uniform compact orange-brown silty sand with no observable bedding and containing well-dispersed stone from gravel to boulders (<0.60 m). Differentiating redeposited subsoil in the form of a primary fill was particularly difficult.

This was an extremely complex area of stratigraphy with the digging, setting, removal and backfilling of two lines of posts (MF and OF) in approximately the same position giving eight episodes of soil movement. Attempts to individually label the multiplicity of soil units thus produced were soon abandoned. Concentrations of stone, which were a useful indicator of secondary fill in other feature groups and a major feature here, also occurred naturally. The best indicator of anthropomorphic soils was the presence of charcoal but this was not evenly distributed through any single soil unit and varied from isolated flecks to dense concentrations and large chunks. The porous nature of this fill led to distinct layers of pale, creamy, leached soil and associated bands of rusty red iron pan. Whereas this could emphasise the interface of different units and apparent cuts, it could also occur within the more homogeneous primary fill and even the natural, so was confusing and could not be relied upon as an indicator. The soil units were further disturbed subsequent to deposition by slumping and compaction.

In common with other feature groups there was a reworking of the primary fill involving the removal of posts but here this activity, and the secondary fill created, was deep and substantial and obliterated much of the evidence for the Middle and Outer Façade posts. Evidence of only five of the presumed nine Middle Façade posts was recovered and only at the lowest levels.

The line of the Outer Façade first became apparent as a spine of concentrated stone rubble, running in a north–south line on either side of the entrance to the Annexe (Fig. 5.34 and Plate 38). Although some stone was found in the primary fill it was concentrated in the core secondary fill of each posthole. Intermixed with the stone was a dark black-grey sandy soil, with discrete areas of charcoal giving the impression of holes or a trench having been rapidly and randomly filled (Plate 39). This central strip was surrounded by a mixed orange-brown, gritty-sandy soil. The size of this feature when it was first revealed prompted the excavation of four box sections (Fig. 5.34). Their positioning was arbitrary, and it was only at the bottom that evidence of individual postholes became apparent resulting in section lines which were not ideally positioned (Fig. 5.36). The sections themselves also became unstable due to the quantity of loose stone and had to be recut.

Section A–B (Fig. 5.36) has caught the southern edge of OF1. Most of the original posthole has been removed with extensive amounts of charcoal, indicating a secondary fill penetrating to the bottom of the hole. It was not

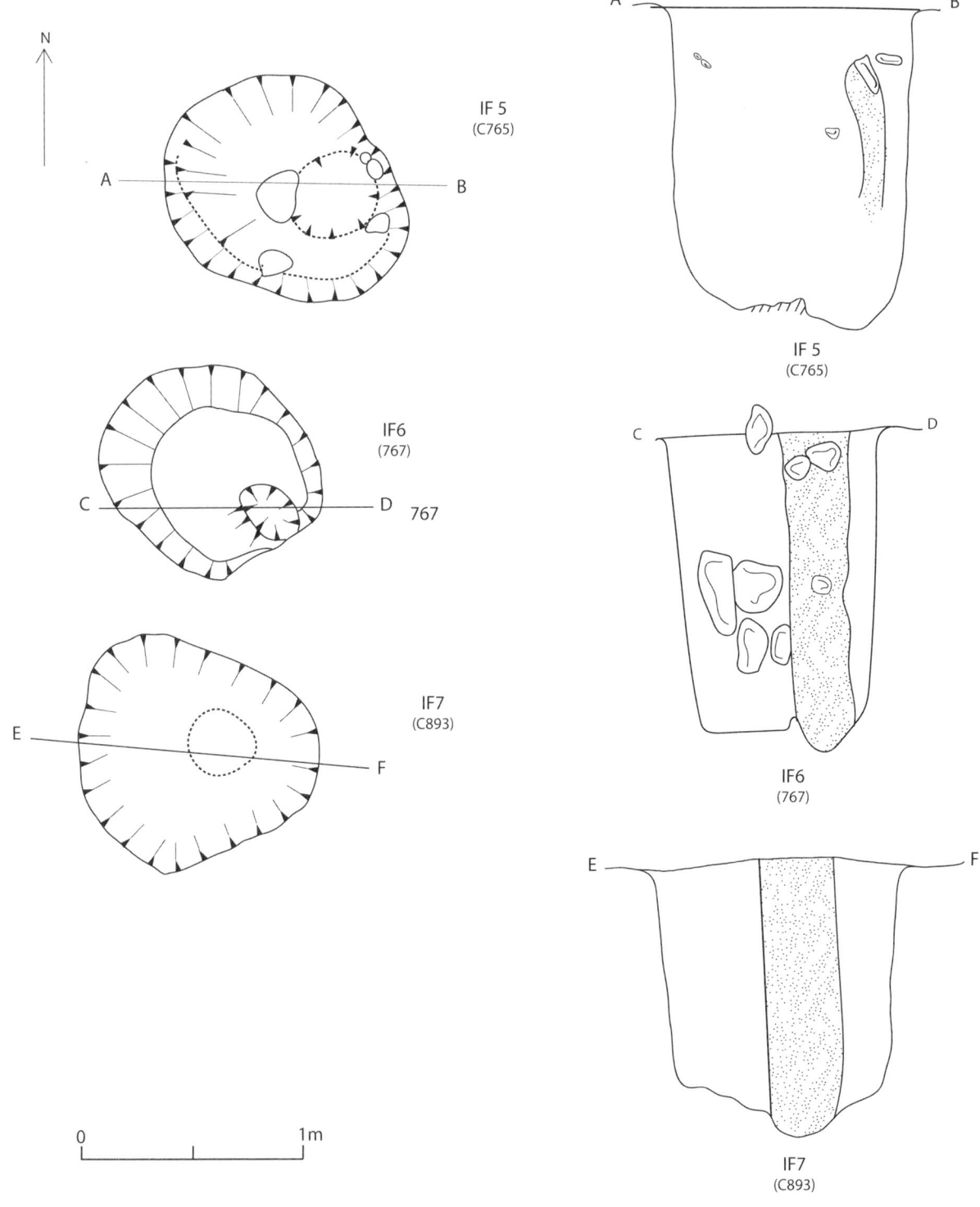

Figure 5.33 Plans and sections of IF5–7.

possible to differentiate the junction of the primary and secondary fills on the west side but this may have started at the large stone at the bottom of the profile. Section C–D, on the north side of box section 3, is the most representative section, apparently showing both primary (west side) and secondary fills (centre). The distinction between them lacks definition in the upper level and on the east side there has been slumping down from the side and the surface as the charcoal-rich core compacted. The bottom of the posthole had a different footprint than the usual concave depression found in other feature groups which was produced by the domed, cut end of a post. In this case, the bottom of OF2 was reversed, with a loose, domed centre surrounded by a hard, depressed ring (Plates

40 and 41). This suggests the selection of large, old trees with heart rot and a central cavity which would have been easier to cut down and lighter to move. Section E–F, on the north side of box section 2, clips the southern side of OF3. There was much slumping here and a concentration of stones in the upper section. The distinction between primary and secondary fills was again difficult, with dense charcoal concentrations penetrating to the bottom of the hole where the edge of a central dome could be seen, suggesting a hollow base to the post. A similar effect can be seen in OF5, shown in section I–J, and in this case, a ring of dense charcoal chunks appeared to be almost continuous, surrounding a softer centre. This could be the remaining outer rim of a post butt, originally charred as a form of preservation. Section G–H cut across OF6 and, like C–D, shows a secondary core of large stones, with charcoal inclusions varying from occasional flecks to dense concentrations and much slumping of the deposits. However, the lower level differed in being devoid of secondary charcoal, and the downward extension of this showed as a brown fill. The lower part of OF6 may have become rotten and broke as the post was being extracted, leaving the stump *in situ*. The west side of the section clips the edge of MF5, which shows as a depression.

The extraction funnels, discreet and clearly seen in other feature groups, were here substantial and erratic, joining together in an irregular trench reflecting the closeness of the posts and the difficulty of removing such large timbers. This process must have been disruptive and disorderly and archaeologically resulted in a problem of distinguishing between some areas of primary and secondary fill. Despite these difficulties of interpretation, there is clear evidence that the Outer Façade to the north of the entrance would have held seven massive posts, approximately 1.00 m in diameter, set less than 0.50 m apart and over 6.00 m in height above ground. The width and height of these posts are considerably greater than the Inner Façade posts and would have created an immensely imposing structure. Rather than a barrier, these posts stood in front of the Inner and Middle Façade posts fulfilling the same function as the stand-alone outer line of the NEW and SEW posts.

The area of the Outer Façade to the south of the entrance was not fully excavated but was cleared down to a level whereby individual postholes could be inferred by concentrations of charcoal and stone and the structure was assumed to be similar to that north of the entrance (Plate 37).

5.6.4 Southern East–West postholes (SEW)

Three rows of postholes were uncovered at the very south of the excavated Annexe area (Fig. 5.37, Plate 42). This feature group runs east–west, with SEW2 of the northern row adjacent to the BNH5 enclosure at OR16 and, thereby, closing the southwest corner of the enlarged Annexe.

At the eastern end, it articulates with the Middle Façade extension at MF4 and 5 at the point where the WNW posts join the Inner Façade. Thirty-nine postholes belonging to this feature group were identified, either on the ground or from aerial photographs and the outline of the 'primary' and 'secondary' fills was planned. Five postholes were excavated at the west end (SEW1–3, 8–9), and seven at the east end (SEW4–7, 10–12) of the paired north and middle rows. The outer, southern row, consisting of eight postholes, though parallel, was set further apart and with a longer interval, recalling the positioning of NEW1–6 and the OF1–7. Burnt human remains were recovered from C1544.

5.6.5 The Entrance Chamber

The Entrance Chamber was uncovered in the Annexe area, linking the entrance through the Main Façade to the entrance into the BNH5 enclosure (Fig. 5.38). The square structure comprised 28 postholes (EC5–31, 36) creating a chamber 9.00 × 9.00 m, with entrance gaps measuring 1.50 m on the east and west sides. In the interior were set four free-standing posts (EC38–41), similar to the Four-Poster setting in BNH6. They differ in that the BNH6 setting was a square defining the platform and the end of a passage with no exit, whereas the four posts in the Entrance Chamber formed a rectangle – a square 'pulled apart' to allow free east–west passage between them. EC40 and 41 in the north are set within EC19–31, 36 whereas EC38 and 39 in the south are set centrally in the area defined by EC5–18. The posts adjacent to EC39/40 are set a little closer together (EC24–27) as are the posts adjacent to EC37/38 (EC10–13) giving the effect of a division of space within the chamber. Possibly these sections of the Chamber wall were blocked, whereas openings either side of EC9, 14, 23 and 28 allowed access to the north and south areas of the Annexe.

This area also saw complex deposition of artefacts within the postholes. When C2 was removed, the Entrance Chamber was identified by a series of oval-shaped, black, circular smudges. On excavation these postholes had near vertical sides and flat bases and were all filled with a distinct 'primary' and 'secondary' fill (Figs 5.39–41). EC5 (Fig. 5.39) survived to a depth of 2.00 m with a charcoal-rich secondary fill (C1414) which penetrated to the bottom of the section. Three postholes at the south of the Entrance Chamber were the exception to this. A trench appears to have been cut to hold three posts (EC10–12), which were all backfilled with the same primary fill (C1447), a red-brown loamy sand. On the north side, EC25 appeared shallower than the other postholes but these all presented difficulties during excavation due to poor definition of the cut. The postholes of the Entrance Chamber measure 1.00 m wide and *c.* 1.80 m deep. It is estimated that they could have held posts of 0.30 m in diameter and 5.40 m above ground. The western exit

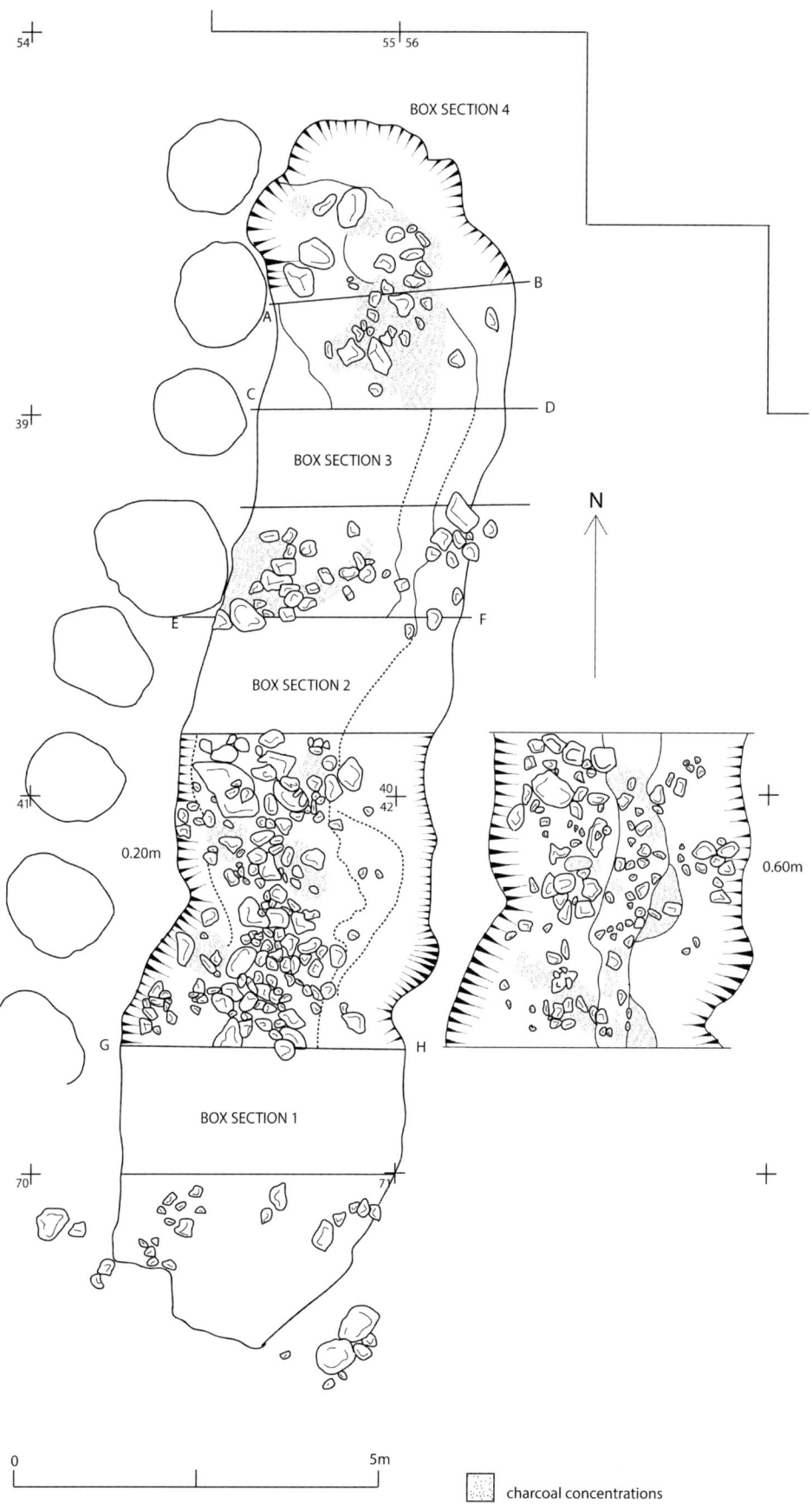

Figure 5.34 Outer Façade at 0.20 m and 0.60 m depth below C2.

5. Ballynahatty 5 and 6: excavating the enclosures

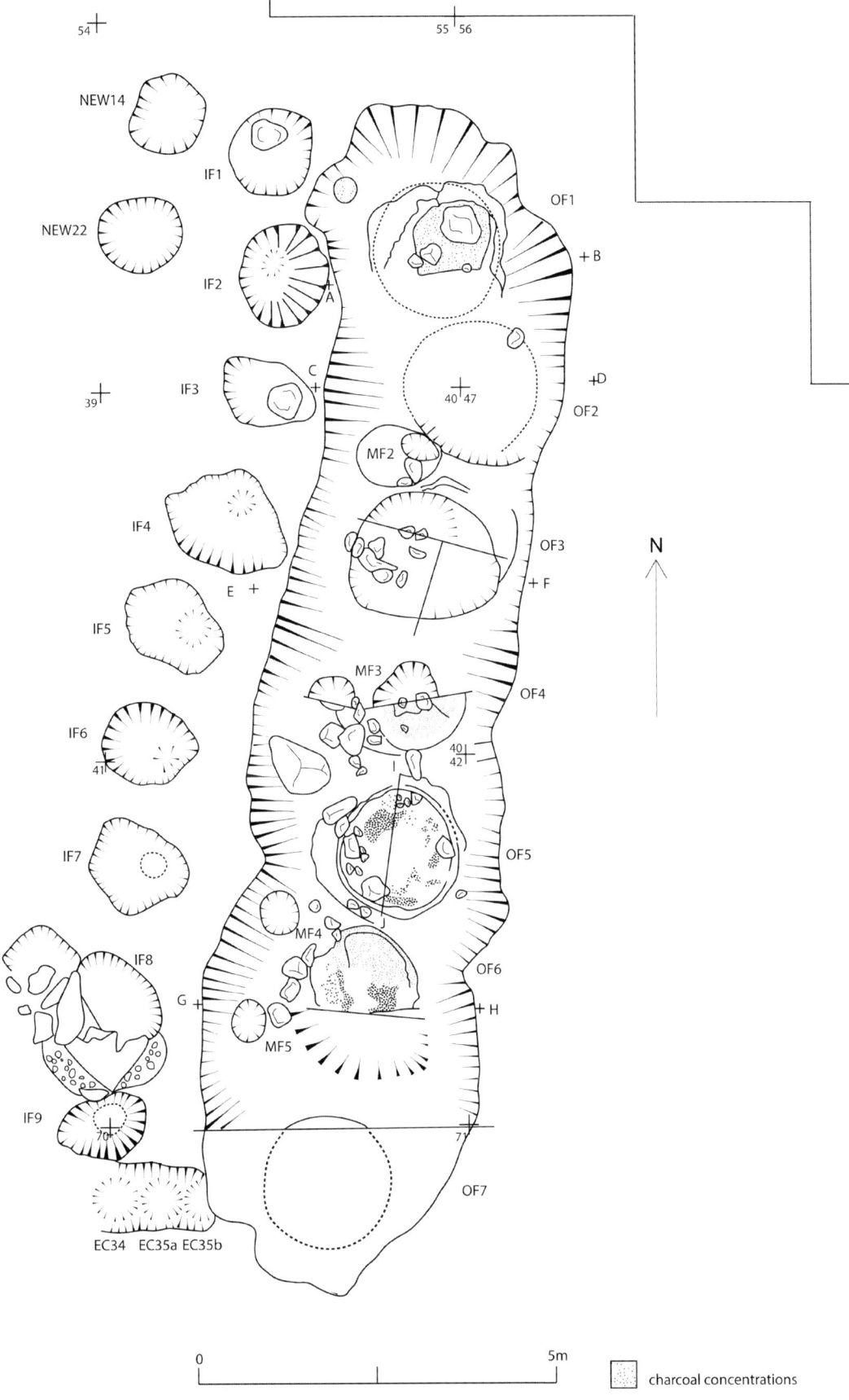

Figure 5.35 Outer Façade at lowest level.

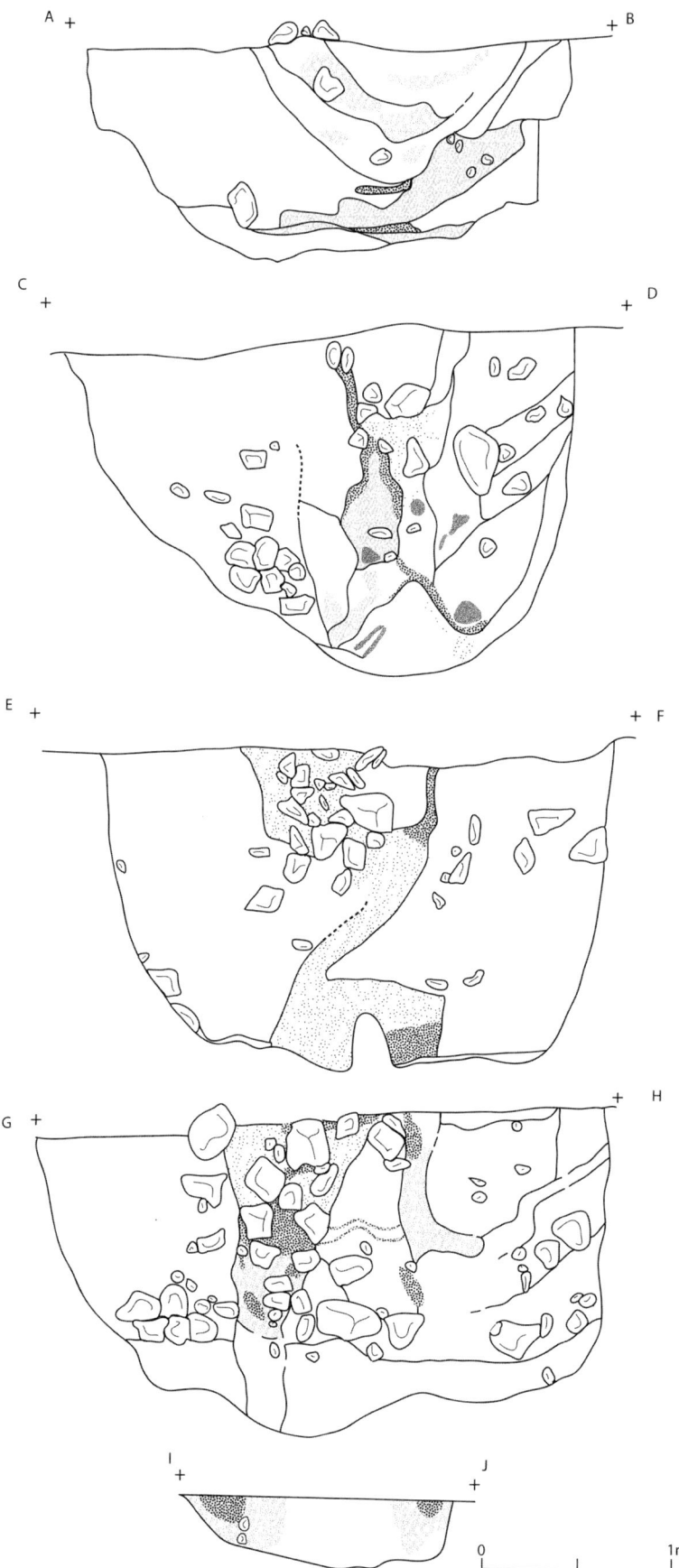

Figure 5.36 Sections across postholes of the Outer Façade. Left is west.

5. *Ballynahatty 5 and 6: excavating the enclosures*

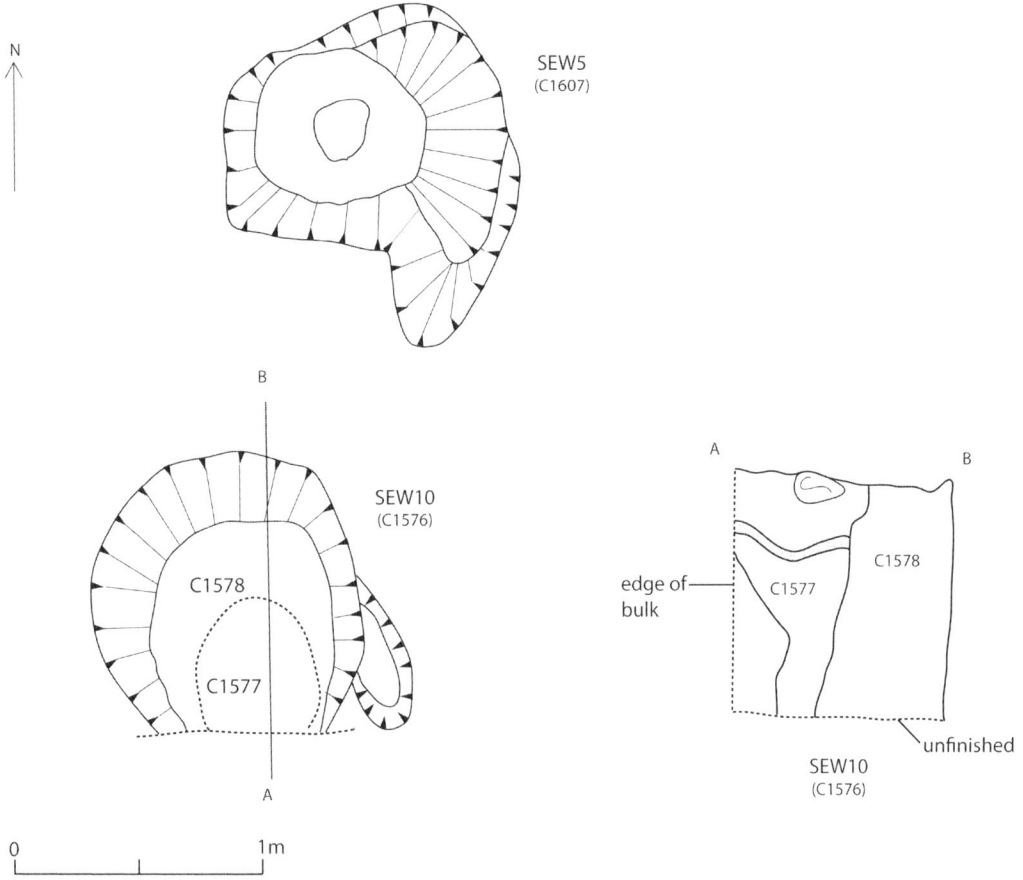

Figure 5.37 Southern East–West postholes (SEW).

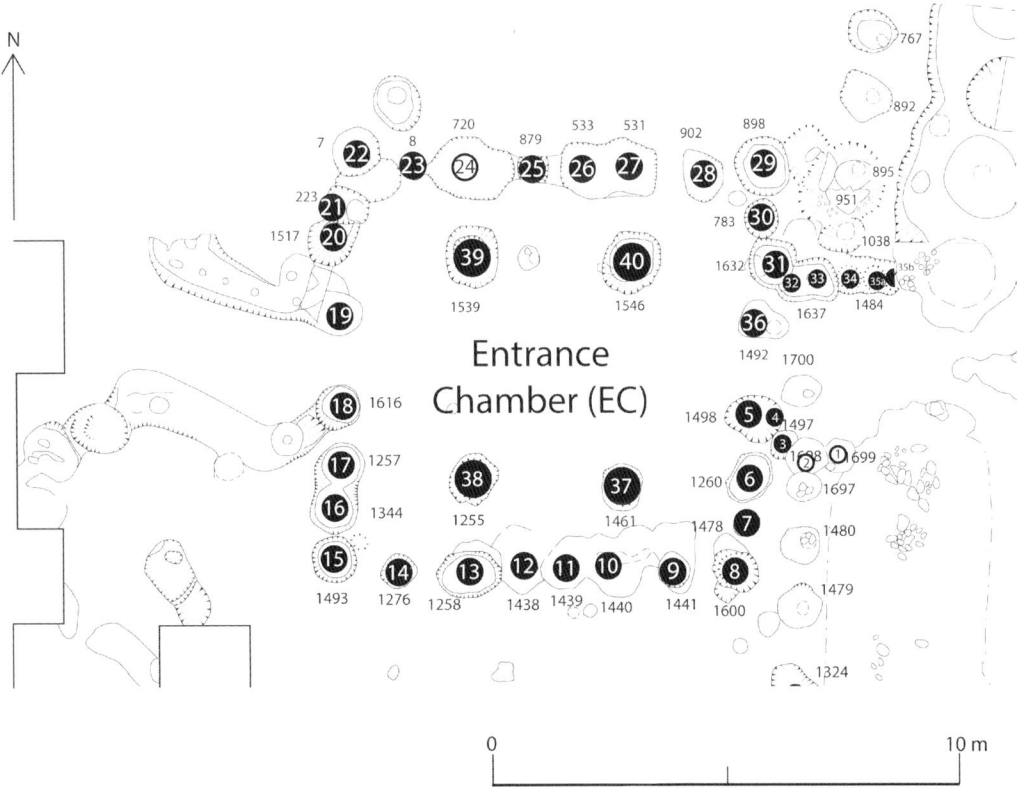

Figure 5.38 Plan of central Annexe with Entrance Chamber.

where it articulates with the entrance to BNH5 has been discussed previously (see Section 5.1.4.2.) and there are similarities with the eastern entrance through the Main Façade. On the north side of the entrance were five posts (EC32–35b) in a truncated slot 0.35 m wide and 0.30 m deep which linked the east side of the entrance chamber at EC31 to the entrance through the Main Façade (Fig. 5.42). Only the north side of the entrance was fully excavated, but the symmetrical form of the entrance allows the widths to be determined. The passage through the façade is estimated as 2.80 m wide which then broadens out to a vestibule, 3.40 m wide, before narrowing to 1.40 m as it enters the Entrance Chamber, the same as the exit. There was a trace of a small posthole alongside EC36 which could have matched EC4 and EC5 and together this may be evidence of a door to close the entrance.

The position of the Entrance Chamber confirmed the re-alignment of the Annexe to the east–west orientation of the ridge, which involved a modification to the larger enclosure BNH5 and the removal of a post in the Outer Ring. The re-alignment pivots on EC19, which is an easterly extension of the north side of the entrance through BNH5. On the south side, another post was inserted to bridge the gap between the entrance and EC18. An additional posthole (EC22) was also set at the northwest corner of the Entrance Chamber so that the line of BNH5 was corrected and the Entrance Chamber remained a square. Despite this, there is a slight distortion to the shape caused by the Main Façade which shows that the Entrance Chamber was inserted after the façade was constructed (Fig. 5.38).

5.7 Early medieval period

A pit (C1453) was discovered in the south of the Annexe area dated to the early medieval period (Figs 5.30 and 5.43, Plates 43 and 44). The pit lay at the meeting point of the SEW posts, the WNW posts and the Inner Façade. The pit ran into the baulk and was not fully excavated; however, the excavated portion measured 1.70 m wide by 2.25 m long, 1.00 m at its deepest and 'U' shaped in section. The pit was filled with a dark, black loam, angular, shattered stones and charcoal (C1452). The sandy natural at the bottom of the hole was reddened by heat and lenses of clean dried sand had drifted from the sides onto the accumulating fill (Plate 43). This shows that the charcoal was in a primary context and subject to repeated fires having taken place within the pit. Despite the black appearance of the fill, only 1.13 kg of charcoal was recovered by wet sieving, most of the particles having passed through the finest sieve – attributable to repeated efficient combustion. C1452 also contained, burnt human bone fragments (jaw and skull) and burnt pig and sheep/goat bones. A quantity of charred grain was present in the lower 0.20 m of the pit and grain from the base produced a radiocarbon date of cal AD 426–576 (UBA-42750).

This does not appear to be a discreet cremation deposit, although it has been suggested that cremated

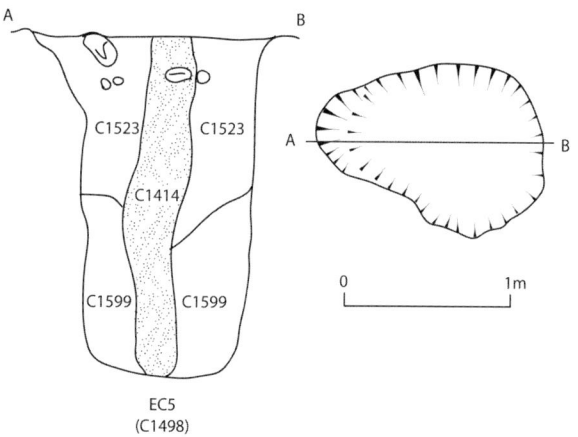

Figure 5.39 Section through EC5.

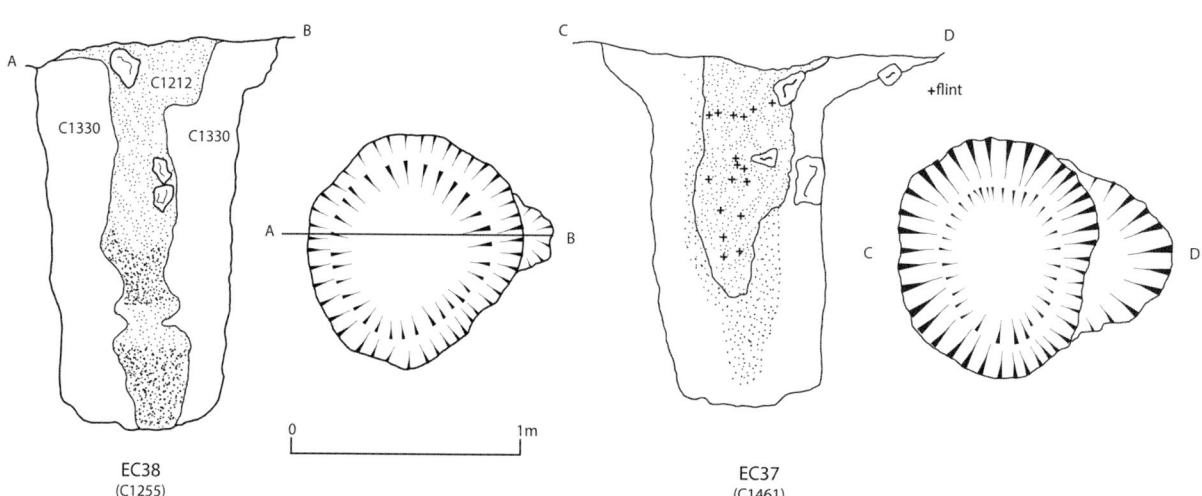

Figure 5.40 Sections through Four-Posters EC37 and 38 showing distinct secondary fill.

5. Ballynahatty 5 and 6: excavating the enclosures

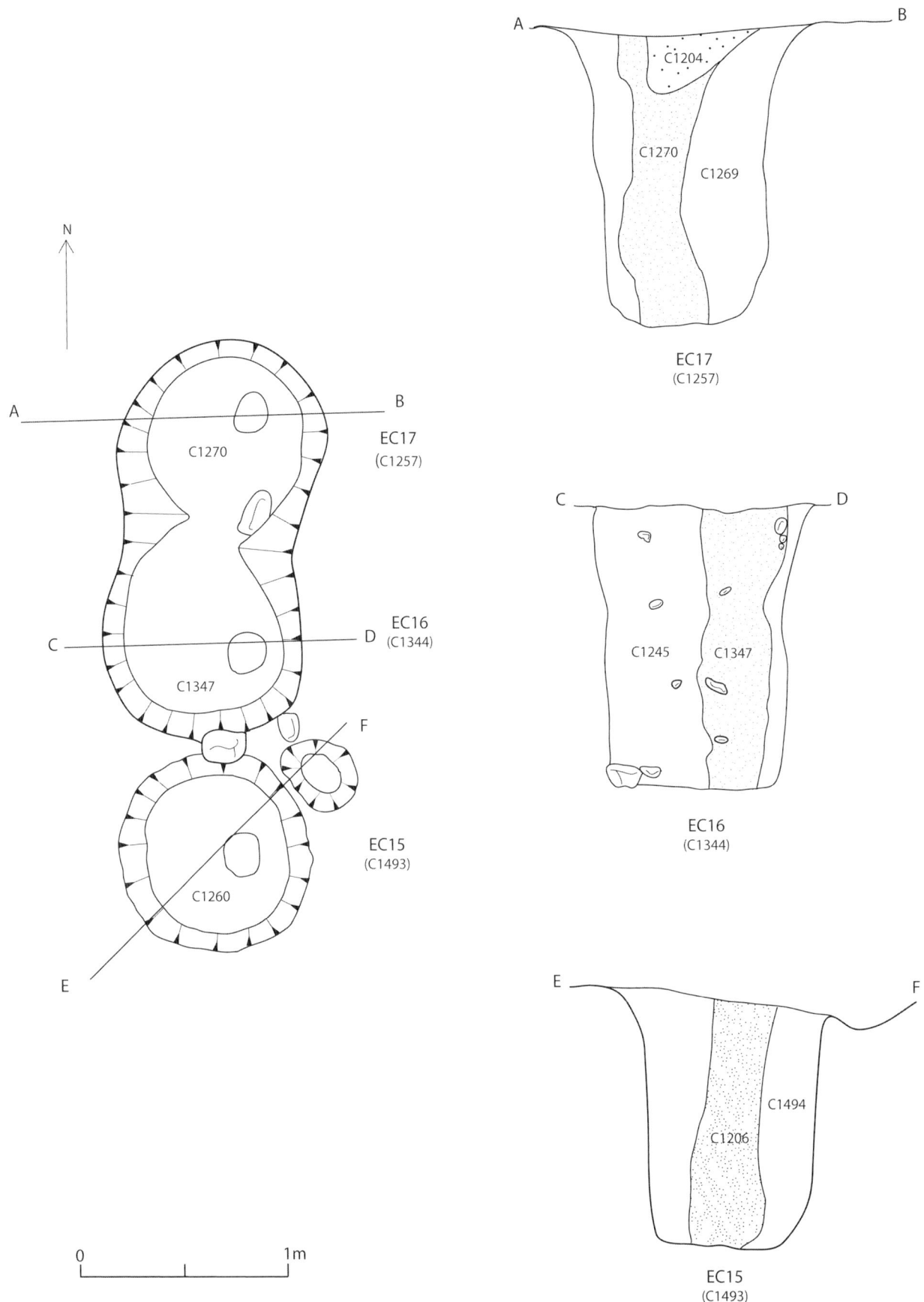

Figure 5.41 Postholes EC15–17 of the Entrance Chamber.

material could exist as spreads or in gullies and postholes and with similar associations (Gleeson and McLaughlin 2021, 395–396). The early date and the cremation deposit could indicate that this is part of a pagan mortuary ritual, although it has been argued that cremation and east–west burials co-existed in Christian commemorative practice into the 7th century AD (Gleeson and McLaughlin 2021, 396). The abundant evidence of burning rules out a simple midden accumulation and together with the presence of cremated human bone suggests that this could be the rarely identified site of a series of pyres. Although of later date (cal AD 710–1000, Beta–150535), a pyre site was identified at Cloghermore Cave, Co. Kerry consisting of a pit and trough containing charcoal and flecks of bone. This site had a timber platform associated with it which seems to have been dismantled (Connolly and Coyne 2005, 30–330).

A circular site (BNH3), situated on the ridge 100 m to the west, was identified by aerial photography. Although little diagnostic cultural material was recovered, the presence of barley in charred grain from the ditch suggests at least an Iron Age/early medieval date. There are several ringforts in the area and as a result, the site has been interpreted as a destroyed ringfort, confirming a strong early medieval presence.

5.8 A note on the archaeobotanical remains from excavated soil samples

Gill Plunkett

Archaeobotanical remains previously extracted from soil samples from the excavations at Ballynahatty were submitted for identification. The remains were examined under a Nikon SMX800 stereomicroscope at ×10 magnification and were identified with the aid of the plant remains reference collection in the Palaeoecology Centre at Queen's University Belfast. Criteria for cereal identification followed Renfrew (1973) and van der Veen (1992). Results are discussed below according to phases outlined in Chapter 11.

5.8.1 Phase 1a

Samples from 19 contexts from Phase 1a contained charred archaeobotanical remains (Table 5.1). Seeds of *Veronica* cf. *hederifolia* (ivy-leafed speedwell) and *Corylus avellana* (hazelnut) shell fragments were the most frequently represented remains, although in only two contexts (174, 102) were both found together. A small quantity of charred grain from the Eastern Setting (ES) was identified as *Avena* spp. (oats). *Chenopodium album* (fat-hen) is well-represented in two samples. Fat-hen is typically associated with arable fields and waste places, although its seeds and leaves are edible. Along with speedwell, the seeds appear to reflect burnt vegetation in cultivated or abandoned land.

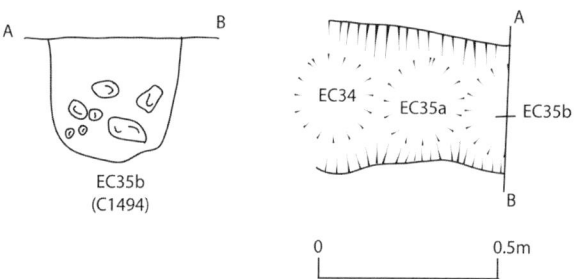

Figure 5.42 Slot on north side of entrance connecting main Facade to Entrance Chamber.

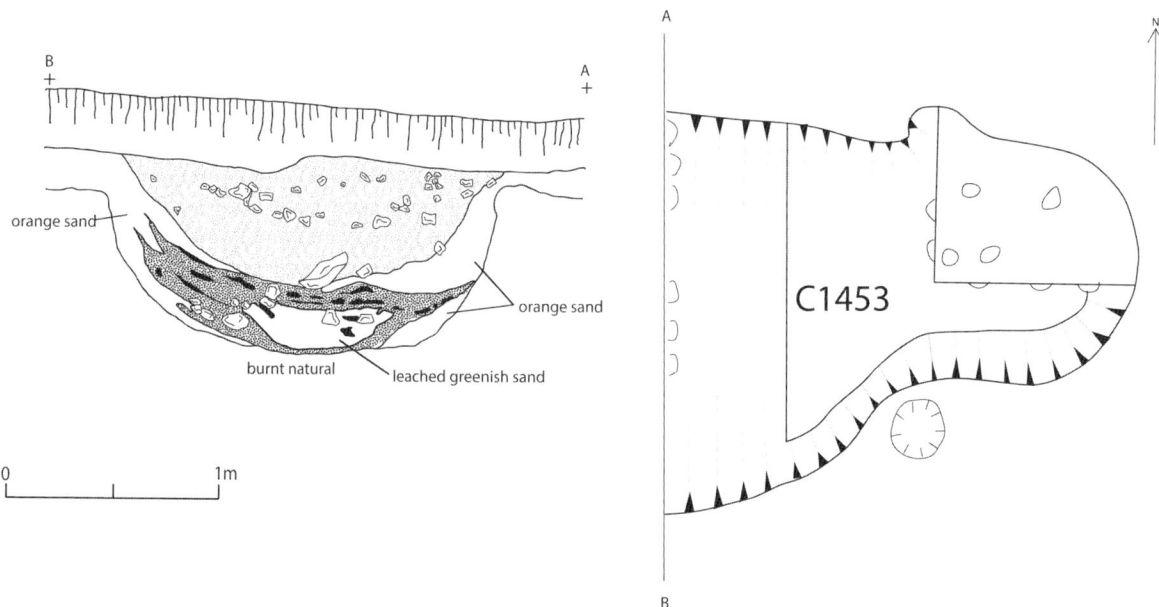

Figure 5.43 Section through the burning pit C1453. For location see Fig. 5.30.

5.8.2 Phase 1b

A small quantity of charred remains was recovered from three samples from BNH5 Inner Ring (5IR) and North–South posts (NS) contexts (Table 5.2). The remains were dominated by *Corylus avellana* shell fragments, but a single *Avena* sp. grain and a seed of *Veronica* cf. *hederifolia* were also present. Like the remains in Phase 1a, these incidental remains possibly include material from activity post-dating the archaeological features.

5.8.3 Phase 3

Four samples from the BNH5 Entrance Chamber yielded charred remains of *Corylus avellana* shell fragments (Table 5.3).

Table 5.1 Archaeobotanical remains (numbers) in samples from Phase 1a, Ballynahatty (NR: Not recorded).

Feature group	Sample identifier	Area	Cut	Context	Soil weight (kg)	Avena spp.	Cerealia undifferentiated	Corylus avellana shell fragments	Chenopodium album	cf. Vicia	Poaceae undifferentiated	Veronica cf. hederifolia
P	34	29	470	482	7.2	–	–	–	15	–	1	–
6 4P	K	29	453	552	NR	–	2	1	–	–	–	–
	56	30	948	644	NR	–	–	1	–	–	–	–
6IR	14,2	19	104	174	181	–	–	3	–	–	–	3
	60,61,63,64,66,67	16	111		NR	–	–	–	–	–	–	30
	49	28	635	635	10.3	–	–	3	–	–	–	–
	62	30	638		5.1	–	–	–	32	–	–	25
	72	30	641	661	5.8	–	–	–	–	–	–	2
6OR	C	17/20	341	347	6.9	–	–	–	–	1	–	19
	47	17	597	527	4.7	–	–	1	–	–	–	–
	55	22	413	377	1.0	–	–	1	–	–	–	–
	36	22	449	447	5.7	–	–	1	–	–	–	–
	71	22		494	NR	–	–	–	–	–	–	32
ES	22	11	57	24	68.0	3	–	–	–	–	–	–
	31	11	58	55	49.8	2	–	–	–	–	–	2
	13	10	31	62	89.4	–	–	–	1	–	–	–
	11	8	35	102	50.5	4	–	9	–	–	–	1
	26	10	30	122	23.1	–	–	4	–	–	–	–
	27	11	25	180	6.1	–	–	1	–	–	–	–

Table 5.2 Archaeobotanical remains (numbers) in samples from Phase 1b, Ballynahatty.

Feature group	Sample identifier	Area	Cut	Context	Soil weight (kg)	Avena spp.	Corylus avellana shell fragments	Veronica cf. hederifolia
5IR	68	2	209	201	8.7	1	13	–
	29	11	244	245	7.4	–	1	1
NS	15	6	53	110	24.4	–	23	–

Table 5.3 Archaeobotanical remains (numbers) in samples from Phase 3, Ballynahatty.

Feature group	Sample identifier	Area	Cut	Context	Soil weight (kg)	Corylus avellana shell fragments
EC	5	15	8	74	64.6	6
	73	37	531	–	4.1	78
	51	37	533	532	6.5	7
	79	41	783	828	31.9	18

5.8.4 Phase 4

Twenty samples from this phase produced a low abundance of charred archaeobotanical remains (Table 5.4). A small assemblage of charred grain was found in one sample (context 613), comprising mainly *Hordeum* spp., including naked barley, and *Avena* spp. Charred seeds of *Persicaria* spp. (knotgrasses) and *Rumex* spp. (docks) in the same sample possibly represent crop weeds, although the seeds are also edible. Small numbers of charred grains were found in three other contexts (219, 318 and 322), of which identifiable specimens were limited to *Avena* spp. (oats). Thirteen samples contained fragments of *Corylus avellana* shell, in most instances amounting to no more than a single nutshell, but half or complete shells were recovered from two contexts (C325 and C451). Charred seeds of *Veronica* cf. *hederifolia* (six samples), *Lathyrus* spp. (vetchlings), Labiatae sp. (mint family) and Poaceae sp. (grass family) were also present, which are likely to represent grassland weeds that were inadvertently burnt.

5.8.5 Early medieval period

A burning pit (context 1452) dating to the early medieval period produced a large quantity of charred remains, from which several samples were collected. A proportion of each sample was analysed but, as their assemblages were comparable, they are considered here as a single sample, altogether representing 30% of the total assemblage from this context (Table 5.5). The remains are significantly dominated by *Hordeum* spp. (barleys), that amount to 88% of the assemblage, of which 94% comprises *Hordeum* var. *nudum* (naked barley). Infrequent occurrences of laterally twisted grains points to two-rowed barley. A small proportion of *Triticum* spp. (wheats, 0.9%) and *Avena* spp. (0.3%) were also recorded. Non-cereal charred remains included a small number of *Corylus avellana* nutshell fragments, and incidental seeds of probably crop and grassland weeds *Persicaria* spp., *Rumex* spp., *Veronica* cf. *hederifolia* and *Viola* spp. (violets). The relative paucity of weed seeds and the excellent condition

Table 5.4 Archaeobotanical remains (numbers) in samples from Phase 4, Ballynahatty.

Feature group	Area	Sample identifier	Cut	Context	Soil weight (kg)	Hordeum var. nudum	Hordeum indeterminate	Avena spp.	Cerealia undifferentiated	Corylus avellana whole shells	Corylus avellana shell fragments	cf. Lathyrus	Labiatae spp.	Poaceae undifferentiated	Persicaria sp.	Persicaria cf. lapithifolia	Rumex sp.	Veronica sp.	Unknown
NS	9	28	216	228	12.2	–	–	–	–	–	–	–	–	–	–	–	–	2	–
EC	15	H	229	224	28.9	–	–	–	–	–	–	–	–	–	–	–	–	2	–
6OR	19	69	326	351	1.3	–	–	–	–	–	5	–	–	–	–	–	–	–	–
6OR	20/19	35	326	325	27.4	–	–	–	–	0.5	20	–	–	–	–	–	–	–	–
6OR	20	52, 57	326	351	17.5	–	–	–	–	–	2	–	–	–	–	–	–	1	–
6OR	20	33	326	451	16.9	–	–	–	–	1.5	6	–	–	–	–	–	–	–	–
6OR	22	59	554	590	19.2	–	–	–	–	–	–	–	–	–	1	–	–	75	–
6OR	46	50	1009	1022	11.7	–	–	–	–	–	–	–	–	1	–	–	–	–	–
6IR	16	65	111	254	32.4	–	–	–	–	–	12	–	–	–	–	–	–	–	–
6IR	19	A	289	296	25.7	–	–	–	–	–	5	–	–	–	–	–	–	7	–
6IR	26	70	583	540	0.5	–	–	–	–	–	18	–	–	–	–	–	–	–	–
6IR	34	41		613	11.4	42	79	33	32	–	–	–	–	–	15	21	9	–	–
6 4P	16	48	582	546	22.6	–	–	–	–	1	10	–	–	–	–	–	–	–	–
5OR	3	25	212	211	6.8	–	–	–	–	–	2	–	–	–	–	–	–	–	–
5OR	12	17	220	219	43.5	–	–	3	2	–	1	–	1	–	–	–	–	–	–
5OR	6	6	284	286	30.7	–	–	–	–	–	20	–	–	–	–	–	–	–	–
5IR	8	10	241	241	2.0	–	–	–	–	–	6	–	–	–	–	–	–	–	1
5IR	11	3	26	318	5.8	–	–	2	–	–	–	–	1	–	–	–	–	–	–
5IR	10	G	32	322	6.4	–	–	1	–	–	1	–	–	–	–	–	–	–	–
5IR	2	12	89		31.6	–	–	–	–	–	–	–	–	–	–	–	–	1	–

Table 5.5 Archaeobotanical remains (numbers) in samples from the early medieval burning pit (Context 1452), Ballynahatty.

Sample identifier	24, 40, 42, 45, P-V
Area	90
Soil weight (kg)	1037
Fraction assemblage counted	0.30%
Hordeum var. *nudum*	1636
Hordeum vulgare	87
Hordeum undifferentiated	813
Triticum spp.	27
Avena spp.	9
Cerealia undifferentiated	316
Corylus avellana shell fragments	11
Persicaria spp.	2
Rumex spp.	2
Veronica cf. *hederifolia*	1
Viola spp.	1

of the grain suggests that the assemblage derives from a fully processed crop.

5.8.6 Discussion

Archaeobotanical remains from samples dating to prehistoric and early medieval periods provide firm evidence that crops were previously grown at this location. Notably, the early medieval Burning Pit featured a large assemblage of naked barley, most likely the two-rowed variety, with a minor component of hulled barley, wheats and oats. A similar assemblage from a Phase 4 context (613), as well as the dominance of oats amongst earlier contexts containing identifiable cereals, raises questions about the integrity or interpretation of the deposits. Oats are generally considered to have been introduced to Ireland in the early medieval period, before which time they are usually interpreted to indicate wild oats growing as a crop weed (Monk 1985/6; McClatchie *et al.* 2016; 2022). In the absence of other cereals in the Phases 1a and 1b samples, it is unlikely that the oat grains represent crop weeds, but they may signify intrusive material from a younger period.

A small range of weed seeds, together with mainly very fragmented hazelnut shell fragments, are found consistently in the Ballynahatty samples. The weed seeds signify open conditions, possibly cultivated or abandoned land, in contrast to the woodland suggested by the hazelnuts. Ivy-leaved speedwell is particularly ubiquitous but is not present in the confirmed early medieval Burning Pit. It is conceivable that the seeds derive from local vegetation that was burnt. Fruits are produced in the summer, which provides a constraint on the time of year during which such burning might have occurred. Although the hazelnut remains may reflect discarded food waste, their generally small quantities are not consistent with a collected food resource. A larger concentration of shell was found in the Entrance Chamber but represents no more than one or two nuts. It is possible that seasoned hazel branches, some bearing nuts, were burnt in the same process that charred the weed seeds.

In conclusion, a large number of samples from Ballynahatty yielded charred archaeobotanical remains. Amongst them, evidence for crop growing very likely relates to activity considerably post-dating the prehistoric period during which the monuments were constructed and used. Weed seeds, on the other hand, may reflect open conditions around the sites at the time of the extensive burning that saw the destruction of the timber circles.

6

The pottery

6.1 Introduction

In total, 490 sherds, ranging in date from the Early Neolithic through to the Early Bronze Age, were found during the excavations at Ballynahatty. The vessels are generally highly fragmented, with only a few sherds from each pot surviving. The only exceptions to this are two badly damaged, but almost complete, Middle Neolithic Coarse Ware bowls recovered from chamber C591, and the sherds making up around 40% of an Early to Middle Neolithic bowl found (as residual material) in the Entrance Chamber. Analysis of the pottery from Ballynahatty has confirmed the presence of Early Neolithic Carinated Bowl pottery; Early to Middle Neolithic 'Globular Bowl' pottery (which might also be described as 'modified Carinated Bowl' pottery since it arguably belongs to that overall tradition); Middle Neolithic Coarse Ware (also known as 'Carrowkeel Ware'); Late Neolithic Grooved Ware; and two Early Bronze Age pots – one probably a Vase Food Vessel (but initially thought to be a Beaker) and the other almost certainly a Collared Urn (Table 6.1). Over half (58%) of the assemblage by sherd count consists of Grooved Ware sherds, which are associated with the timber monument complex, and a further 17% consists of Middle Neolithic Coarse Ware sherds, which are mostly associated with chamber C591 and with other deposits of cremated remains. Around 20% of the pottery could not be attributed to a particular tradition or period. Overall, the assemblage provides intriguing glimpses into activities spanning around two millennia and an important and relatively well-dated assemblage of Grooved Ware.

The Grooved Ware was examined macroscopically and reported by Helen Roche as part of her MA research, University College Dublin (Roche 1995), while the rest of the ceramic assemblage was reported by Sarah Gormley, with additional input from Catriona Brogan. The reports were prepared for publication by Catriona Brogan, with additional comments being provided by Alison Sheridan in 2023 (albeit without first-hand sight of the assemblage).

This chapter starts by reviewing the spatial distribution of the pottery before describing and discussing the material in chronological order.

Table 6.1 Breakdown of the different pottery traditions recovered during the excavation at Ballynahatty.

Pottery tradition	No. sherds
Carinated Bowl	4
Globular Bowl	14
Coarse Ware	83
Grooved Ware	282 • 194 Knowth Style 1 • 67 Knowth Style 2 • 4 Knowth Style 3 • 17 unknown
Vase Food Vessel	5
Collared Urn	3
Indeterminate	99
Total	490

6.2 The distribution of the pottery at Ballynahatty

6.2.1 Topsoil and ploughsoil (contexts C1 and C2)

A substantial portion (147 sherds: 30%) of the pottery assemblage was recovered from the topsoil and ploughsoil layers at Ballynahatty. This number also includes pottery which has no recorded context information. Fourteen of those sherds also lacked information on their findspot area and the remaining 134 were recovered from 15 different area squares. The distribution of the pottery recovered

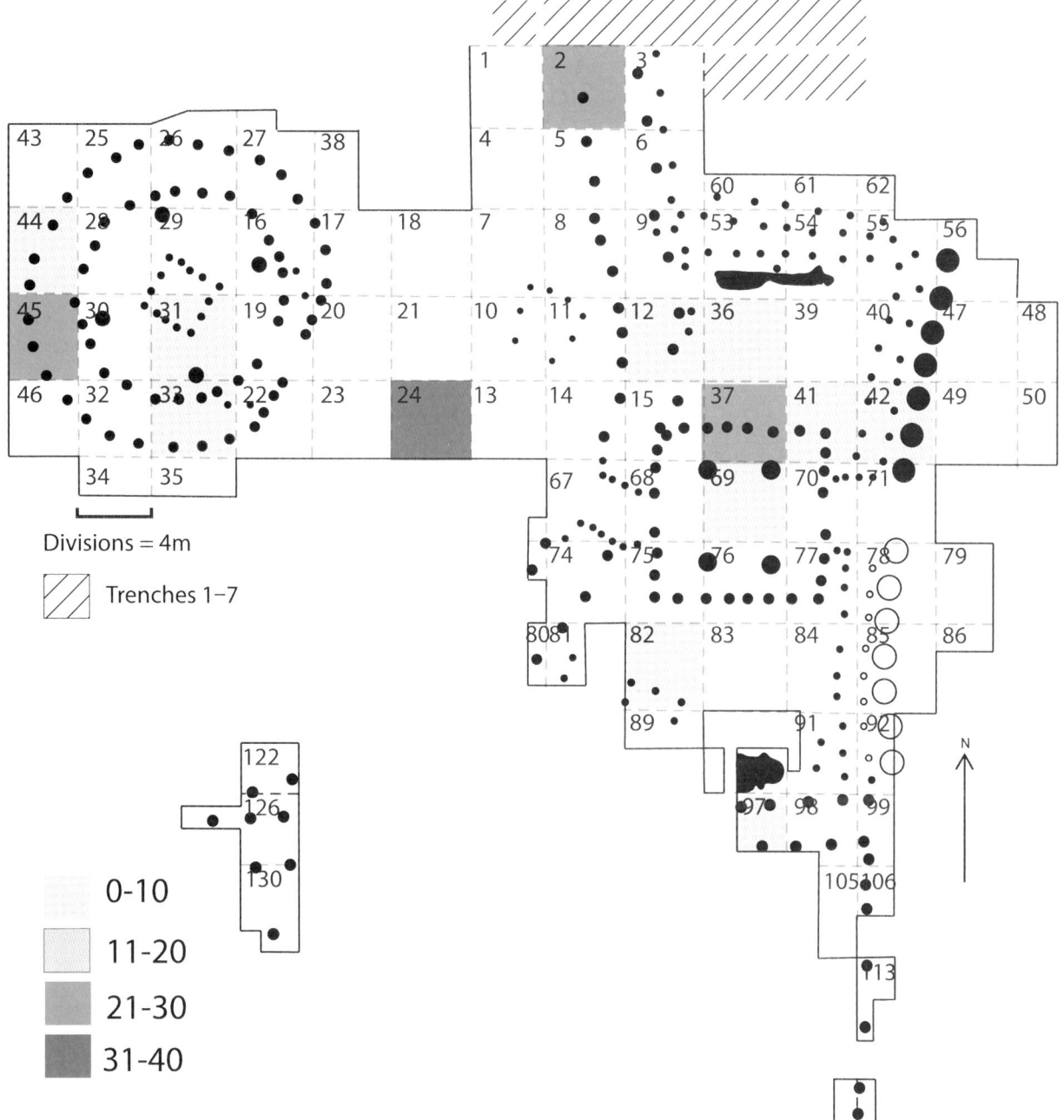

Figure 6.1 Distribution of the pottery recovered from the topsoil and ploughsoil contexts.

from the topsoil (Fig 6.1) correlates well with that recovered from the underlying features, and it is proposed that, generally speaking, the material in the topsoil is likely to be derived from these features.

One hundred and three sherds from the topsoil are of Grooved Ware (89 of which are of 'Knowth Style 1'), 36 sherds are of Coarse Ware, and there are also three possible Carinated Bowl sherds from the topsoil. The remaining five sherds are too abraded and fragmented to identify. The distribution of these pottery types in the topsoil and ploughsoil is shown in Figure 6.2. This highlights an interesting concentration of Coarse Ware in areas 44, 45 and 46, rather than in areas 41, 42 and 50, where a series of funerary deposits including chamber C591 and 'cist' C1014 are located. The Grooved Ware pottery is concentrated in areas 2, 24 and 37. The distribution of the pottery recovered from the topsoil is detailed by area square in Figure 6.2 and Table 6.2.

6.2.2 Features and contexts

A total of 269 pottery sherds were recovered from a total of 49 different features underlying the topsoil and ploughsoil layers (Table 6.3). This is only 55% of the total assemblage recovered from the site. The distribution of the pottery is detailed below under the various 'feature group' headings (Fig. 6.3).

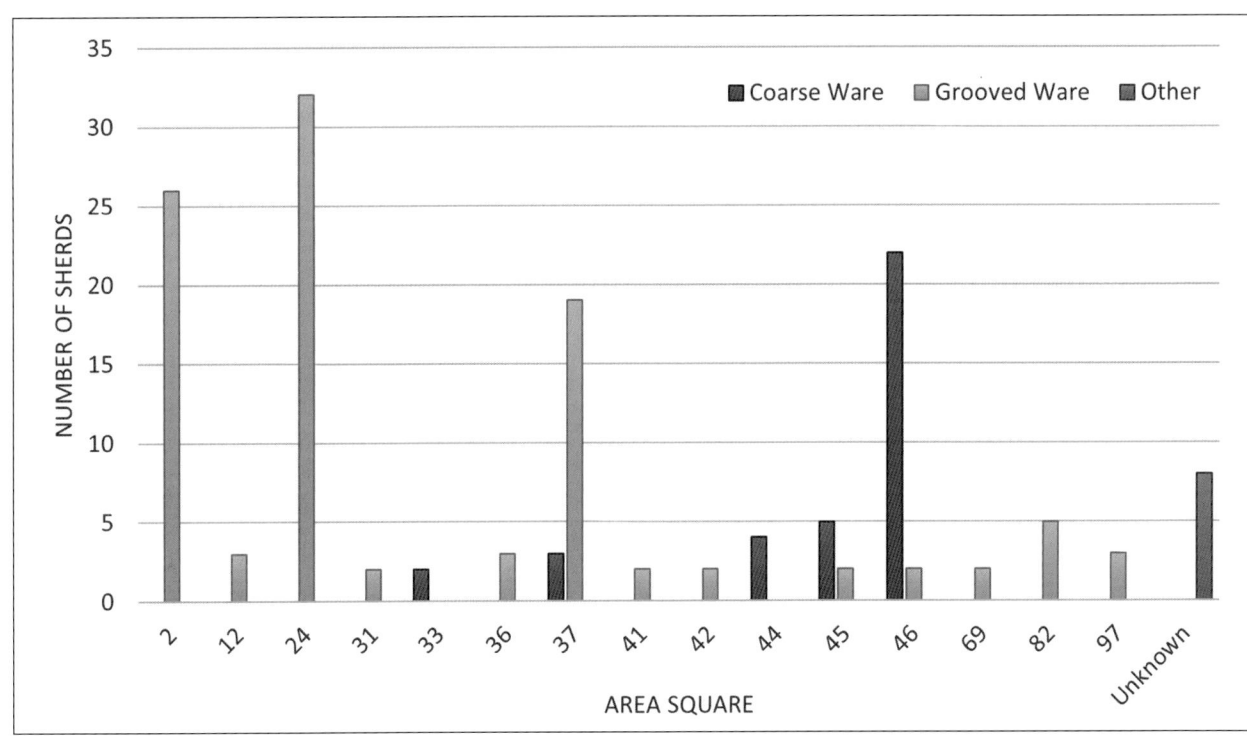

Figure 6.2 Distribution of pottery by type in the topsoil and ploughsoil layers of each area square.

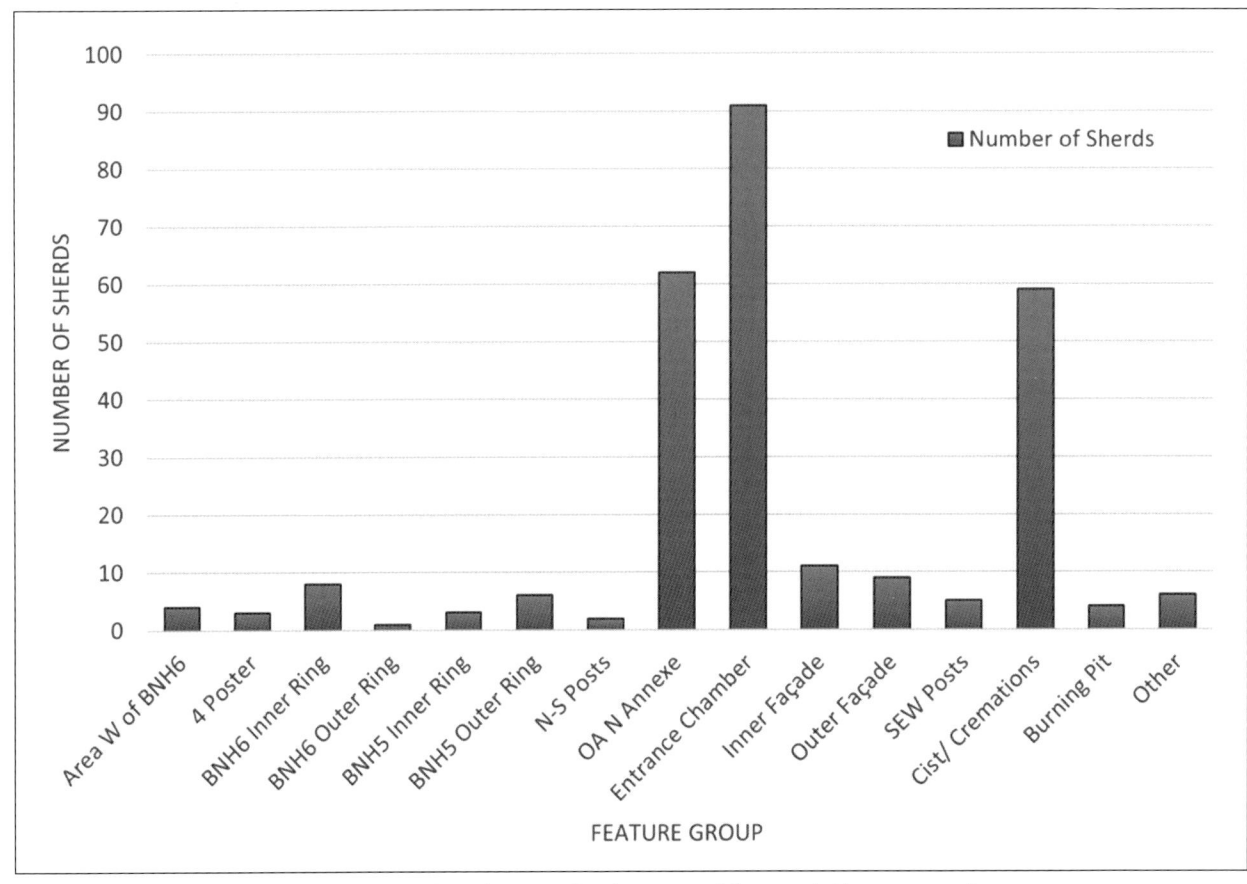

Figure 6.3 Number of pottery sherds recovered from each 'feature group'.

The pottery was recovered from both the primary and secondary fills of the posthole features. Of a total of 106 pottery sherds which could be assigned to either the primary or secondary fill of a posthole, over half (59 sherds) were found in the primary fill (Fig. 6.4). This finding is at odds with the results from the lithics,

Table 6.2 Detail of the distribution of the pottery sherds from the topsoil and ploughsoil layers by area.

Area	Context no.(s)	Pottery tradition	Pottery sub-classification	Vessel number	No. sherds	Overlies posthole(s)
2	209	Grooved Ware	Knowth Style 1	19.1	26	BNH5 IR1
12	664, 735 & 26; 220 & 622	Grooved Ware	Knowth Style 1	19.2	3	BNH5IR7, 8 & 9; BNH5 OR6 & 7
24	–	Grooved Ware	Knowth style 1	18	32	No features
31	490, 949, 926, 958, 657, 830 & 493	Grooved Ware	Knowth Style 3	–	2	EP1–5, 13 & 14
33	630; 507, 853, 813, 501; 503 & 504	Coarse Ware	–	–	1	4P1; BNH6 IR3–6; BNH6 OR6 & 7
36	–	Grooved Ware	–	12	3	NW corner of Northern Annexe
37	720, 879, 533 & 531	Coarse Ware	–	–	3	EC24–27
		Grooved Ware	Knowth Style 1	12	4	
			Knowth Style 1	17	10	
			Knowth Style 1	27	1	
			–	unknown	4	
41	902, 898, 783; 951; 895	Grooved Ware	–	–	1	EC28–30; Cist 951; IF8
42	951; 895	Grooved Ware	–	–	1	Cist 951; IF8; Outer Façade
44	941, 938 & 802; 943 & 931	Coarse Ware	–	5	1	OR15–17; Stone filled pits 943 & 931
	–	–	–	6	3	
45	921 & 918	Coarse Ware	–	4, 6 & 7	22	BNH6 OR13 & 14
	–	Grooved Ware	–	–	1	
46	960 & 1024	Coarse Ware	–	4	5	BNH6 11 & 12
	–	Grooved Ware	–	1	1	
69	1539 & 1546	Grooved Ware	–	–	1	EC39 & 40
82	1503, 1401, 1508, 1505 & 1589	Grooved Ware	Knowth Style 1	30	5	WNW2–4, 12 & 13
97	1574, 1544 & 1576	Grooved Ware	Knowth Style 2	–	3	SEW4, 5 & 10
Unknown	–	–	–	–	14	–
Total					147	

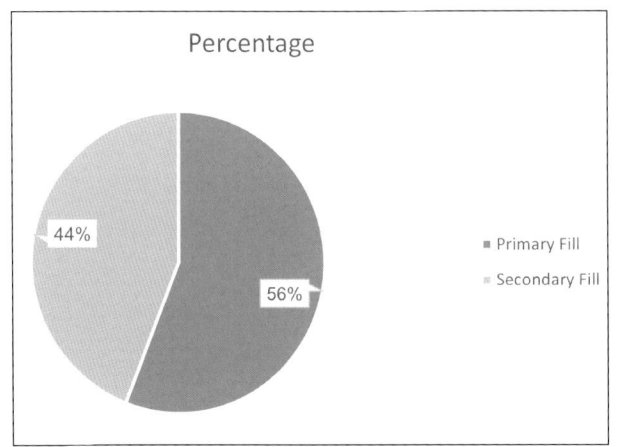

Figure 6.4 Percentage of pottery sherds recovered from the primary and secondary fills of postholes.

bone and stone assemblages, which showed that the majority of those assemblages were recovered from the secondary fill contexts. The primary fill of the postholes was established when the post was raised and so the pottery recovered from these contexts is associated with the erection of the post. The secondary fill was deposited within the posthole when the post was destroyed/extracted and the pottery recovered from these contexts was evidently involved in the destruction phase of the site. Since the creation of the Late Neolithic timber monument complex involved cutting through the existing land surface, with traces of earlier activities present, there is some scope for some residuality, as will be seen below in the discussion of the Early to Middle Neolithic Globular Bowl pottery from the Entrance Chamber area.

Table 6.3 Detail of the distribution of the pottery sherds from the feature groups.

Area	Associated posthole(s)	Context no. (s)	Primary/secondary	Pottery tradition	Pottery sub-classification	No. sherds
West of BNH6	–	931	–	Coarse Ware	–	3
	OR12	936	–	Grooved Ware	–	1
BNH6 OR	OR2	449	Secondary	Grooved Ware	Knowth Style 2	1
BNH6 IR	IR1	104	105: depression at edge of IR1	Grooved Ware	Knowth Style 1	1
	IR6	446	Context overlies IR6	Grooved Ware	Knowth Style 3	1
		578	Context overlies IR6	Grooved Ware	Knowth Style 3	1
	IR17	519	519 overlies IR17	Grooved Ware	Knowth Style 1	1
	IR22	186	Secondary	Grooved Ware	Knowth Style 2	2
		291	Secondary			2
	IR25	578	Context overlies IR6	Grooved Ware	Knowth Style 3	1
Four-Poster	4P2	948	Secondary	Miniature Vessel	–	1
	4P3	552	Secondary	Grooved Ware	–	1
BNH5 IR	IR1	209	Secondary(?)	Grooved Ware	Knowth Style 1	1
	IR10	287	Primary	Unknown	–	1
BNH5 OR	–	64	–	Carinated Bowl	–	1
	OR10	1596	Primary	Unknown	–	1
	OR13	1711	–	Beaker	–	5
Entrance to BNH5	–	1343	–	Grooved Ware	Knowth Style 1	1
N–S posts	NS9	–	–	Grooved Ware	Knowth Style 1	1
		127	Secondary	Grooved Ware	Knowth Style 1	1
North Annexe	EC28 & EC29	876	876 overlies EC28 & EC29	Coarse Ware	–	2
		876	876 overlies EC28 & EC29	Grooved Ware	Knowth Style 1	11
	–	777	777: oval scoop	Coarse Ware	–	1
	–	779	779 covers chamber C591	Coarse Ware	–	1
	–	779	779 covers chamber C591	Grooved Ware	–	1
	–	500	C500: possible floor layer	Grooved Ware	–	1
	–	555	Below C500	Grooved Ware	Knowth Style 1	32
	–	605	Below C555	Grooved Ware	Knowth Style 1	12
	–	612	Below C605	Grooved Ware	Knowth Style 1	1
	–	518	Charcoal spread	Grooved Ware	Knowth Style 1	1
	–	786	Shallow scoop	Grooved Ware	Knowth Style 1	1

(Continued)

Table 6.3 Detail of the distribution of the pottery sherds from the feature groups. (Continued)

Area	Associated posthole(s)	Context no. (s)	Primary/ secondary	Pottery tradition	Pottery sub-classification	No. sherds
Entrance Chamber	EC14	1272	Primary fill	Grooved Ware	–	6
		1280	Primary fill	Grooved Ware	Knowth Style 1	1
	EC15	1494	Primary fill	Grooved Ware	Knowth Style 2	3
		1494	Primary fill	Globular Bowl	–	1
	EC16	1345 & 1347	1345 primary; 1347 secondary	Grooved Ware	Knowth Style 2	14
		1345	Primary fill	Globular Bowl	–	2
	EC17	1269	Primary fill	Grooved Ware	Knowth style 2	17
		1270 & 1204	Secondary fills	Grooved Ware	Knowth Style 1 & 2	24
		1269	Primary fill	Globular Bowl	–	2
		1270	Secondary fill	Globular Bowl	–	3
	EC18	1618 & 1619	Primary fill	Grooved Ware	Knowth Style 1	4
	EC24	742	–	Grooved Ware	Knowth Style 1	1
	EC27	587	Primary fill	Grooved Ware	Knowth Style 1	3
	EC28	903	–	Grooved Ware	Knowth Style 1	7
	EC29	–	–	Grooved Ware	Knowth Style 1	1
	EC30	827	Primary fill	Grooved Ware	Knowth Style 1	6
	EC31	1417	Secondary fill	Grooved Ware	Knowth Style 1	1
	EC40	1430	Secondary fill	Unknown	–	1
Inner Façade	–	896	Primary fill: IF8 cuts through cist	Coarse Ware	–	10
Middle Façade	MF13	–	Secondary fill	Grooved Ware	Knowth Style 1	1
Outer Façade	–	–	Area 42	Coarse Ware	–	9
Chamber	C591		–	Coarse Ware	–	24
Stone Lined Cremation C1012	C1012	–		Coarse Ware	–	25
Stone Lined Cremation C1014	C1014	–		Coarse Ware	–	1
SE–W posts	SEW4	1574	–	Grooved Ware	Knowth Style 1	1
	SEW5	1544	–	Grooved Ware	Knowth Style 1 & 2	4
Burning pit	–	–	–	Unknown	–	4
Total						268

The recovery of pottery from the primary fills appears to be concentrated within the Entrance Chamber (Fig. 6.5 and Table 6.3). The material from the primary fill of posthole C895 is residual from the disturbance of the cist.

Analysis of the distribution of the Grooved Ware pottery by sherd type also highlights some distinct patterns (Plate 45). Within the timber circle BNH6, the Grooved Ware is represented solely by body sherds. However, within the postholes of the Entrance Chamber and in the northern Annexe area there is a greater variety of sherd types with rim, body and base sherds all represented. The number of sherds recovered from these two feature areas is also much greater than that found in the timber circle.

6.2.3 Interpretation of the pottery distribution

The distribution pattern confirms the close association of the Late Neolithic Grooved Ware assemblage with the timber monument complex, and that of the Middle Neolithic Coarse Ware with the funerary activities of that period. The spatial proximity of the various pottery styles, spanning some two millennia, reflects the repeated use of the area for different purposes, and this is a phenomenon noted at other sites associated with Grooved Ware in Ireland.

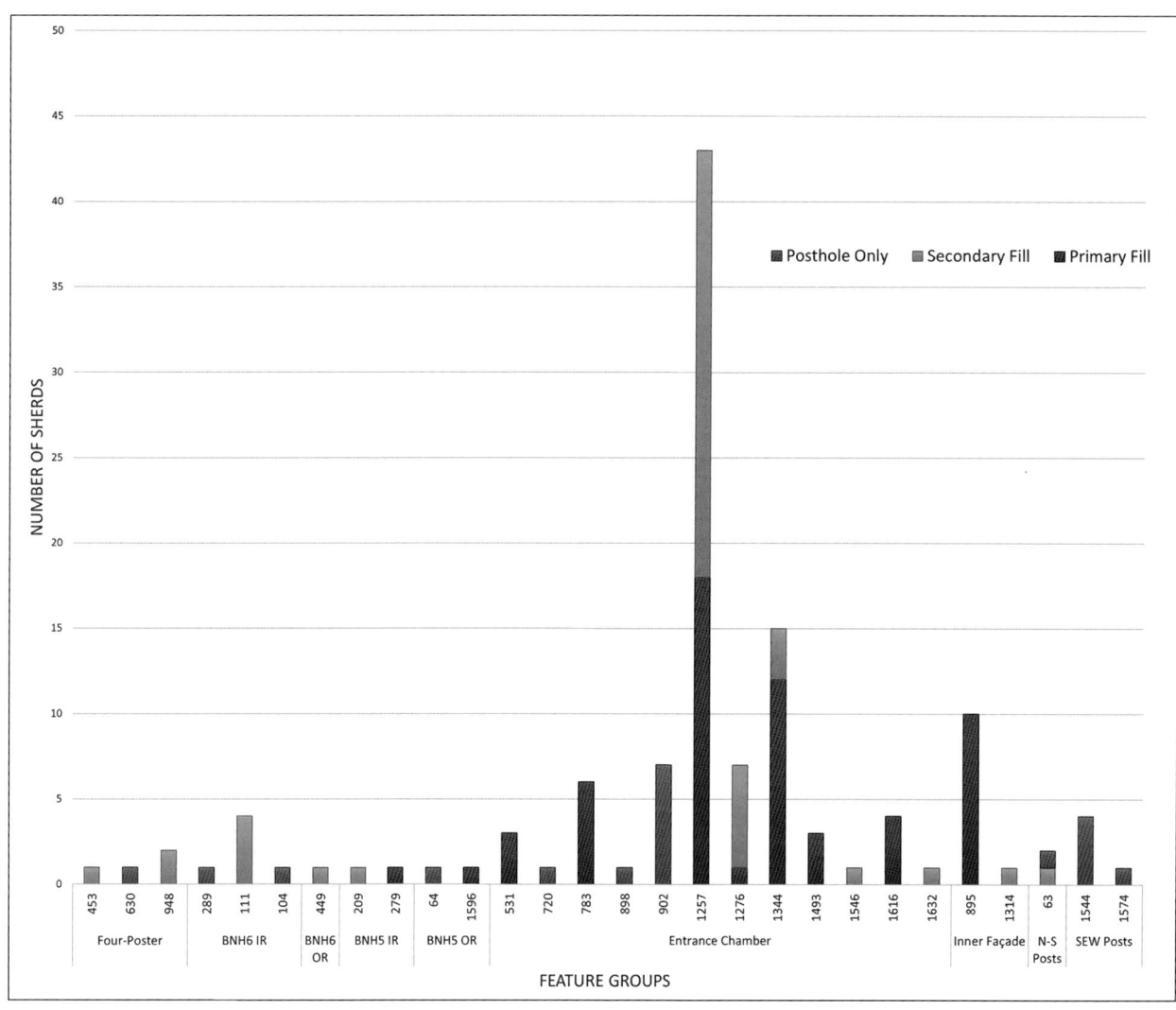

Figure 6.5 Number of pottery sherds from the post holes of each 'feature group', indicating the number of sherds recovered from the primary and secondary fills.

6.3 Pottery traditions present at Ballynahatty

6.3.1 Carinated Bowl pottery

Just four sherds of Carinated Bowl pottery (excluding the modified Carinated Bowl Globular Bowl sherds, described below) were found, and three of these come from unrecorded locations in the topsoil/ploughsoil. Only one sherd, which probably comes from the rounded base of a bowl, 10 mm thick, was associated with a feature: the plough-damaged upper fill of posthole OR1 (C64) in the Outer Ring of the BNH5 Enclosure. Its position there must be as residual material.

The fabric is hard, well-finished and dark brown in colour. Two of the small sherds found in the ploughsoil are featureless body sherds, but the third shows the carinated profile that is characteristic of this ceramic tradition (Fig. 6.6). The assemblage is too small to determine whether it belongs to the earliest, 'traditional Carinated Bowl' variant of this kind of pottery or to the subsequent 'modified Carinated Bowl' variant (as defined by Sheridan; 1995; 2007); but it may well be that it was broadly contemporary with the residual charcoal sample from the centre of BNH6 that produced a radiocarbon date of 3785–3639 cal BC (UB-4296; see Chapter 10 for details). It is not possible to tell whether the presence of this pottery, and of other Early Neolithic material (discussed in Chapter 11), relates to a phase of domestic occupation on the Ballynahatty ridge.

Carinated Bowl pottery is the earliest type of pottery to have been found in Ireland, and its origins can be traced in one of the regional groups of users of Chasséo-Michelsberg pottery in the Nord-Pas de Calais region of northern France (Sheridan 1995; 2007; 2016; Grogan and Roche 2010). The know-how to make it arrived with groups of immigrant farmers in the centuries around 4000 BC; however, here is not the place to rehearse arguments about the 'when' and 'how' of its arrival in Ireland.

To date, Carinated Bowl pottery has been found at or near seven other timber circles across Ireland, attesting to

6. The pottery

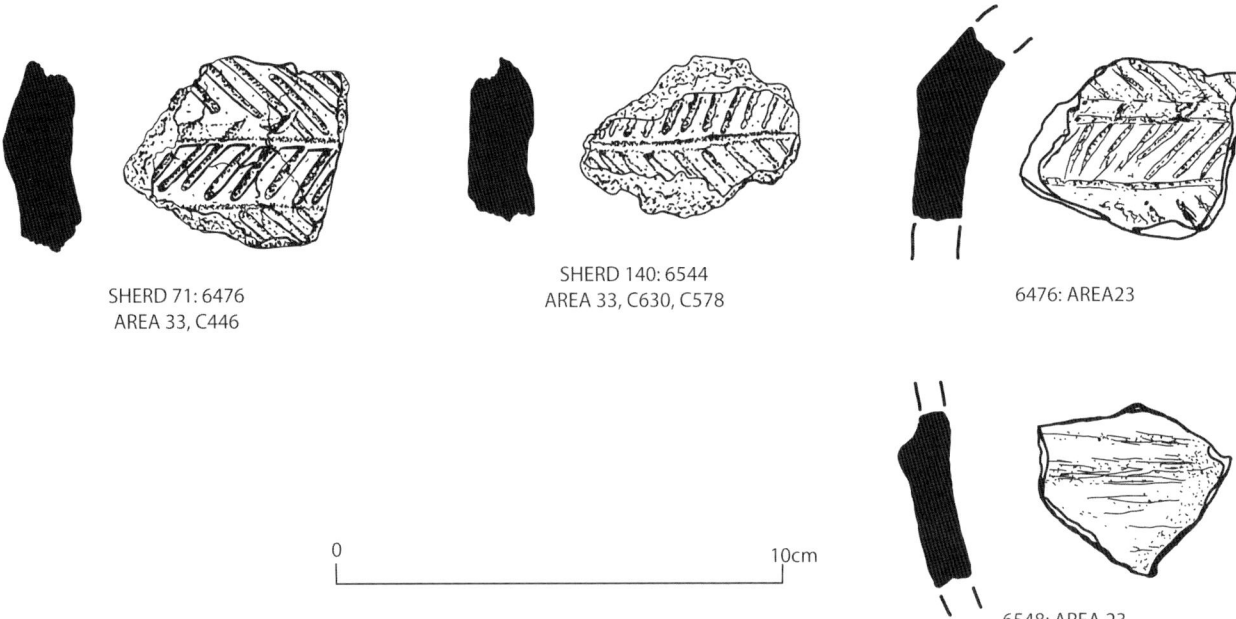

Figure 6.6 The Carinated Bowl sherds. Note the base and carination sherds.

pre-circle activities during the Early Neolithic. Sherds of Carinated Bowl were found at Ask, Co. Wexford (Grogan and Roche 2011a, 161) and Scart, Co. Kilkenny (Roche and Grogan 2010a), but they came from unsecured contexts and can tell us little about the nature of this early activity at these sites. With some other timber circles, it was found in pits. At Donnydeade, Co. Tyrone, Carinated Bowl pottery appears to have been deliberately deposited in a pit in which a single sherd of later Grooved Ware pottery was also found; however, it is not clear how (if at all) this pit relates to the timber circle, or whether the Grooved Ware was intrusive in an Early Neolithic pit (*Unearthed* 2019, 13). Five pits containing 172 Carinated Bowl sherds were excavated approximately 40 m to the east of the timber circle at Lowpark, Co. Mayo (Cleary n.d., 54). Much of this evidence for pre-circle activity was, however, badly disturbed or destroyed by later Bronze Age activity (Gillespie n.d., 15). A number of pits containing Carinated Bowl pottery were also excavated at Lagavooren in Co. Meath, with the majority (103 sherds) found in a single pit (265) (Grogan and Roche 2012, cxiv). There is no direct link between the Early Neolithic activity and the timber circle site. A single sherd of Carinated Bowl was also recovered from a gully to the north of the timber circle at Glenshane, Co. Derry. It is believed that this may relate to early funerary activity at the site, but this cannot be confirmed (Nicol and Donaghy 2020, 133).

Knowth, Co. Meath is the only site to have produced Carinated Bowl pottery from features other than pits. Here it was recovered in association with unambiguous evidence for domestic activity, including the remains of a rectangular house (Eogan and Roche 1997, 39). Located at the site of the great passage tomb, this so-called 'Western Neolithic Complex' was the earliest activity uncovered at Knowth (*ibid.*, 5). Radiocarbon dates of 4324–4045 cal BC, 3975–3789 cal BC and 3948–3784 cal BC were obtained from the excavation of the area where the Carinated Bowl pottery was recovered, although the dated charcoal was not directly associated with the pottery, and the publication does not state its species (*ibid.*, 39).

6.3.2 Early to Middle Neolithic 'Globular Bowl' pottery

Fourteen sherds identified as belonging to this kind of pottery were found. Nine, found in the primary and secondary fills of postholes EC15–17 in Area 75, the south-west corner of the Entrance Chamber, appear to belong to a single vessel (Plate 47), while the remainder come from pit C16 in Area 45. This pit, whose fill was rich in charcoal, is located at the rear of the timber circle BNH6, opposite the entrance. Unfortunately, it is not possible to determine the chronological relationship between this pit and the timber circle, as the pit is not dated.

The vessel whose sherds were found in the three neighbouring pits EC15–17 is undecorated, with a simple rounded, slightly inturned rim, wall thickness of *c.* 9 mm, and an estimated rim diameter of 160 mm. Abundant fragments of a black and white speckled igneous rock are present as a filler; further investigation would be necessary to determine whether this is the microdiorite noted in some of the Grooved Ware sherds (discussed below). Only the upper part of the vessel survives. During examination of the assemblage, these sherds were not regarded as belonging to the Grooved Ware tradition and, if their identification as Globular Bowl sherds is correct, the vessel will have had a rounded, rather than a flat base. One cannot, however, rule out the possibility that the pot was in fact an undecorated, flat-based Grooved Ware pot. The presence of its sherds in three neighbouring postholes suggests that they may have been deposited deliberately – an interpretation that would favour a Grooved Ware attribution – although if the vessel is a Globular Bowl, then the sherds will be residual from an earlier phase of activity. It should be noted that abundant sherds of Grooved Ware were found in the fills of the postholes in question.

If one accepts the identification as Globular Bowl, then an Early to Middle Neolithic date seems likely – possibly *c*. 3600–3300 BC (Sheridan 1995; Grogan and Roche 2010). The term 'Globular Bowl' is a catch-all term to refer to various kinds of bowls, dating to the Early to Middle Neolithic, that have a globular profile; most are decorated. At least some may be regarded as having evolved, through a process of style drift, from the uncarinated vessels within the repertoire of the Carinated Bowl tradition (hence the use of the term 'modified Carinated Bowl' to describe it above). A possible comparandum for the vessel shown in Plate 46 comes from Audleystown court tomb, Co. Down (Collins 1954, fig. 6.1). However, the absence of the lower part of the vessel shown in Plate 46 leaves some doubt about the reliability of its identification as a round-based pot.

6.3.3 Middle Neolithic Coarse Ware

Seventy-six sherds of Coarse Ware pottery (also known as 'Carrowkeel Ware'), from at least eight vessels, were recovered during the excavations, and these add to the three or (more likely) four such vessels found in the 'subterranean chamber' found in 1855 (described in Chapter 1). The pottery was recovered from three distinct areas within the site (Plate 46). The greatest concentration was recovered from chamber C591 and the immediate surrounding area (Plate 46: area a). The chamber was disturbed during the erection of posthole IF8, which was driven through the western edge, shattering both the capstone and base stone. A number of residual sherds, probably originally from the chamber, were recovered from the fill of the posthole IF8 and also from several spreads in areas 37 and 41, which lie just to the west of the chamber, covering areas of the Annexe and also the northern section of the Entrance Chamber. The apparent regard for the pottery vessels and their contents by the builders of the Inner Façade is interesting to note (as discussed in Chapter 5, Section 5.2). Large parts of two pots were recovered from chamber C591. The first (Vessel **1**: Plate 48) varied from 12–18 mm in thickness and is of a very sandy fabric. It is very poorly made, and the decoration is roughly applied in the form of stab and drag ornament covering the surface of the pot. The rim is pointed and inverted, and the vessel is globular in shape with a round base. The second pot (Vessel **2**: Plate 49) from the chamber ranges from 11–16 mm in wall thickness and is completely covered with stab and drag decoration, arranged in nested swags. The fabric of the second pot is not as sandy, and some small angular lithic inclusions are visible. The rim of this pot is flat and inverted and has lines faintly incised along the top of the rim. The diameters of both pots are estimated to be around 160 mm, although the accuracy of these measurements is compromised somewhat by the pots' poor condition. This was the only part of the site where Coarse Ware was recovered in association with postholes.

The second concentration of Coarse Ware was associated with the deposits of cremated remains outside (and pre-dating) the Façade (Plate 47: area b), including 'cist' C1014. The pottery included the fragmented remains of two vessels. Both vessels were of similar fabric and decoration to the Coarse Ware pots found in chamber C591. The third concentration of Coarse Ware was identified to the west of the BNH6 structure (Plate 47: area c) and comprised the fragmented and abraded remains of at least four Coarse Ware vessels. The first is represented by a portion of a rim with an estimated rim diameter of *c*. 160 mm. The wall thickness varies between 9 mm and 19 mm and the fabric is harder and better finished than that recovered from the chamber. The rim is flat and inverted. The top of the rim has been ornamented with finger- or thumbnail impressions, and below the rim are horizontal rows of heavily impressed 'dots', which are almost rectangular in shape. Body fragments of another vessel, with stab and drag decoration. were also recovered from this area. The fabric is coarse with large inclusions and is 10–13 mm thick. Another group of body sherds, 10–16 mm thick and roughly ornamented with stab and drag decoration, make up another Coarse Ware pot. A final group of sherds are from the base of a round-bottomed vessel. The sherds are very abraded, ranging from 6 mm to 13 mm in thickness, and are undecorated. These sherds pre-date the erection of the timber circle; they are from sealed contexts. No Coarse Ware sherds were found in the postholes that lay in the same area. The sherds recovered from the topsoil to the west of BNH6 are, therefore, likely to have come from sealed contexts which are likely to have been disturbed relatively recently by ploughing.

The types of Coarse Ware fabric recovered from Ballynahatty fall into two groups. The first comprises four vessels of poor quality, two of which were associated with the subterranean chamber C591 on the line of the Inner Facade (Plate 46 area a), and the other two come from the line of four deposits of cremated remains outside the entrance through the Outer Façade (Plate 46: area b). The vessels are all of similar flaky, thick sandy fabric (with varying amounts of sand), adorned with roughly executed stab and drag decoration. No Coarse Ware was recovered in association with the deposits of cremated remains in the BNH6 area.

The second fabric group was located within C931, a stone-packed cut to the west of BNH6, with some sherds in the overlying topsoil (Plate 46: area c). These features were recovered at the limit of the excavated area and so may be indicative of remains lying further within the BNH5 enclosure. The pottery in this area was of a harder and better-finished fabric than the types recovered from chamber C591 and the deposits of cremated remains in stone-lined hollows. It may be that the difference in fabric and finish is indicative of the function that the pottery was required to perform. It is unlikely, for example, that the pottery recovered from the cist could ever have been handled to any degree, so rough is the construction of the vessel. The pot is likely to have been prone to breakage, and it is difficult to see how it could have held liquid or foodstuffs or been heated over a fire. The pottery recovered from the area to the west of BNH6, in contrast, is of a harder, better-finished fabric. The decoration on this pottery is also better executed. It may be that the pottery found within chamber C591 was made specifically to be deposited within that tomb. The better-finished pottery may have been required for more quotidian usage, for example, storage or for heating foodstuffs over a fire. This need not necessarily mean a 'domestic' function, with food and drink undoubtedly forming part of ritual ceremonies.

The decoration adorning the Coarse Ware from Ballynahatty has parallels elsewhere (Fig. 6.7). Both the pots from chamber C591 have stab and drag decoration, roughly horizontal in the case of the pot featured in Plate 48 and in swags in the case of the pot in Plate 49. Vertical lines of stab and drag decoration are present on the vessel recovered from Millin Bay, Co. Down (Collins and Waterman 1955). Vessels with all-over stab and drag lines were recovered from the Mound of the Hostages at Tara, Co. Meath; Altanagh, Co. Tyrone; and under peat at Lislea, Co. Monaghan (Herity 1974). A sherd with lines of impressed marks made with a small point was recovered from Ballynoe stone circle in Co. Down (Groenman-van Waateringe and Butler 1976), comparable with the decoration found on one sherd from Ballynahatty. A rimsherd decorated with finger- or thumbnail impressions has also been recovered from Knowth, Co. Meath (Eogan and Roche 1997), although is not as neatly executed as the example from Ballynahatty. Rims similar to the types found at Ballynahatty (*i.e.* pointed and inverted or flat and inverted) are typical of this type of pottery and have been recovered, for example, from Altanagh, Co. Tyrone and Ballynoe, Co. Down.

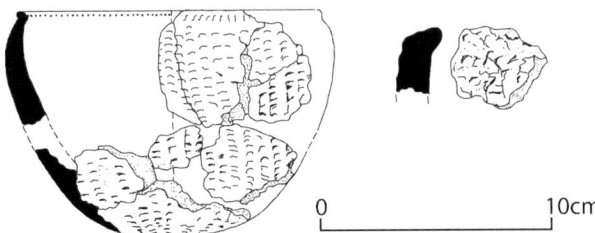

Figure 6.7 Coarse Ware pottery from Millin Bay and Ballynoe, Co. Down (from Herity 1974, fig. 138).

Coarse Ware ('Carrowkeel Ware') has most frequently been found in passage tombs but, as pointed out by Sheridan (1995), it has also been found in a variety of non-passage tomb contexts (as noted above), including at the court tomb at Audleystown, Co. Down (Collins 1954, fig. 7.4) and the porphyry extraction site at Eagle's Nest, Lambay, Co. Dublin (Sheridan 2022). The main currency of its use appears to lie between *c.* 3300 BC and *c.* 2900 BC; the radiocarbon dates obtained for associated excavated material, and for human remains from the 1855 'subterranean chamber', are consistent with this. Elsewhere, at Knowth, it has been suggested that the 'Decorated Pottery Complex' featuring this kind of pottery came into existence around 3500 cal BC and continued for a number of centuries (Eogan and Roche 1997, 96).

The choice of the term 'Coarse Ware', rather than 'Carrowkeel Ware', to describe this pottery follows Sheridan's recommendation (1995, 12; *cf.* Brindley and Lanting 1990, 6; Cooney 2000, 124), which sought to break down the assumption that this ceramic tradition is solely associated with passage tombs (as reflected, for example, in Herity's assertion that it is 'found exclusively with passage grave burials … and was clearly the normal funerary ware of the passage grave builders' (1974, 142). However, the term 'Carrowkeel Ware' remains in common use. The key point to note is that the same type of pottery is being referenced in the use of both terms.

6.3.4 Grooved Ware

6.3.4.1 Introduction

A total of 282 Grooved Ware sherds, comprising 212 body sherds and fragments, 32 base and base angle sherds and 38 rim sherds, were recovered during the course of the excavation. It is estimated that the sherds from Ballynahatty represent a minimum of 30 vessels (labelled **BNH1–30**), all flat-based. (One further pot – a possible 'miniature' vessel from the secondary fill of posthole 4P2, C948, in the Four-Poster setting inside BNH6 – may well be of Grooved Ware, but it was not allocated a pot number.) The vessels were subdivided into three styles, based on a classification scheme that had been developed for the Grooved Ware assemblage at Knowth (Roche 1995). Most of the assemblage (194 sherds, 69%; 22 out of 31 pots) belong to Knowth Style 1. These pots (**BNH 1–22**) are small, thin-walled (2–8 mm thick), finely made pots of tub and barrel shape, with carefully smoothed exterior surfaces and estimated rim diameters of *c.* 145–255 mm. Decoration is minimal and takes the form of one or up to four horizontal incised lines on the interior, immediately below the rim (Figs 6.8–6.13). One Knowth Style 1 pot, **BNH 18**, has a horizontal applied cordon on its exterior (Fig. 6.14). The fabric of Knowth Style 1 pots contains moderate to high quantities of finely crushed microdiorite and quartzite inclusions.

Knowth Style 2 pots (**BNH23–26**) are also small and probably have the same shape as Knowth Style 1 pots,

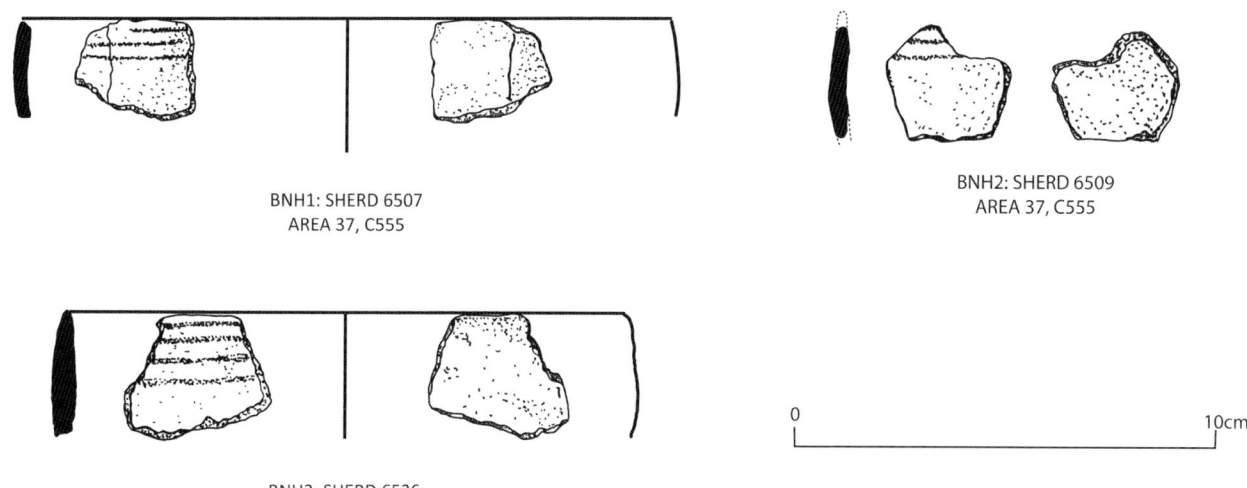

Figure 6.8 Knowth Style 1 vessels **BNH1–3** *(after Roche 1995, fig. 22).*

but they are generally thicker (with wall thickness ranging between 4 mm and 14 mm) and of a coarser fabric, with large quartzite inclusions that range from 3.7 mm to 4.6 mm. This style is represented by 67 sherds (24%) from four vessels. None of the Knowth Style 2 vessels has horizontal grooving on the interior of the rim. The only evidence of decoration comes in the form of a narrow, applied cordon and a narrow, applied V-shaped cordon on the exterior of vessel **BNH23** (Fig. 6.14).

Knowth Style 3 is represented by just four sherds (1% of the assemblage), from four pots (**BNH27–30**, Fig. 6.15). The Knowth Style 3 vessels are thick, at 10–11 mm. The sherds are well-finished and hard, with an external applied slip. The exterior of the sherds is covered all over with finely incised decoration, which takes the form of oblique lines and, in one case, forms a herringbone pattern. Two of the sherds have evidence of an applied cordon, creating a ridge on the surface. It is not possible to tell from the small sherds what overall form the decoration may have taken or whether the cordon ran horizontally or vertically. The Knowth Style 3 fabric can be either fine or coarse and contains microdiorite and quartz inclusions. The inclusions range in size from <=1 mm to <=5.5 mm.

In addition, there were a further 17 sherds (6%) were too fragmented and abraded to classify. The principal characteristics of all the identifiable Grooved Ware pots from Ballynahatty are summarised in Table 6.4.

6.3.4.2 Fabric, use, condition

The use of igneous rock inclusions, as seen in the Knowth Style 1 and 3 vessels, is a common feature in Irish Grooved Ware. Dolerite was the main inclusion in the Grooved Ware pottery from Scart, Lagavooren (Grogan and Roche 2012, cxv), Ask (Grogan and Roche 2011a), Kilmainham 3 (Grogan and Roche 2011b) and Newgrange (Williams and Jenkins 2020, 29), but there were also a number of other inclusion types including quartzite (Lagavooren and Ask), fine shale (Ask), and granite (Kilmainham 3).

Grooved Ware from the pits at Fourknocks Ridge, Co. Meath contained large microdiorite inclusions (King 1999. 173), while that from Lowpark appears to have been tempered with diorite and porphyry (Cleary n.d.). The Grooved Ware pottery at Glenshane, Co. Derry does not contain any igneous rock inclusions; instead, it is noted as being tempered with quartz and 'gravel' (Barkley 2020). It is possible that the igneous rock and quartzite inclusions were deliberately added as temper, as they both increase thermal shock resistance, which is desirable for cooking vessels or vessels that would have been exposed to thermal stresses (Quinn 2013, 158).

Examination of the exposed cross-sections of the Grooved Ware sherds from Ballynahatty revealed that the fabric ranged in colour from black, through brown to orange. While a few sherds were completely black throughout, which is indicative of a reduced firing atmosphere, the majority of the sherds displayed chromatic variation across their core and margins, with orange and brown hues common. This would suggest that the sherds were fired in an oxidising atmosphere, which is indicative of open firing, where firing temperature and atmosphere can be highly variable.

Regarding the use of the vessels, a striking 85% of the Grooved Ware assemblage from Ballynahatty has sooting on the surfaces (Plate 50). It is likely that the vessels became charred prior to fragmentation rather than as broken sherds, as no sherd profiles are charred as would be expected if broken potsherds were thrown onto a fire. This is a common feature of Grooved Ware assemblages recovered from timber circles. Evidence of carbonised accretions has been observed on Grooved Ware sherds from sites including Lowpark, Co. Mayo (Cleary n.d., 57), Kilbride (Roche and Grogan 2008, 29–30), Scart (Monteith 2008, 118), Balgatheran (Ó Drisceoil 2009, 95), Ballynacarriga 3 (Roche and Grogan 2010b, 247) and Lagavooren (Grogan and Roche 2012, cxv). This suggests that the Grooved Ware deposited at timber

6. The pottery

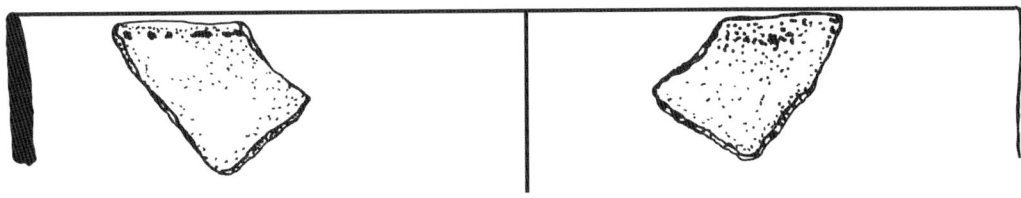

BNH4: SHERD 6480
AREA 36

BNH4: SHERD 6508
AREA 37

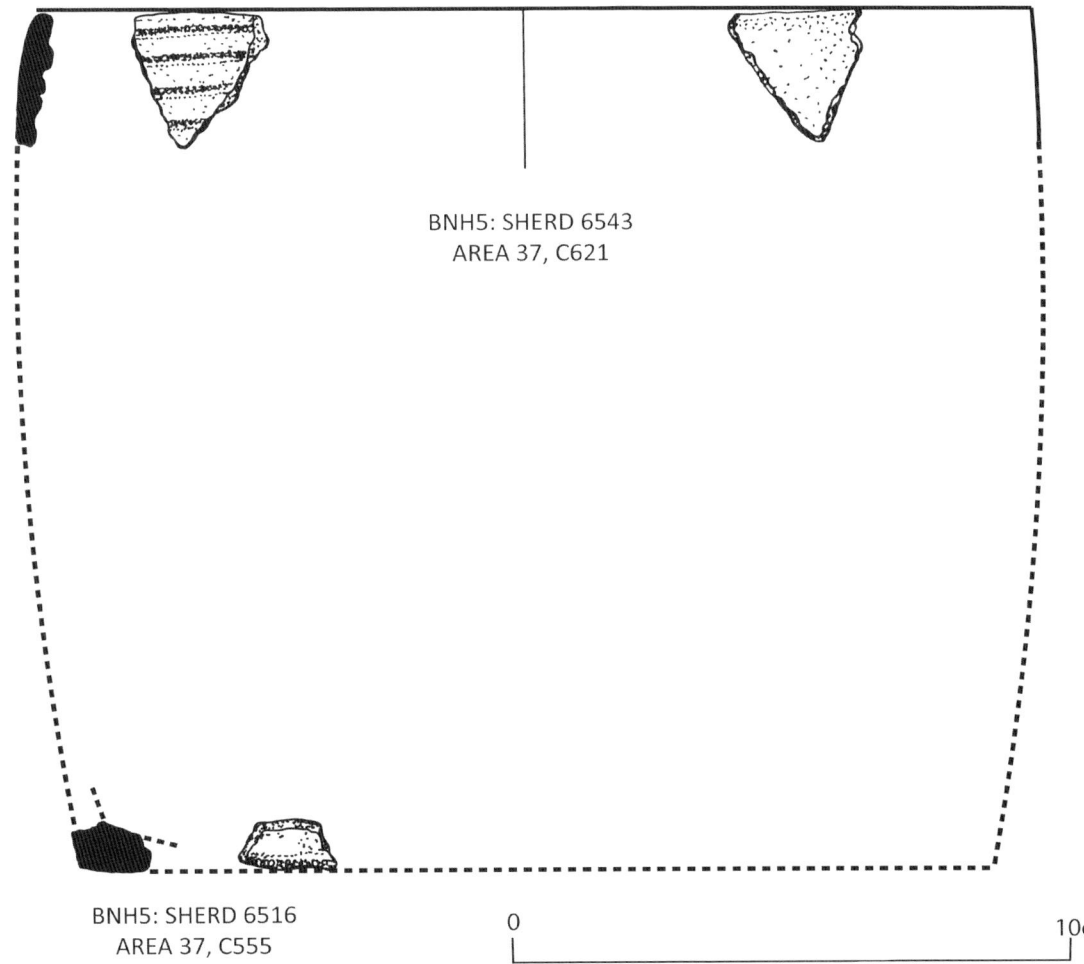

Figure 6.9 Knowth Style 1 vessels **BNH4** *and* **BNH5** *(after Roche 1995, fig. 22).*

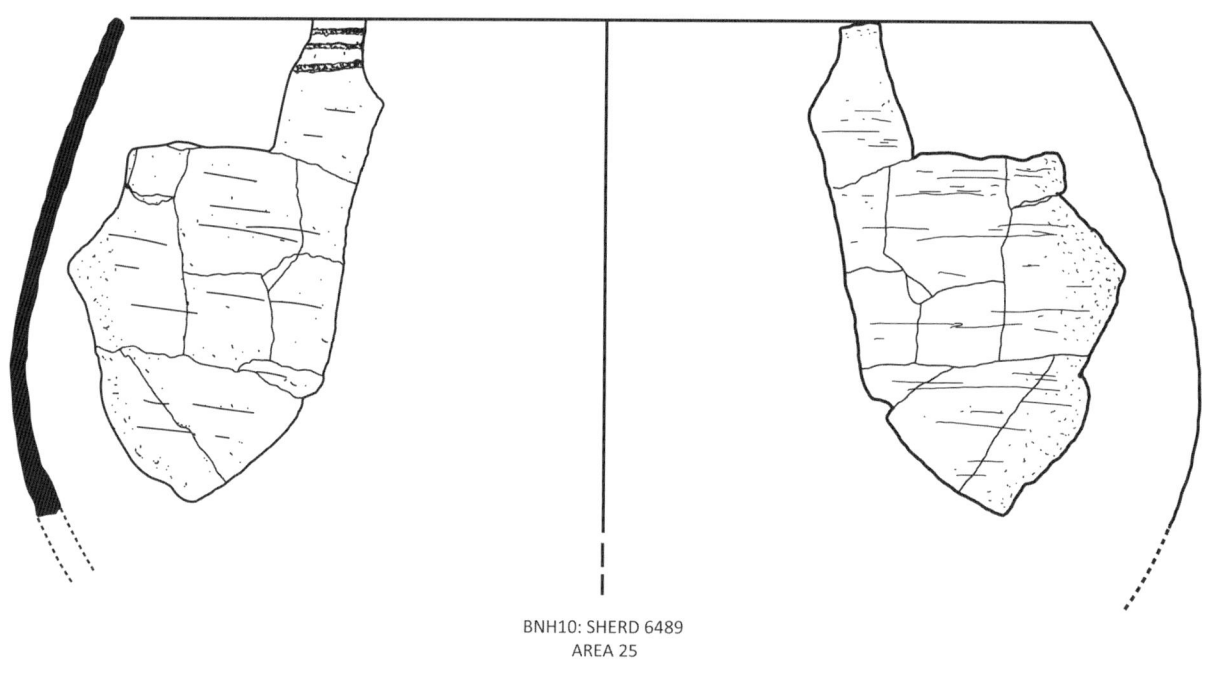
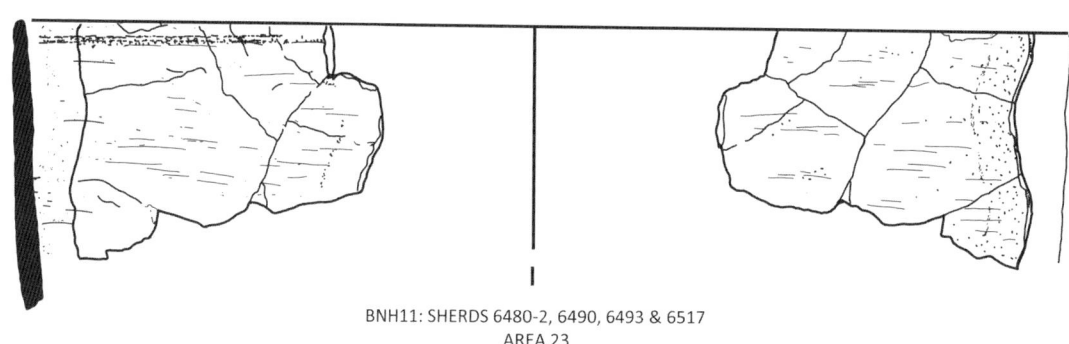
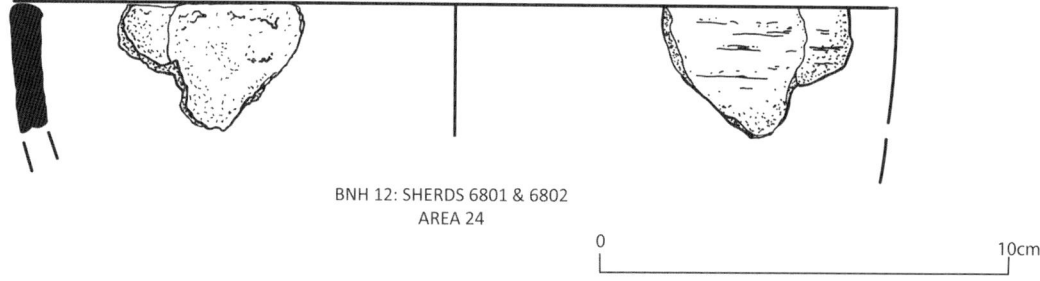

Figure 6.10 Knowth Style 1 vessels **BNH10–12** *(after Roche 1995, fig. 22).*

circles was often used for the preparation of foodstuffs. The deposition of these used vessels may represent an aspect of the 'ritualising' of everyday domestic activities to build and maintain group identity based on 'the idea of the household' (Carlin and Cooney 2017, 46).

The Grooved Ware from Ballynahatty is not particularly abraded; only a small number of sherds are weathered, suggesting that most of the assemblage was sealed into a context shortly after breaking rather than lying around on the ground surface. This is also the case at Knowth

6. *The pottery*

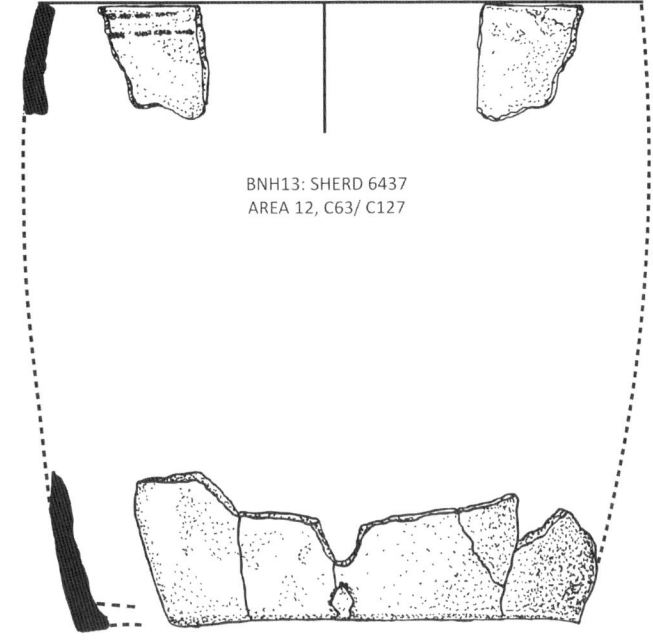

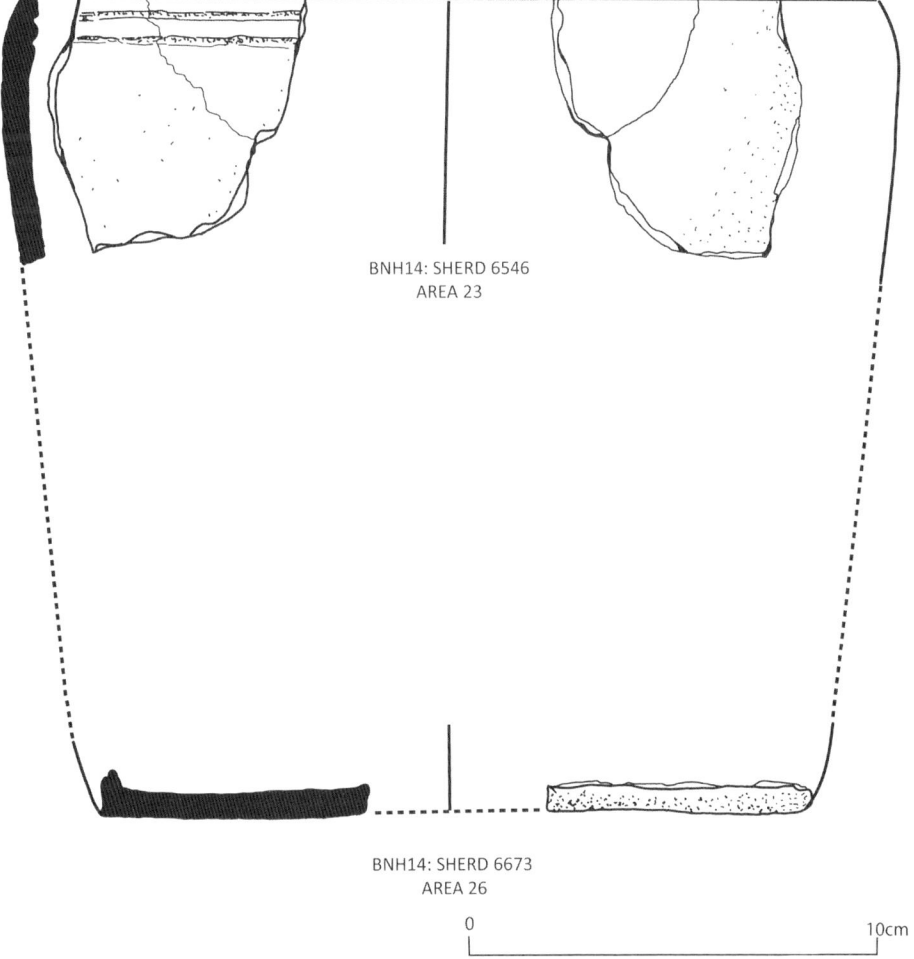

Figure 6.11 Knowth Style 1 vessels **BNH13** *and* **BNH14** *(after Roche 1995, figs 22 and 23).*

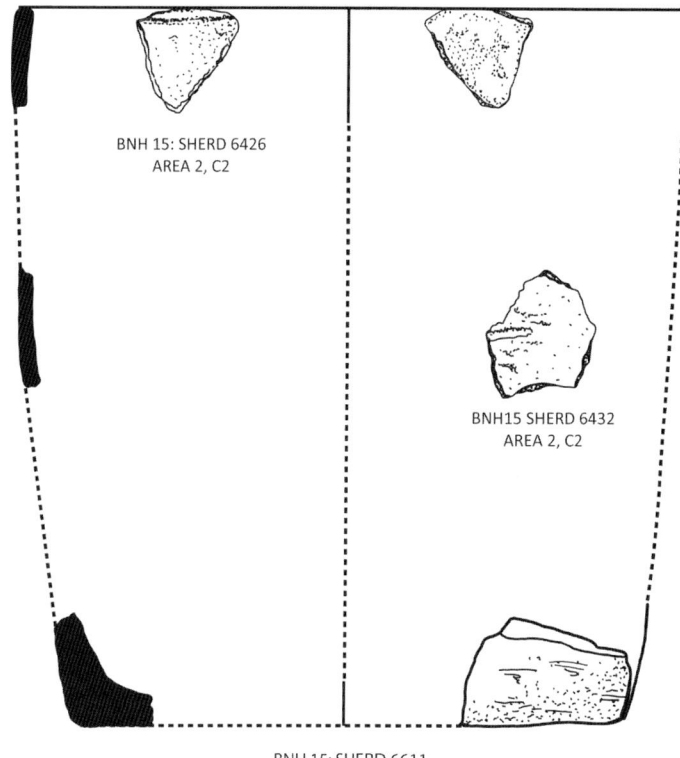
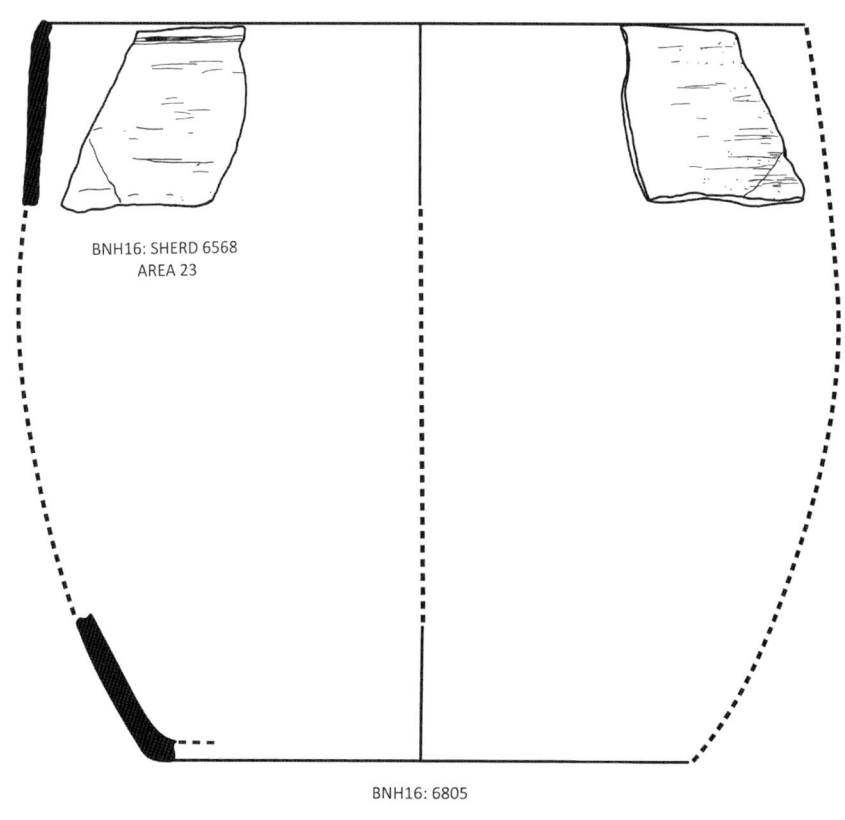

Figure 6.12 Knowth Style 1 vessels **BNH15** *and* **BNH16** *(after Roche 1995, figs 23 and 24).*

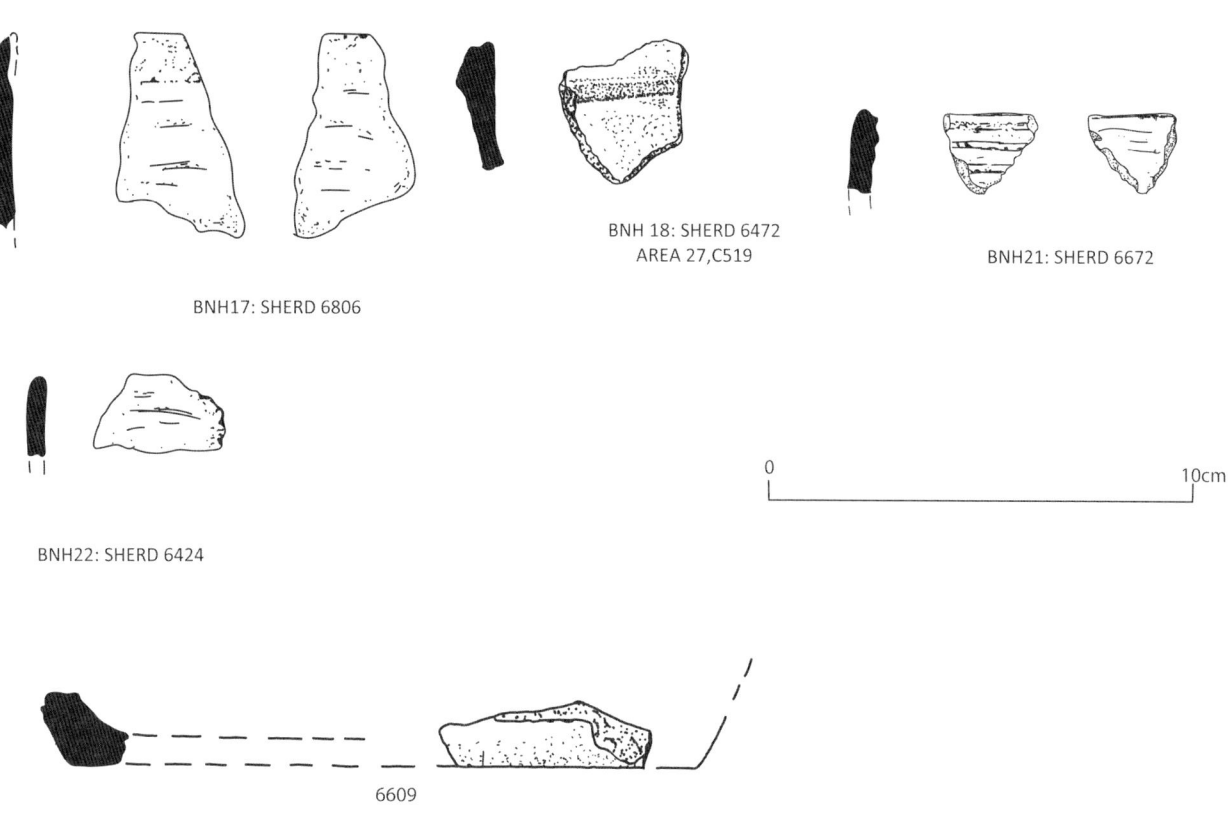

Figure 6.13 Knowth Style 1 vessels **BNH17**, **18**, **21** *and* **22** *(after Roche 1995, fig. 24).*

Figure 6.14 Knowth Style 2 pot **BNH23**, *with narrow applied V-shaped cordon (after Roche 1995, fig. 24).*

(Eogan and Roche 1997, 101), Lowpark (Cleary n.d., 58) and Kilmainham 3 (Grogan and Roche 2011b, xlii), where there is little evidence for erosion on either the sherd surface or edge break. This may suggest that there was only a short period of time between the use of the pots and their subsequent deposition in the post pits. There are, however, a number of sites, including Lagavooren (Grogan and Roche 2011b, cxv), Kilbride (Roche and Grogan 2008), Ballynacarriga 3 (Lehane and Leigh 2010, 246), Scart (Roche and Grogan 2010a), where the worn condition of some of the Grooved Ware sherds suggests that the vessels may have been used or exposed prior to their deposition, or that they may even have been derived from midden material (Carlin and Cooney 2017, 42).

6.3.4.3 Spatial and contextual distribution of the Grooved Ware

An area-by-area review of the distribution of the Grooved Ware pottery is offered below. As regards the distribution of the different styles of Grooved Ware, while the Knowth Style 1 sherds were found widely distributed across the entire site, they are mainly found in BNH6, the northern Annexe and areas of the Entrance Chamber. Knowth Style 2 sherds were recovered mainly from the Entrance Chamber, and the Knowth Style 3 sherds were only found within the timber circle BNH6 (Plate 51).

Grooved Ware pottery was excavated from both primary and secondary contexts within the postholes of the Ballynahatty complex. The sherds recovered from the primary fill of postholes in the Annexe presumably date from the construction event. The problem lies in confirming whether or not any of the sherds are residual. It is

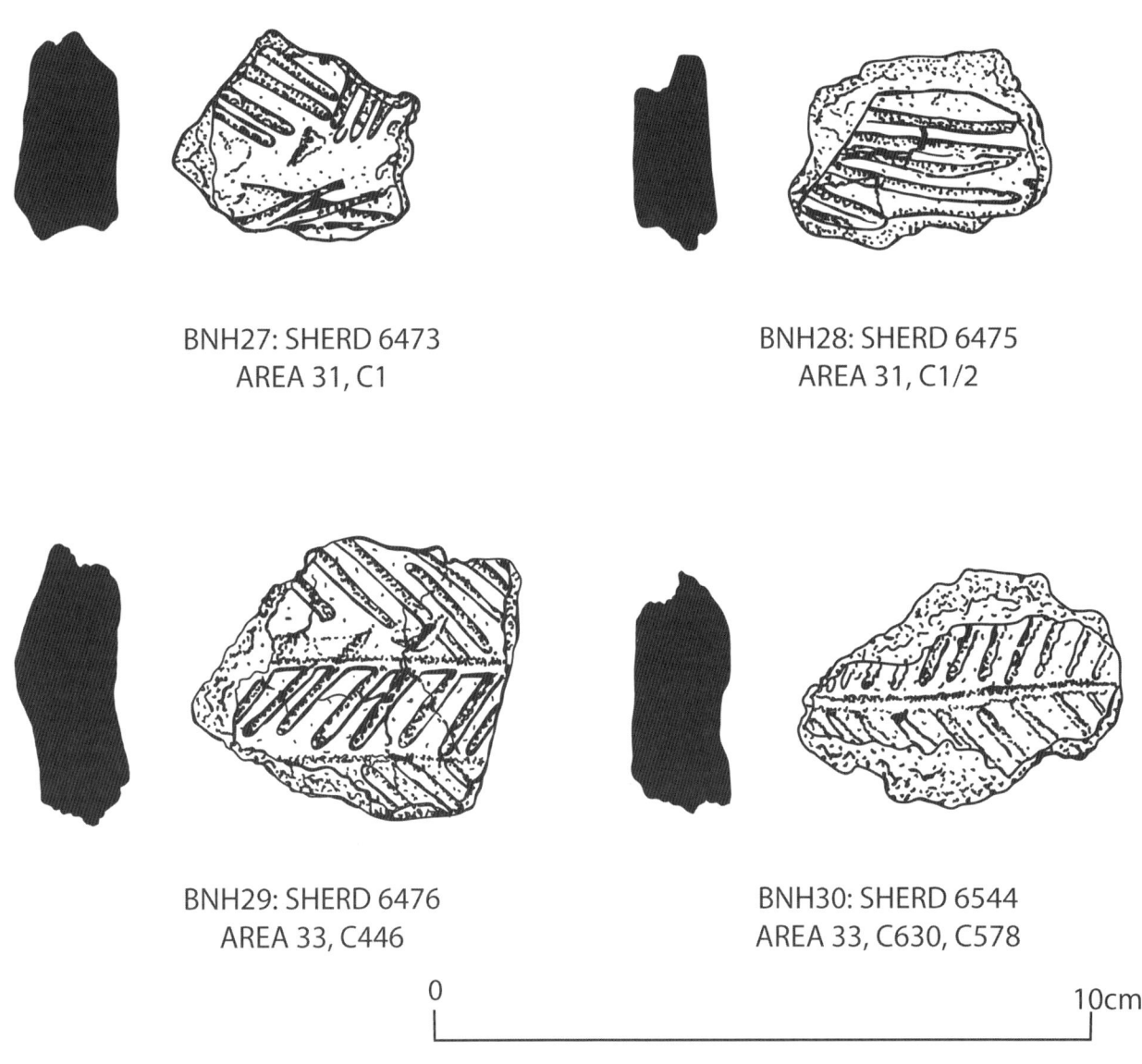

Figure 6.15 Knowth Style 3 vessels **BNH27–30**.

considered that the pottery recovered from the primary fills in the southwest of the Entrance Chamber is not residual but rather was incorporated within the primary fills when the posts were erected. The majority of the pottery in this area is of a type not recovered in the surrounding area. It may be that one or two sherds from the northern Annexe did get incorporated into the adjacent postholes of the north wall of the Entrance Chamber. It is proposed, however, that the assemblage is not one characterised by the presence of residual material. Patterns are detectable in the distribution, and the pottery is extremely restricted in its concentration, making it unlikely, for example, that there was sufficient pottery lying around the site, which might become incorporated into the holes of newly erected posts. If the material had been distributed on the ground surface prior to the construction of the Annexe area, it would surely be detectable in the ploughsoil layers. It is conceded that a small number of sherds in the primary fill of the northern Entrance Chamber may be residual. It is felt, however, that this is not to such an extent that it affects the overall distribution picture. This is particularly apparent when the distribution of the lithic material is also taken into consideration. Lithics were not recovered from the northern Annexe area to the same degree as the pottery; however, it was found that the lithics were also recovered in considerable quantities from the primary fill of the Entrance Chamber. The inclusion of pottery sherds in the primary fill of the postholes in the Entrance Chamber and Annexe area is therefore considered to represent deliberate placement, not the result of residuality.

Grooved Ware pottery was also recovered from the secondary fills of postholes. This is indicative of activity carried out during the destruction of the monument. These contexts are considered fairly reliable as it appears that great effort was expended in their formation (detailed below). The sherds within these features are considered to have been deliberately placed. It should be noted, however, that in the southwestern corner of the Annexe,

Table 6.4 Summary of the characteristics of the Ballynahatty Grooved Ware pottery (after Roche 1995).

Type	Vessel	Sherds	Vessel shape	Rim	Fabric	Grit content	Inclusion size (mm)	Colour
Knowth Style 1	BNH1	6432, 6440, 6499, 6501–03, 6507	Barrel	Unexpanded, almost pointed	Hard & compact	High	<=1.9	Orange-brown/ black/black
	BNH2	6498, 6509, 6512	Barrel	–	Hard & compact	High	<=1.5	Black/black/ black
	BNH3	6526	Barrel	Unexpanded rounded rim	Hard & compact	High	<=5.5	Black-brown/ orange/orange
	BNH4	6449, 6508	Barrel	Unexpanded, rounded rim	Hard & compact	High	<=1	Brown-orange/ black/black
	BNH5	6496, 6497, 6510, 6511, 6513, 6516, 6517, 6518, 6520, 6522, 6543	Barrel	Unexpanded, rounded rim	Hard & compact	Moderate	<=1.6	Brown/brown/ orange- brown-black
	BNH6	6514, 6519, 6521, 6523, 6524, 6525	–	–	Hard & gritty	High	<=4.7	Orange/orange/ orange
	BNH7	6529, 6530, 6531	–	–	Brittle & flakey	Moderate	<=3.5	Orange/black/ black
	BNH8	6532, 6533, 6534, 6535, 6536, 6537, 6538, 6539, 6540, 6541, 6542	–	–	Hard & chalky	Moderate	<=2.5	Orange/orange/ orange-black
	BNH9	6528	–	–	Hard & compact	Moderate	<=1	Orange/black/ black
	BNH10	6485, 6486, 6489, 6494	Barrel	Unexpanded, rounded rim	Hard & compact	Moderate	<=2	Brown-orange/ brown-black/black
	BNH11	6480, 6481, 6482, 6490, 6493	Barrel	Unexpanded, rounded rim	Hard & compact with soapy texture	Moderate	<=1.5	Dark orange/black /black
	BNH12	6446–6470	Barrel	Unexpanded, rounded rim	Hard & slightly friable	Moderate	<=3.5	Brown orange throughout
	BNH13	6425, 6431, 6434–6437	Barrel	Unexpanded, rounded rim	Hard & compact	High	<=1.2	Black throughout
	BNH14	827, 903	Barrel	Unexpanded, rounded rim	Hard to compact	High	<=2.4	Orange/black/ black
	BNH15	6412–6414, 6418, 6424, 6426, 6432	Barrel	Unexpanded rounded rim	Hard & compact	High	<=2.4	Orange/grey/ orange
	BNH16	–	Barrel	Unexpanded, rounded rim	Hard & compact	High	<=1.4	Brown throughout
	BNH17	6421	Barrel	Unexpanded, rounded rim	Hard & compact	High	<=2.9	Orange/black/ black
	BNH18	6472	–	–	Hard & compact	Moderate	<=2.9	Brown throughout
	BNH19	6444	–	–	Hard & compact	Moderate	<=3.1	Brown/Brown/ black
	BNH20	6410–26	–	–	Hard & compact	High	<=2.4	Brown-orange
	BNH21	6414	Barrel	Unexpanded, rounded rim	Hard & compact	Moderate	<=2	Brown
	BNH22	–	Barrel	Unexpanded, rounded rim	Hard & compact	High	<=1.2	Brown-orange/ black/ black

(Continued)

Table 6.4 Summary of the characteristics of the Ballynahatty Grooved Ware pottery. (Continued)

Type	Vessel	Sherds	Vessel shape	Rim	Fabric	Grit content	Inclusion size (mm)	Colour
Knowth Style 2	BNH23	6439–40, 6442–43	–	–	Coarse & friable	High	<=3.9	Orange/black/black
	BNH24	6415–16	–	–	Rough & chalky	High	<4.6	Orange/orange/orange
	BNH25	6535	–	–	Coarse & hard	High	<3.7	Orange/black/orange
	BNH26	6445	–	–	Hard & rough	High	<=4.2	Orange/black/orange
Knowth Style 3	BNH27	6473	–	–	Hard & rough	High	<=5.5	Orange/orange/black
	BNH28	6475	–	–	Hard & rough	High	<=3	Orange/orange/black
	BNH29	6476	–	–	Hard & rough	High	<=3.6	Orange/orange/black
	BNH30	6544	–	–	Hard & rough	High	<=3.4	Orange/orange/orange

sherds of a vessel were recovered from the secondary fill, and sherds of the same vessel were found in the primary fill. Although it may be the case, it seems unlikely that fragments of vessels were saved in order to be deposited during the destruction of the site, particularly as the sherds recovered from the primary and secondary fills are in a similar state of preservation. It is possible that in the cases of the postholes at the southwest of the Entrance Chamber, the pottery was originally thrown in between the post and the primary fill when the post was erected. When the post was extracted, some of the sherds may have become dislodged from their position at the interface and have fallen into the cavity left by the post and so have subsequently been incorporated into the secondary fill.

6.3.4.3.1 Grooved Ware and the BNH6 Timber Circle

Construction: Grooved Ware was not recovered from the primary fill of the BNH6 structure and so cannot be directly linked to the construction of the monument.
Use: One Grooved Ware sherd was recovered from a feature which cuts posthole IR25 (C289), and another came from a small depression on the edge of posthole IR1 (C104). These are the first postholes on either side of the entrance through the BNH6 Inner Ring. The Knowth Style 1 sherds were not recovered directly from the postholes but appear to be from related contexts. It could be envisaged that the pottery was placed into cuts located at the edge of the entrance posts, which were intended to receive offerings. Sherds from a single Knowth Style 1 vessel were also deposited at an apparently important point just outside BNH6, where the line of sight from the Eastern Setting and its opposite and the BNH6 entrance intersect. Evidently, this pottery was placed, either as a whole vessel or as a group of sherds, to mark this important spot.
Destruction: Four sherds of Knowth Style 1 pottery were recovered from the secondary fill of posthole IR22 (C111). A single sherd was recovered from the secondary fill of one Outer Ring posthole (OR2: C449). The Grooved Ware sherds, recovered from the burnt secondary fill, were evidently inserted into the posthole with the destruction of this part of the structure. Only two Grooved Ware pottery sherds were recovered from the large Four-Poster postholes. One sherd was recovered from the secondary fill of posthole 4P3 (C453). A base sherd from a possible 'miniature' vessel was recovered from the secondary fill of posthole 4P2 (C948). Again, it is reasonable to assume that these pottery pieces were included with the secondary fill when the postholes were backfilled. No Grooved Ware pottery was associated with the postholes of the 'Excarnation Platform' or the Eastern Setting.

Two sherds of Knowth Style 3 pottery were recovered from the layers above IR6 (C501) and 4P1 (C630), and another two from the topsoil above the centre of BNH6. It appears that the pottery sherds had been disturbed from their original location by the plough and have come to rest in these features overlying the posthole of the Inner Ring (IR6) and the Four-Poster (4P1). The recovery of this highly decorated type of Grooved Ware at Ballynahatty is significant. Brindley (1999a) assumes the extensively decorated Grooved Ware to be the earliest type recovered in Ireland, and this accords with the current typo-chronological evidence. It could be assumed, therefore, that the use of this pottery type at Ballynahatty followed the Middle Neolithic Coarse Ware phase of activity and pre-dated the construction of the timber circle. Its distribution at Ballynahatty is interesting, close to the Coarse Ware activity to the west of BNH6.

6.3.4.3.2 GROOVED WARE AND THE BNH5 OUTER ENCLOSURE

A single Knowth Style 1 sherd was found in the secondary fill of posthole IR1 (C209) of the Inner Ring of BNH5. One further small, abraded fragment of pottery was recovered from BNH5 Inner Ring, from the primary fill of posthole IR10 (C279). A small fragment of Grooved Ware was recovered from the primary fill of the BNH5 Outer Ring posthole OR12 (C1596). A single fragment of Knowth Style 1 pottery was also found in the BNH5 entrance area. The sherd came from a charcoal layer within a shallow trench which had been dug at the north side of the entrance through the BNH5 inner and outer enclosures to hold a row of posts.

6.3.4.3.3 GROOVED WARE AND THE NORTH–SOUTH POSTS

The North–South posts cut the primary fill of the postholes of the BNH5 Outer Ring, suggesting that the postholes of BNH5 were still in place whenever the new postholes were dug (as discussed above). Two Knowth Style 1 sherds were recovered from the southernmost of these postholes, NS9 (C63).

6.3.4.3.4 GROOVED WARE AND THE ENTRANCE CHAMBER

The pottery recovered from the Entrance Chamber was concentrated in the southwest corner, within four postholes. The sherds were largely from a single Knowth Style 2 vessel and, as detailed above, there are a number of cross-context joins between the four postholes EC15 (C1493), EC16 (C1344), EC17 (C1257) and EC18 (C1616). The rim sherds were spread throughout all four postholes, and the base fragments were found in postholes EC17 and EC18. A whole vessel could not be reconstructed from these sherds; however, a substantial portion of the vessel remains. The sherds were recovered from the primary fill of postholes EC15 and EC18 but were present in both the primary and secondary fills of postholes EC16 and EC17. With EC16, there were 12 sherds in the primary fill and three sherds in the secondary, and with EC17, there were 18 from the primary and 25 from the secondary fill. It appears that initially, the broken sherds of this vessel were included in the primary fills of the four postholes, perhaps thrown into the post-pit cutting before the primary fill was packed in around the post. The recovery of a substantial number of the sherds from the secondary fill of posthole EC17 would suggest that during the destruction process, the sherds from the primary fill became incorporated into the secondary fill. A body sherd from this vessel was also recovered from one of the Southern East–West posts, SEW5 (C544), perhaps indicating that the Southern East–West posts and the Entrance Chamber were erected at the same time, or that the sherd within the posthole SEW5 (C1544) is residual and so this post was erected at a later stage than the Entrance Chamber. Again, this sherd is not in significantly different condition from the sherds of a pot recovered from the Entrance Chamber. It may be that the sherds of this vessel were also incorporated into other as yet unexcavated features. Sherds of three different Grooved Ware vessels were recovered from spreads in the northern Annexe and also in the postholes of the Entrance Chamber. The sherds were found in the primary fills of EC24 (C720 and EC30(C783) and in postholes EC28 (C902) and EC29 (C898), as well as in the secondary fill of posthole EC17(C1257). It may be that the sherds of these vessels within the postholes are residual, the Entrance Chamber having cut through the northern Annexe deposits. It may also be the case, however, that the sherds were broken in the northern Annexe and were then placed within the postholes of the Entrance Chamber.

6.3.4.3.5 GROOVED WARE AND THE FAÇADE

Only a single sherd of Grooved Ware was recovered from the Façade area; a Knowth Style 1 sherd from the secondary fill. The remaining pottery sherds from this feature are Coarse Ware from the Middle Neolithic chamber C591. As might be expected, a quantity of the chamber sherds was recovered from the primary fill of the postholes which cut it. A quantity of the chamber material was also recovered from the primary fill of the adjacent Outer Façade area.

6.3.4.3.6 GROOVED WARE AND THE NORTHERN ANNEXE

A substantial amount of Grooved Ware (almost 20%) recovered during the excavation came from the northern Annexe area. The northern Annexe appears to have had a floor surface at one time, and from this, over 50 sherds (representing at least seven different vessels) of Knowth Style 1 were recovered. This floor surface appears to have been made up of a layer of fine yellow sandy soil covered with a stony deposit, which also had associated spreads of charcoal. The northern Annexe appears to have been the location for a particular kind of activity not observed elsewhere within the site. The build-up of sherds here may be the result of ceremonial activity carried out within the confines of the structure, but prior to entering into the interior of BNH5 and BNH6. It is unlikely that this pattern is simply due to survival. If pottery had been within surface spreads elsewhere on the site, there is likely to have been evidence of this within the ploughsoil. The distribution of the pottery in the ploughsoil at Ballyhahatty overlies features which contained pottery, and so it is proposed that the pottery was disturbed from the contexts but generally did not move very far from its original position. The lack of Grooved Ware in the topsoil of other areas would indicate that the concentration in surface spreads in the northern Annexe is a real phenomenon rather than a product of survival.

6.3.4.3.7 Structured Deposition in Timber Circles

Although only a few Grooved Ware sherds were associated with the timber circle (BNH6), commonalities can be found between the distribution patterns of the Grooved Ware assemblage from Ballynahatty and those from other Irish timber circles. At Ballynahatty, it is considered that the deposition of the Grooved Ware sherds is suggestive of highly structured practices and that the position of the sherds should be viewed as significant. The variation in distribution patterns between the BNH6 structure and the Entrance Chamber is particularly notable and is perhaps suggestive of a chronological separation. It has been suggested elsewhere that the deposition of Grooved Ware in timber circle contexts is a relatively late phenomenon in the currency of Grooved Ware (although that is contradicted by the evidence from Machrie Moor, Arran, where early Grooved Ware has been found). Paul Garwood highlights the paucity of Grooved Ware associated with the Phase 1 southern circle and northern circle at Durrington Walls, Wiltshire, and at Mount Pleasant, Dorset, and suggests that the deposition of Grooved Ware at these monuments is a relatively late practice within the overall currency of Grooved Ware use (Garwood 1999, 152). In southern Britain, it seems that the early use of Grooved Ware is related to isolated pits or pit groups (*ibid.*, 157). Alex Gibson has suggested that Grooved Ware is associated with the climax of timber circle use (1999, 78). Perhaps the deposition of Grooved Ware became an important element at Ballynahatty in its later stages, that is, with the insertion of the Entrance Chamber and possibly the Southern East–West posts.

The association between the Grooved Ware deposits and the later stages or end of the timber circle at Ballynahatty is a trend that is replicated at other timber circles across Ireland. At Bettystown, Co. Meath, the posts were removed before a charcoal-rich fill containing Grooved Ware pottery, struck flint and burnt and unburnt bone, was backfilled into the voids (Eogan 2000). There is evidence of similar activity at Balgatheran, Co. Louth, where the postholes from Buildings 1 and 3 appear to have been deliberately removed/destroyed before being infilled with deposits containing lithics, polished stone axehead fragments and Grooved Ware (Ó Drisceoil 2009, 80, 84). At the site of Lagavooren, Co. Meath, the position of the sherds within the upper layers of the postholes indicates that the Grooved Ware was deposited after the construction of the site (Carlin and Cooney 2017, 45). Although the partial section of the timber circle at Lowpark, Co. Mayo, only produced seven sherds of Grooved Ware from two postholes (C657 and C633), these were deposited after the posts had been removed (Gillespie n.d., 31). The deposition of Grooved Ware sherds at two of the timber circles (structures 3 and 4) at Scart, Co. Kilkenny, occurred after the timbers had been removed in what appears to have been a deliberate act of destruction (Laidlaw 2017, 47). Substantial quantities of Grooved Ware were also recovered from the timber circle at Armalaughey, Co. Tyrone. The Grooved Ware at this site appears in a secondary context within the postholes, indicating that the sherds were deposited after the posts had been removed (Carlin 2016, 205). There are only two examples of Grooved Ware within primary contexts at Irish timber circles. At both Knowth, Co. Meath (Eogan and Roche 1997, X) and Glenshane, Co. Derry (Barkely 2020, v), the Grooved Ware sherds appear to have been deliberately incorporated into the packing material surrounding the posts. This suggests that in both cases, the sherds had been deposited into the postholes during the construction of the timber circle.

The evidence from the majority of the excavated Irish timber circles indicates that Grooved Ware was deposited after the construction and primary phase of activity at the sites had come to an end. Many of the sites show evidence of the postholes being deliberately removed through either extraction or burning or that they had been left to rot *in situ*. These voids were then filled with artefact-rich deposits, which often contained sherds of Grooved Ware pottery alongside lithics, animal and charcoal. The significance of these deposits is still not fully understood, but Carlin has proposed that the placement of Grooved Ware sherds within secondary contexts may be regarded as 'ritualised acts of abandonment and/or commemoration [of the site's past history]' (Carlin 2016, 206). Laidlaw has also suggested that these deposits may also commemorate individual postholes, which may have held architectural or symbolic significance (Laidlaw 2017, 47).

6.3.4.4 Discussion: Irish Grooved Ware studies and the place of the Ballynahatty assemblage within the wider picture of Grooved Ware use in Ireland and Britain

The earliest occurrence of Grooved Ware is in the Orkney Islands, where this type of pottery seems to have originated and from where it seems to have spread throughout the rest of Britain and Ireland (Schulting *et al.* 2010; MacSween *et al.* 2015; Richards *et al.* 2016). The pottery is flat-based, with straight sides, giving it a bucket or barrel-shaped profile. The surface of the pots is decorated with a series of incised lines. While some Grooved Ware has extensive decoration covering the whole surface, the Irish examples (with a few exceptions) have much more restricted decoration, often confined to a few incised lines located on the interior surface of the rim. Generally, Irish Grooved Ware is of a fine fabric, with thin or medium-thickness walls, and well-fired.

Grooved Ware, first defined in Britain in 1936 (Warren *et al.* 1936; Wainwright and Longworth 1971), did not receive the same recognition in Ireland until around 60 years later (Roche 1995; Brindley 1999a;

1999b). References to Grooved Ware pottery in Ireland are found in publications as early as 1951 and 1954 (Ó Ríordáin 1954) and again in 1959 (Collins 1959, 7). There was also recognition of the type at Newgrange (O'Kelly *et al.* 1983) and Knowth (Gibson 1982; Brindley 1984), but it was not until the discovery of the 'Grooved Ware Timber Circle' at Knowth that inroads were made into the understanding of this pottery type and its date in Ireland. Writing in 1995, Helen Roche argued that

> with the discovery of a sealed Grooved Ware assemblage in a typical British Grooved Ware type context at Knowth, Co. Meath and with the discovery of Grooved Ware associated with a timber circle complex at Ballynahatty, Co. Down, the importance of Grooved Ware in Ireland can no longer be ignored. (1995, 3)

Her thesis addressed this problem, and all assemblages which were known to contain both Neolithic and Beaker pottery were examined for the work. Roche's re-analysis of the pottery assemblages led to the identification of Grooved Ware at sites from which it had previously not been recognised. A total of 13 sites were identified in all, with a further four identified as possible Grooved Ware sites (*ibid.*, 7). Roche classified the pottery into three groups, termed Knowth Style 1, 2 and 3 (*ibid.*, 14) and that is the classification system used here.

A further study of the Grooved Ware in Ireland was put forward by Anna Brindley (1999a). Brindley identified Grooved Ware sites and classified them into five different groups based on their fabrics and decoration, naming them according to their widest geographical distribution. These groups are known as the Dundrum-Longstone type, the Knowth type, the Kiltierney type, the Grange-Geroid Island type and the Donegore-Duntryleague type (*ibid.*, 24, 30). Brindley suggests a date of 3000 cal BC for the earliest, Kiltierney type and a date of 2600–2450 cal BC for the later types (*e.g.* Dundrum-Longstone; 1999a, 30–31). Her work suggests that Grooved Ware in Ireland began with the use of highly decorated wares, such as those recovered from Kiltierney, Co. Fermanagh (Daniells and Williams 1977, fig. 7) and evolved to the more sparsely decorated examples, such as those recovered from Ballynahatty and Knowth. Note, however, that since Brindley's study was published, a radiocarbon date of 1874–1617 cal BC [3415±35 BP], obtained from cremated bone associated with the highly decorated type-sherd from Kiltierney, indicates that the pot in question is not Grooved Ware but is instead an Irish-Scottish Vase Food Vessel: Alison Sheridan pers. comm. Nevertheless, it is the case that the earliest Grooved Ware in Ireland is relatively highly decorated, as the dates of *c.* 3080–2920 cal BC for human remains associated with the vessel from Knowth tomb 6 demonstrates (Schulting *et al.* 2017).

In her study, Brindley classes all the Grooved Ware material from Ballynahatty as belonging to the 'Dundrum-Longstone' type. This is certainly valid and, in general, the pottery from Ballynahatty is homogeneous. For the purposes of this study, however, we have adopted Roche's scheme, as it allows for more detailed analysis within the assemblage and recognises the three classes which are present within the group from Ballynahatty.

These seminal studies by Roche and Brindley still form the basis of the many subsequent studies of Grooved Ware material excavated across Ireland. In recent years there has been significant growth in the number of known Grooved Ware sites due to development-led excavations, with over 50 Grooved Ware sites now identified in Ireland (Carlin 2016, 198). The current evidence suggests that Grooved Ware is confined to a number of restricted contexts in Ireland, namely 'pits, spreads, timber circles and [secondary contexts from] developed passage tombs' (Carlin 2017, 155). Grooved Ware is often found deposited in association with a number of other artefacts, including 'worked flint, polished stone axes, charred hazelnut shells, animal bone and marine molluscs' (Laidlaw 2017, 48).

6.3.5 Early Bronze Age pottery

Sherds of two Early Bronze Age pots were found during the excavations at Ballynahatty.

6.3.5.1 Vase Food Vessel

Five sherds of a pot with a complex geometric incised design (Fig. 6.16) are recorded as being broadly associated with posthole BNH5 OR13 (C1711) in the outer ring of the timber enclosure at the point where it joins the WNW posts. Initially tentatively identified as a Beaker by the late Derek Simpson, this pot has recently been re-classified by Alison Sheridan as belonging to a specific type of Vase Food Vessel which Arthur ApSimon (1969, 40–44, figs 4 and 5) described as an 'Irish-Scottish Vase'. These distinctive vessels, found mostly in Cos Antrim and Tyrone but with a few examples in southwest Scotland, fall within Brindley's 'Stage 3' of her Vase typo-chronology, which she dates to *c.* 1830–*c.* 1740 cal BC (Brindley 2007, 262–264, 328, fig. 106, table 69).

While such pots are normally found in funerary contexts, the sherds at Ballynahatty were not accompanied by any human remains, although one cannot altogether rule out the possibility that a grave containing unburnt human remains had once existed at the findspot. What can be stated with reasonable confidence is that this pot was deposited at least 700 years after the timber monument complex was in use. It is unlikely that many, if any, traces of that Late Neolithic set of structures would still be visible during the Early Bronze Age and so it is not necessary to invoke notions of commemorating the ancestors to account for its presence.

104 *Ballynahatty*

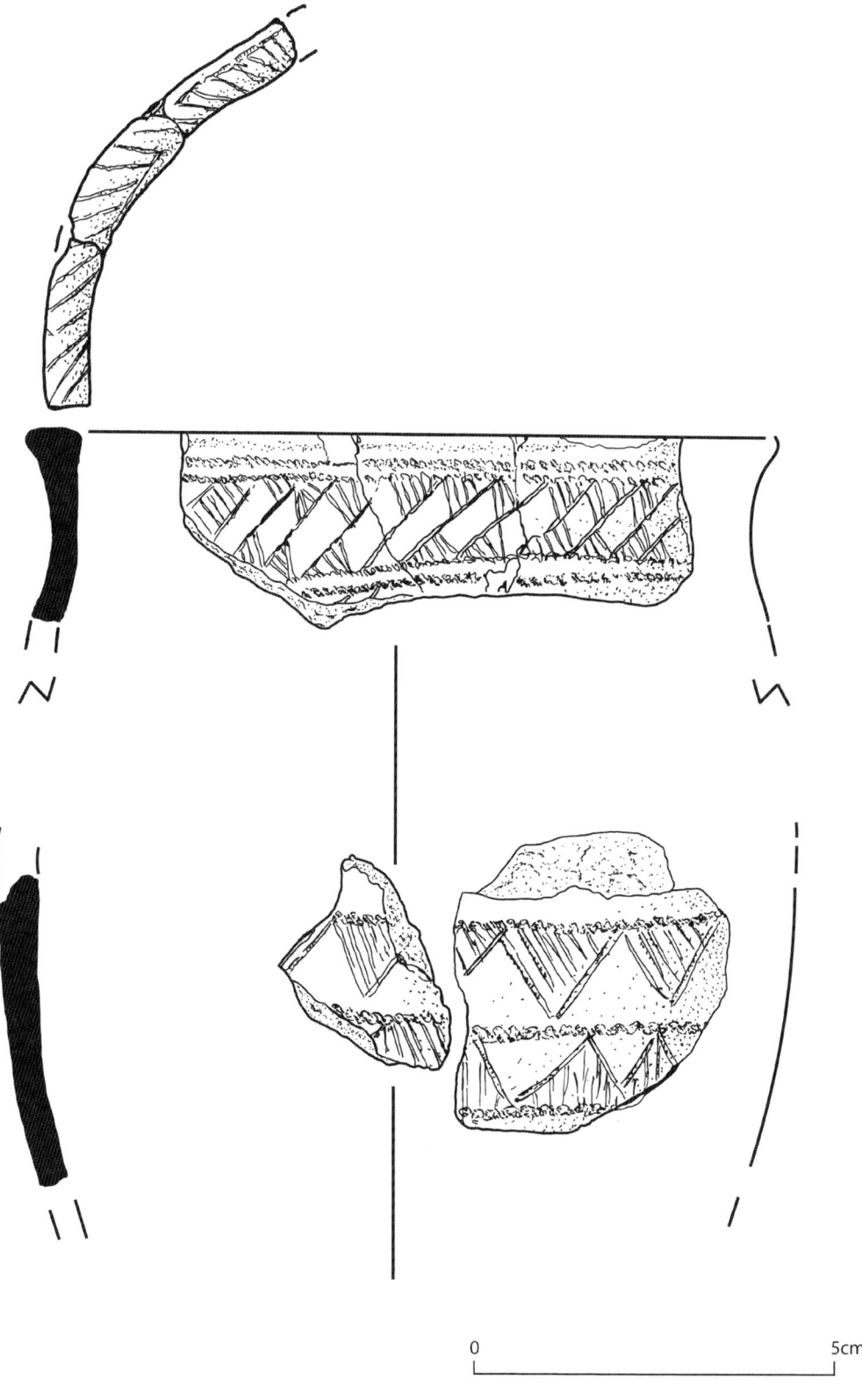

SHERD 6603
AREA 23, C81

Figure 6.16 Reconstruction of the Vase Food Vessel.

The absence of Beaker pottery from Ballynahatty is interesting since elsewhere this kind of pottery has been found in a few Irish timber circles, where its pattern of deposition seems to replicate that noted with Grooved Ware (Carlin 2011). At Paulstown, Co. Kilkenny, for example, Beaker sherds were found in contexts associated with the deconstruction of the timber circle (*ibid.*, 179), while at the complex 'square-in-circle' monument at Armalughey, Co. Tyrone (Dunlop and Barkley 2016, 36–47), Beaker deposition appears to have focused on the central area where the initial four-post structure had been, which may mark an attempt by the Beaker users to forge a link with the earliest phase of the site's history (Carlin 2011, 183).

6.3.5.2 Collared Urn

Three sherds that are almost certain to be from a Collared Urn (Fig. 6.17) were found in the topsoil/ploughsoil overlying the timber circle (areas 31 and 33). The rimsherd illustrated in Figure 6.17 is decorated with crude incised designs on the internal rim bevel and on the exterior. The sherds are not associated with any features from the site but there is abundant evidence of Bronze Age activity in the surrounding fields. Ballynahatty lies within the main distribution area for Collared Urns in Ireland, which encompasses Co. Antrim and the northern section of Co. Down; examples have been found in the Belfast area (Brindley 2007, 132, fig. 34). Collared Urns are usually found containing cremated human remains in simple pits or cists and it is highly likely that this pot had been used in that way. According to Brindley's typo-chronology of Irish Early Bronze Age pottery, the currency for Collared Urns in Ireland is estimated to be 1850/1830–1700 cal BC.

As with the 'Irish-Scottish' Vase described above, and in contrast to the Chalcolithic and Early Bronze Age deposits of Beaker pottery seen at a few timber circle sites in Ireland, there is no reason to believe that the Collared Urn was deposited here because a major Late Neolithic monument complex had existed at the spot.

Early Bronze Age pottery is not a common find at Irish timber circle sites but there are parallels for the re-use of the landscape long after the timber monuments had ceased to exist. At Lagavooren, Co. Meath, for example, two sherds from a poorly made Vase Food Vessel were recovered from a pit in the centre of the timber circle; Grooved Ware was also present in the pit (Grogan and Roche 2012, cxvi). It is, however, unclear whether the pit was an original feature of the timber circle or a later addition (*ibid.*, cxviii, n. 7). At Lowpark, Co. Mayo, a deposit of cremated remains accompanied by sherds of three crudely made Food Vessels was found in a burial pit located *c.* 40 m from the site of the earlier timber circle (Cleary n.d., 64). A large assemblage of Bronze Age pottery, including Food Vessels, Cordoned Urns and coarse domestic vessels, was also recovered from a number of deposits of cremated remains at Ask, Co. Wexford (Grogan and Roche 2011a, 168).

At other timber circles, including Glenshane, Co. Derry and Lagavooren, the sites appear to have taken on more domestic roles during the Early Bronze Age. Such activity at Glenshane is attested by the presence of a roundhouse and associated features. Three flat-based urns were identified from the ceramic assemblage from this phase (Barkley 2020, iv). At Lagavooren, 54 sherds of Early Bronze Age Cordoned Urn pottery, believed to be of a domestic nature, were found in a large pit to the south of the timber circle (Grogan and Roche 2012, cxviii), while a Late Bronze Age vessel found in a pit at the northern edge of the site is believed to represent domestic waste (Grogan and Roche 2012, cxix).

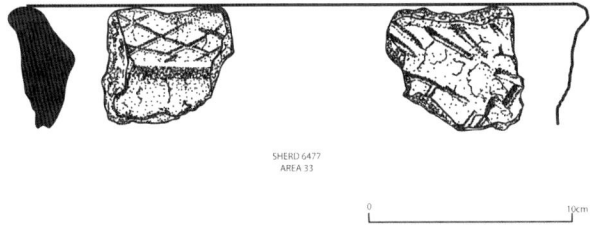

Figure 6.17 Rim sherd from a Bronze Age Collared Urn. Note: the rimsherd is too small to allow a reliable estimate of rim diameter; the diameter shown in the illustration is likely to be too small.

7

The lithic assemblage: chipped stone

Eiméar Nelis, abridged by Caroline Malone

7.1 Introduction

Lithic material is commonly found on British and Irish timber circle sites, with some sites producing a vast number of artefacts. Ballynahatty produced a very substantial assemblage of over 5000 recorded lithics, the bulk of which are mainly associated with the great timber structures of the monument. The assemblage makes a significant contribution to an understanding of lithics from similar sites of monumental ritual activity in Ireland in the later Neolithic period. This chapter presents the study of the composition and results of the technological analysis of the lithic assemblage from Ballynahatty undertaken by Eiméar Nelis as part of her PhD studies (Nelis 2003). The tables and figures derive from the original thesis and are renumbered in this chapter. The thesis may be consulted in the University Library of Queen's University Belfast. The comprehensive study outlines the composition of the assemblage with regard to the raw materials, production techniques and condition of the lithics, and it includes a detailed analysis of the distribution of lithics across the site. These distribution patterns help to define how the timber circle complex was used and provide intriguing insights into the lifecycle of the sites and aspects of ritual activity. The analysis below is divided into two sections. The first presents an assessment of the primary technology of the assemblage and the worked assemblage. The second section examines the distribution patterns of the lithic material at Ballynahatty and compares the assemblage to other broadly contemporaneous site assemblages in Ireland.

7.2 Composition: flint and non-flint

Of the 5036 chipped stone artefacts recovered during excavations at Ballynahatty, the overwhelming majority (5007 pieces/99.0%) are of flint. Chert (11/0.2%), quartz (17/0.3%) and one piece of porcellanite (a single flake) (0.1%) make up the remaining 1% (Table 7.1). The majority of quartz pieces are unworked lumps (12 of 17 artefacts), flakes (two pieces) or angular debitage (three pieces). Some of the chert is unworked (two pieces), but the rest are retouched tools (five out of nine worked chert

Table 7.1 Classification of flint and non-flint assemblage and types of material present (Nelis 2003, 723).

Type	Flint	Porcellanite	Quartz	Chert	No.	%
Unworked	664	–	12	2	678	13.5
Possible hammerstone	1	–	–	–	1	0.0
Cores	61	–	–	–	61	1.2
Flakes and blades	1089	–	2	1	1092	21.7
Flake and blade shatter	1763	1	–	–	1764	35.0
Angular shatter	1218	–	3	3	1224	24.3
Retouched tools	194	–	–	5	199	3.9
Core tools	17	–	–	–	17	0.4
Total	5007	1	17	11	5.36	100
%	99.4	0.1	0.3	0.2	100	

artefacts), or else angular shatter: (three pieces) or flake (one piece).

7.2.1 The flint assemblage

A substantial element of the 5007 pieces of flint in the assemblage is unworked material (664 pieces/7.3%), and one possible hammerstone has been identified. The assemblage is mainly composed of knapping debitage; cores account for a small element (61/1.2%), but flakes and blades, both complete and shattered, constitute over half of all artefacts (2852/56.9% of flint assemblage), and angular shatter, occurring as a result of knapping, accounts for almost one-quarter (1218/24.3%) (Plate 52). The tool assemblage is composed of retouched tools (retouched or utilised: 194/3.8%) and a substantial number of core tools (17/0.3%).

7.2.1.1 Raw material, cortex and condition

In most cases, the source of the flint could not be determined, but a reliance on beach pebble material was not evident, with only 31 artefacts being identified as being made from beach pebbles in addition to a small number made from possible river pebbles (5 pieces: Table 7.2).

The presence or lack of cortex is almost evenly divided within the assemblage, with just over half of the artefacts having no cortex (2827 pieces/56.5%: Table 7.3). Most corticated material retains only partial coverage (2080/41.5%), with only a small number being fully corticated (100/2%). In addition to the unworked material (32), the bulk of fully corticated artefacts are flakes and blades (both complete and shattered: 48 pieces). The condition of the assemblage is interesting since almost half has been burnt (2321/46.4%). Of the remainder, over one-third is in a fresh or patinated condition (1872/37.4%), and less than one-fifth of the assemblage is abraded (814/16.3%).

Table 7.2 Inferred source of flint utilised (Nelis 2003, 724).

Possible source Character	Indet. flint source	Beach pebbles	Poss. river pebbles	Total
Unworked	657	5	–	664
Cores	59	2	–	61
Complete flakes	900	9	1	910
Complete blades	179	–	–	179
Flake/blade shatter	1756	5	2	1763
Angular shatter	1212	4	1	1217
Retouched	189	5	–	194
Hammerstone	1	–	–	1
Core tools	15	1	1	17
Total	4971	31	5	5007
%	99.3	0.6	0.1	100

Table 7.3 Basic composition of flint material recovered, showing the extent of the cortex (Nelis 2003, 725).

Character	Cortex			Total	%
	Primary	Secondary	Tertiary		
Unworked	32	414	218	664	13.3
Hammerstone	–	1	–	1	0.0
Cores	–	9	52	61	1.2
Complete flakes & blades	18	570	501	1089	21.7
Flake/blade shatter	30	522	1211	1763	35.2
Angular shatter	13	411	794	1218	24.3
Retouched	4	99	91	194	3.8
Core tools	1	13	3	17	0.3
Total	100	2080	2827	5007	
%	2.00	41.5	56.5	100	100

Table 7.4 Basic composition of flint assemblage showing condition (Nelis 2003, 725).

Character	Fresh	Patinated	Abraded	Burnt	Total
Unworked	124	117	322	101	664
Hammerstone	–	–	1	–	–
Cores	15	14	24	8	61
Complete flakes & blades	633	157	138	161	1089
Flake/blade shatter	360	128	101	1174	1763
Angular shatter	97	77	191	853	1218
Retouched	78	58	34	24	194
Core tools	9	5	3	–	17
Totals	1316	556	814	2321	5007
%	26.3	11.1	16.3	46.4	100

In general, all artefact types display abrasion and shatter. Almost half of the unworked material is abraded, which may be a factor in this material not being selected for use (322/48.5% abraded unworked pieces). The burnt material accounts for most of the angular shatter and is damaged beyond further identification (853/70% burnt angular).

Notably, most of the complete flakes and blades are fresh or patinated (790/72.5%), in contrast with the burnt flake and blade shatter (1174/66.6%). The contrast indicates that burning was a major factor in creating the broken flake and blade assemblage, rather than this material occurring primarily as a result of knapping failure or subsequent trample. A study of the mass of the flint assemblage (Table 7.5) shows that burnt material constitutes less than one-fifth (15.8%) of the assemblage, whereas numerically, these artefacts account for almost half of

Table 7.5 The condition of the flint assemblage by count and mass (Nelis 2003, 726).

Type	Burnt proportions	Total (no.)	Burnt (%)	Burnt (mass)	Total (mass) (g)	Burnt (%)
Unworked	101	664	15.2	196.7	6328.9	31
Hammerstone	–	1	–	–	476.4	–
Core	8	61	13.1	340.9	3816.1	8.9
Complete flakes and blades	161	1089	14.7	519.24	5895.2	8.8
Flake/blade shatter	1174	1763	66.6	2102.9	4364.8	48.1
Angular shatter	853	1218	70	1949.7	10,440.4	18.7
Retouched	23	194	11.9	310.9	2524.9	12.2
Core Tools	1	17	5.9	17.3	482.13	36
Total	2321	5007	46.4	5437.9	34,346.8	25.8

the assemblage. Therefore, while the burnt assemblage is abundant, it evidently includes much of the small scale, lighter debitage, and its disproportional numerical strength is partly a factor of the friability of such artefacts, leading to their proliferation within the assemblage. This is borne out by a comparison of the numbers and mass of the burnt proportion of the assemblage: for example, numerically, burnt pieces account for two-thirds of the flake/blade shatter but, in terms of mass, constitute less than one-half. Just one-fifth of the mass of angular shatter consists of burnt material (18.7%), but numerically, burnt material accounts for over two-thirds (70%) of the angular shatter.

7.2.2 Primary technology

A total of 5036 chipped stone artefacts representing knapping material were recovered during excavations at Ballynahatty, of which, as already stated, the overwhelming majority are of flint, with the remainder of chert, quartz and (in one case, a flake) porcellanite. Some of the chert is unworked, but most pieces are retouched tools, and the remaining material is angular or flake shatter. In most cases, the source of the flint could not be determined, but reliance on beach pebble material was not evident, with only a small number of artefacts being identified as such.

The flint assemblage is mainly composed of knapping debitage. Cores are a small element within the assemblage and comprise randomly flaked pieces/bipolar cores, platform cores, and a small number which appear to combine bipolar and platform techniques. A component of the core assemblage, most of which is bipolar or scalar reduced, may have served as wedges.

Flakes and blades, both complete and shattered, constitute over half of the assemblage. Of these, most are complete, constituting over one-fifth of the flint assemblage. Most of the complete pieces are a product of platform technology, with the remainder being bipolar examples. The platform types present on flakes and blades are highly variable, perhaps reflecting the variability of techniques used during reduction, as well as, perhaps, in the use of various hammer types. In the main, planar platforms are dominant; however, faceting is also common, especially among blades. The vast majority of flakes and blades have feathered terminations with most of the remainder having either light hinges (especially flakes) or plunging distals (more common among blades). Flake and blade shatter is also abundant, constituting over one-third of the flint assemblage, and most pieces are fairly undiagnostic as a result of burning. Of the remainder of the shattered material, most pieces could be specified as either being broken flakes or blades. Distal fragments are most commonly found, and many of the remainder are proximal fragments, with relatively small numbers of medial fragments being evident, and this is the case with both flakes and blades. Platform and termination types found on proximal and distal flake and blade shatter reflect the patterns evident within the complete assemblage. Angular shatter, the result of knapping debris, accounts for almost one-quarter of the assemblage and confirms that knapping was undertaken on the site. Most angular shatter is burnt material which has been damaged beyond further identification.

7.2.2.1 Unworked material

Over 10% of the flint assemblage (495 pieces) has not been worked, the bulk being thermal flakes (Table 7.6). Some have been edge-damaged in such a way to suggest working but are so heavily abraded that they are considered to be pseudoliths (*i.e.* naturally occurring pieces resembling retouched tools, such as edge-retouched pieces or sometimes, in the case of thermal flakes, scrapers).

Most of the remainder of the unworked material appears to be regolithic material (*i.e.* glacial drift), unsuitable for use since such heavily abraded, derived material can often be difficult to work and is usually overlooked when an alternative is available. Some of the regolithic material has further suffered thermal fracturing (47 of 151 regolithic lumps). The remainder of the unworked pieces are flint pebbles, most of which have been thermally damaged (15 of 18 pebbles); approximately one-third of these are beach pebbles. This burnt, unworked flint is useful as an indicator of areas of burning.

Table 7.6 Basic composition of unworked flint (Nelis 2003, 728).

Unworked flint	Regoliths/ amorphic lumps	Pebbles	Thermal flakes	Total
Intact	104	3	489	596
Thermally split	47	15	–	62
Pseudoliths	–	–	6	6
Total	151	18	495	664

7.2.2.2 Hammerstones

The assemblage includes just one possible hammerstone: a flint pebble, measuring over 90 mm in maximum length, and weighing over 465 g, which exhibits multiple areas of percussion.

7.2.2.3 Cores and flaked material

Cores and flaked pieces account for just over 1% of the flint assemblage (61 pieces/1.2%) and comprise randomly flaked pieces (10), bipolar cores (13), platform cores (25), and a small number which appear to combine bipolar and platform techniques (4). In addition, there are nine fragments of cores which eluded clearer classification (Table 7.7).

In general, the randomly flaked pieces exhibit an effort to utilise existing material, most commonly thermally-damaged lumps, with the removal of one or two flakes without preparing a platform but, instead, using an existing planar surface. As such they represent an expedient core technology, tending not to be exhaustively reduced, being already inherently limited in their potential. For the most part, they exhibit flake scars with feathered terminations.

Most commonly, cores have prepared platforms but such examples still constitute less than half the core assemblage (25 pieces). These tend to be single-platform cores, partially flaked (14 pieces: Table 7.7), with no fully flaked examples being found. Some of the single-platform cores were formed on quartered pebbles, and of note, some also appear to have deployed the anvil technique rather than hand-held reduction, with resultant damage at the base of the core suggesting that care was not taken to protect this area. Most of the remainder are dual-platform cores, one of which is formed on a large flake. Multi-platform cores are relatively rare (3), and the assemblage includes one keeled core. Most of the platforms have been formed by the removal of one flake (single facetted), with small numbers having dihedral platforms (or two facets).

The remaining complete cores comprise pieces exhibiting bipolar technology (13, with an additional four examples perhaps combining bipolar with platform methods). These bipolar examples are in addition to the bipolar cores and flakes that may have been used, or created for use, as wedges or *pièces escaillées*, and are discussed below.

Table 7.7 Composition of core assemblage (Nelis 2003, 729).

Flaked pieces and cores	Flint
Flaked pieces	10
Split pebble	1
Flaked pebbles	1
Flaked chunks	8
Platform cores	25
Single-platform, fully flakes (Type A1)	–
Single-platform, partially flaked (Type A2)	14
Dual-platform, parallel (Type B1)	1
Dual-platform, parallel (Type B2)	2
Dual-platform, parallel (Type B3)	3
Multi-platform (Type C)	3
Keeled (type d)	1
Keeled & 1 platform (type e)	–
On flake	–
Dual-platform, parallel (Type B1)	1
Bipolar and Platform	4
Bipolar	13
Split pebble	1
Flaked pebble	1
Flaked chunks	1
On flake	10
Unclassifiable	9
Possible fragment	7
Thermally damaged	2
Total	61

Most bipolar cores appear to have re-used flakes (10 of 13 examples) and are the only core type predominantly to do so. Of interest, those exhibiting a combination of bipolar and platform technology tended to re-use thermally-damaged material and, similar to the flaked pieces discussed above, may be seen as expedient re-use of existing material. It appears, then, that the re-use of thermal-damaged material deployed a kind of 'pseudo-' platform technique, whereas the re-use of existing flakes required a primarily bipolar technique; however, it was not unheard of to combine both of these methods during the course of reduction.

In general, all cores tend to have provided flakes more frequently than blades, although rarely to the total exclusion of blades; only a small number of single-platform examples show a dominance of blade scars. Most of the removals seem to have had feathered terminations with occasional hinge scars being evident. Only one of the single-platform cores was dominated by hinge scars and another seems to have provided mostly overshot removals.

A reason for the abandonment of most of the complete cores could be proposed: in most cases and all types,

Table 7.8 Complete flakes and blades, indicating extent of cortex and burning (Nelis 2003, 732).

Classification	Level of cortex			Total	Burnt no.	Burnt %
	Primary	Secondary	Inner			
Platform	13	458	418	889	124	13.69
Flakes	12	396	349	757	105	13.75
Blades	1	62	69	132	19	15.45
% total	1.5	51.5	47	100	–	–
Bipolar	5	112	83	200	34	17
Flakes	4	84	62	150	23	15.33
Blades	1	28	21	50	11	22
% total	2.5	56	41.5	100	–	–

the cores appeared exhausted (27). Of the remainder, some were abandoned early, possibly due to limited potential for creating good flakes (12); others were abandoned due to raw material flaws, such as cavities (4) or massive internal fractures (2), or due to knapping flaws, with some having large step/hinge fractures (2).

The cores tend to be quite diminutive, with the largest example measuring less than 80 mm in maximum length, whilst 75% of cores measure less than 45 mm in maximum length. With a few exceptions, the platform cores tend to dominate the assemblage. The bipolar cores are the least frequent, and flaked pieces and those exhibiting a combination of bipolar and platform flaking are of similar dimensions to the majority of the bipolar pieces (Plate 53).

7.2.2.4 Flakes and blades

A total of 1089 complete flakes and blades were recovered, constituting over one-fifth of the flint assemblage. Of these, most are a product of platform technology (889 pieces), with the remainder being bipolar examples (200 pieces). The length: breadth ratio for platform flakes and blades ranged from 0.2:1 to 4.5:1, with the same ratio for bipolar flakes and blades ranging from 0.5:1 to 3.5:1 (Table 7.8).

Of flakes and blades resulting from platform technology, flakes dominate (757 pieces/85.5% of complete platform removals) over blades (132/14.5% of complete platform removals). Fully corticated examples are rare, and when they do occur, tend to have flake proportions (12 flakes, 1 blade). In general, partially corticated flakes and blades are only slightly more common than those retaining no cortex (Table 7.9). Just over 10% of both complete flakes and blades are burnt, with this percentage being marginally higher with regard to blades.

Complete bipolar flakes and blades (200/18.4%) are a substantial component of the flake and blade assemblage, comprising almost one-fifth of this assemblage. Bipolar removals tend to shatter more frequently than platform removals; in addition, some removals created by bipolar techniques can carry what are essentially platform features (*i.e.* intact 'platforms' and terminations). Therefore, it is

Table 7.9 Length, breadth, thickness (mm) and mass (g) range for platform and bipolar flakes and blades (Nelis 2003, 728).

Measurement (mm)	Platform		Bipolar	
	Flake	Blade	Flake	Blade
Length (range)	3–68	4–71	8–58	17–78
Breadth (range)	3–65	2–18	8–42	5–35
Thickness (range)	1–40	1–51	8–42	2–14
Mass (range)	0.01–153.2	0.29–113.1	0.18–58.56	0.18–18.84

plausible that some of the 'platform' flakes and blades may have been produced by the bipolar technique. This is all the more feasible within this particular assemblage since some of the cores appear to have combined bipolar and platform techniques. Bipolar technology, therefore, is a major element within the reduction strategy at Ballynahatty.

Exactly 75% of the complete bipolar removals have flake proportions (150/75%), and the remainder are blades (50/25%). Again, fully corticated removals are rare, more commonly having flake proportions (4 flakes, 1 blade); partially corticated examples are most common (112/56%), and uncorticated pieces constitute just over two-thirds of bipolar flakes and blades (83/41.5%). A substantial proportion of the bipolar flake and blade assemblage is burnt, including just over one-sixth of flakes (34/15.3%) and over one-fifth of blades (11/22%).

Assessment of the length, breadth and thickness ranges for platform and bipolar flakes and blades showed that platform blades tend to be marginally longer but substantially narrower than flakes, although the blades are just as thick, if not thicker, than the flakes. Bipolar flakes and blades tend to be larger in terms of length and breadth than platform examples but are generally not as thick and are substantially lighter than platform examples (Table 7.9; Plate 54). Likewise, pressure flakes tend to be the smallest among the assemblage, whilst bipolar removals range in size and are more given to blade-like dimensions than platform removals.

7.2.2.5 Platform flakes and blades

Various types of platform flake and blade were identified within the assemblage, with regular pieces (*i.e.* reduction flakes not concerned with preparing the core for flaking) the most common. Examples include morphologically variable flakes from all stages of the knapping process (433 regular flakes, 57.2% of flakes; 104 blades, 78.8% of blades). These include a number of core platform, platform edge and core face rejuvenation flakes (Table 7.10) and large numbers of core trimming flakes (as well as a few with blade proportions: 195 core trimming flakes; 5 core trimming blades). The recognition of flakes caused by pressure flaking can be more problematic than the recognition of percussion removals, but there appears to be a substantial number within the assemblage (129/14.5% of flakes and blades). In addition, are a small number of double ventral flakes, none of which is particularly fine or appears very controlled; the assemblage also includes a single possible bifacial thinning flake.

Little can be said of the types of bipolar flakes and blades present within the assemblage; in general, they are simply described as being regular. However, the combination of platform and bipolar techniques during the knapping process is endorsed by the presence of a number of platform core preparation flakes removed by bipolar strikes: a total of nine artefacts could be described as bipolar flakes that appear to be aimed at trimming a platform core, and two platform edge rejuvenation flakes (or single crested flakes) were struck at a perpendicular angle to the core face, removing the platform edge, apparently using bipolar methods (Table 7.11).

Table 7.10 Types of complete flakes and blades (Nelis 2003, 279).

Types	Flakes	Blades	Total
Platform rejuvenation	10	–	10
Platform edge rejuvenation	2	–	2
Core face rejuvenation	6	–	6
Core trimming	195	5	200
Pressure flake	106	23	129
Regular flake	433	104	497
Double ventral	4	–	4
Bifacial thinning	1	–	1
Total	757	132	889

Table 7.11 Types of bipolar flakes and blades.

Types	Flakes	Blades	Total
Regular	139	50	189
Core trimming	9	–	9
Platform edge rejuvenation	2	–	2
Total	150	50	200

7.2.2.6 Complete flakes and blades: platforms

The platform types present on flakes and blades are highly variable (Plate 55), perhaps reflecting the variability of techniques used during reduction, as well as, perhaps, a variability in hammer types. In the main, flakes tend to have planar platforms, both medium (5 mm+ in thickness) and small (<5 mm in thickness), some of which are accompanied by edge preparation; these include winged and double platforms. However, faceting is also common, and there are substantial numbers of dihedral examples. Platform variability is reflected among blades, although faceting is more commonly found. The bipolar technique is evidenced by the crushed and shattered platforms present in both assemblages (Plates 55 and 56).

The use of indirect percussion might be implied by the presence of substantial numbers of punctiform platforms (and occasional Hertzian cones), but many of these are also a consequence of high-impact direct percussion directed at or near the edge of the platform. However, these tend to be found on the smaller flakes (although not exclusively so) and particularly on those inferred to be pressure flakes. In general, however, the type of platform does not appear to be related to the size of the removal.

7.2.2.7 Complete flakes and blades: terminations

Most flakes and blades have feathered terminations (Plate 57), with the remainder having either light hinges or plunging distal ends. Plunging terminations are more commonly found on blades than flakes and hinged terminations are mostly associated with removals of flake proportions. Stepped terminations, though generally infrequent, are less common on blades.

Such results are to be expected: the very nature of a plunging termination, which extends the length of the removal, tends to encourage blade proportions; conversely, hinging and stepping are abrupt, truncating terminations, which tend to create removals of flake proportions (Plate 58). In general, this trend is endorsed by the length/breadth scatter plot of flakes and blades, showing termination types: plunging terminations are more commonly found on blades or blade-like flakes, and hinging terminations tend to be found on short, broad flakes.

7.2.2.8 Flake and blade shatter

Flint flake and blade shatter fragments number 1763 artefacts, constituting over one-third of the flint assemblage (35.3%) with only one indeterminate flake fragment of chert (Table 7.12). Of the flake and blade shatter assemblage, the main components are those pieces damaged through burning; based on the recognition of ventral features, these artefacts were still identifiable as flake (or possibly blade) removals but could not be further classified (633/36.5% of flake shatter). In addition, are lateral fragments (77/4.4%), quartile fragments (167/9.5%) and thermally-damaged flake shatter (158/8.9%).

Table 7.12 Showing the classification of flake shatter, indicating extent of cortex and burning.

Classification	Cortex			Total	Burnt	
	Primary	Secondary	Inner		No	%
Flint shatter	11	179	273	463	229	49.5
Proximal	4	63	107	174	96	55.2
Medial	–	25	30	55	31	56.4
Distal	7	91	136	234	102	43.6
Blade shatter	3	97	165	265	105	39.6
Proximal	3	36	63	102	38	37.3
Medial	–	12	28	40	16	40.9
Distal	–	49	74	123	51	41.5
Lateral fragments	4	20	53	77	51	71.8
Indeterminate fragments	4	52	111	167	104	62.3
Indeterminate fragments, due to burning	6	117	510	633	633	100
Thermally damaged	2	57	99	158	47	29.7
Total	30	522	1211	1763	1169	66.3

Table 7.13 Showing fragment size of flake and blade shatter.

Fragment size (mm)	Flake shatter			Blade shatter			Total	%
	Proximal	Medial	Distal	Proximal	Medial	Distal		
<5	1	2	9	–	–	1	13	1.8
5–9	12	5	30	5	13	5	70	9.6
10–19	68	20	97	20	13	22	240	32.9
20–29	53	17	54	41	11	39	215	29.5
30–39	27	6	28	22	3	30	116	15
40–49	11	4	13	9	–	16	53	7.3
50–59	–	1	2	5	–	9	17	2.3
60–69	2	–	1	–	–	1	4	0.5
70–79	–	–	–	–	–	–	–	–
80–89	–	–	–	–	–	–	–	–
Total	174	55	234	102	40	123	728	100

The remainder are broken flakes (463/26.3%) or blades (165/9.4%) whose orientation could be more clearly established and some flake fragments may have had blade proportions when complete. Almost one-half of the determinate flake shatter are burnt (229/49.5% of flake shatter) but fewer blades are recognisable when burnt (105/39.6% of blade shatter).

In the case of both flakes and blade, distal fragments are most commonly found, accounting for half of flake and blade fragments (234/50.5% distal flake shatter; 123/46.4% distal blade shatter). Most of the remainder are proximal fragments, accounting for around two-fifths of both flakes and blades (174/40.4% proximal flake shatter; 102/38.5% proximal blade shatter). Both flake and blade shatter assemblages only include small numbers of medial fragments (55/12.5% medial flake shatter; 40/15.1% of medial blade shatter). These results suggest that similar fracture processes impacted upon both flakes and blades, and that, in general, they suffered one major fracture, resulting in a proximal and distal fragment.

Fragment sizes range from less than 5 mm to 80–89 mm in maximum length, with the bulk of most types of fragments measuring 10–29 mm (1079/61.2% of shatter, Tables 7.13 and 7.14). Indeterminate fragments, being too damaged to be further identified, tend (unsurprisingly) to be small, with burnt pieces mostly measuring 5–19 mm (456/72% of burnt indeterminate pieces), and unburnt indeterminate pieces mainly falling between 5 mm and 29 mm (126/78.4% of indeterminate shatter).

From an observation of the proximal fragments of flakes and blades, in general, platform percussion flaking accounts for the vast majority of knapping techniques (236/85.5%), with small numbers representing bipolar reduction (26/9.4%) and pressure flaking (14/5.1%) (Table 7.15).

Plate 38 The Outer Façade, looking north from the entrance, showing the massive stone core of the secondary fill (box section 1).

Plate 39 The Outer Façade, looking north from the entrance, showing persistent stone concentrations in secondary fill at a lower level.

Plate 41 Showing leeched ring and central hump of OF5. During excavation the stones formed a distinct ring, defining the diameter of the post butt at about 1.00 m.

Plate 40 Façade, looking south at the lowest level. The width is the combination of the Outer and Middle Façade postholes.

Plate 42 Truncated postholes of the SEW group: inner (foreground), middle and outer.

Plate 43 Section through the Burning Pit, C1453.

Plate 44 Section through the Burning Pit, C1453.

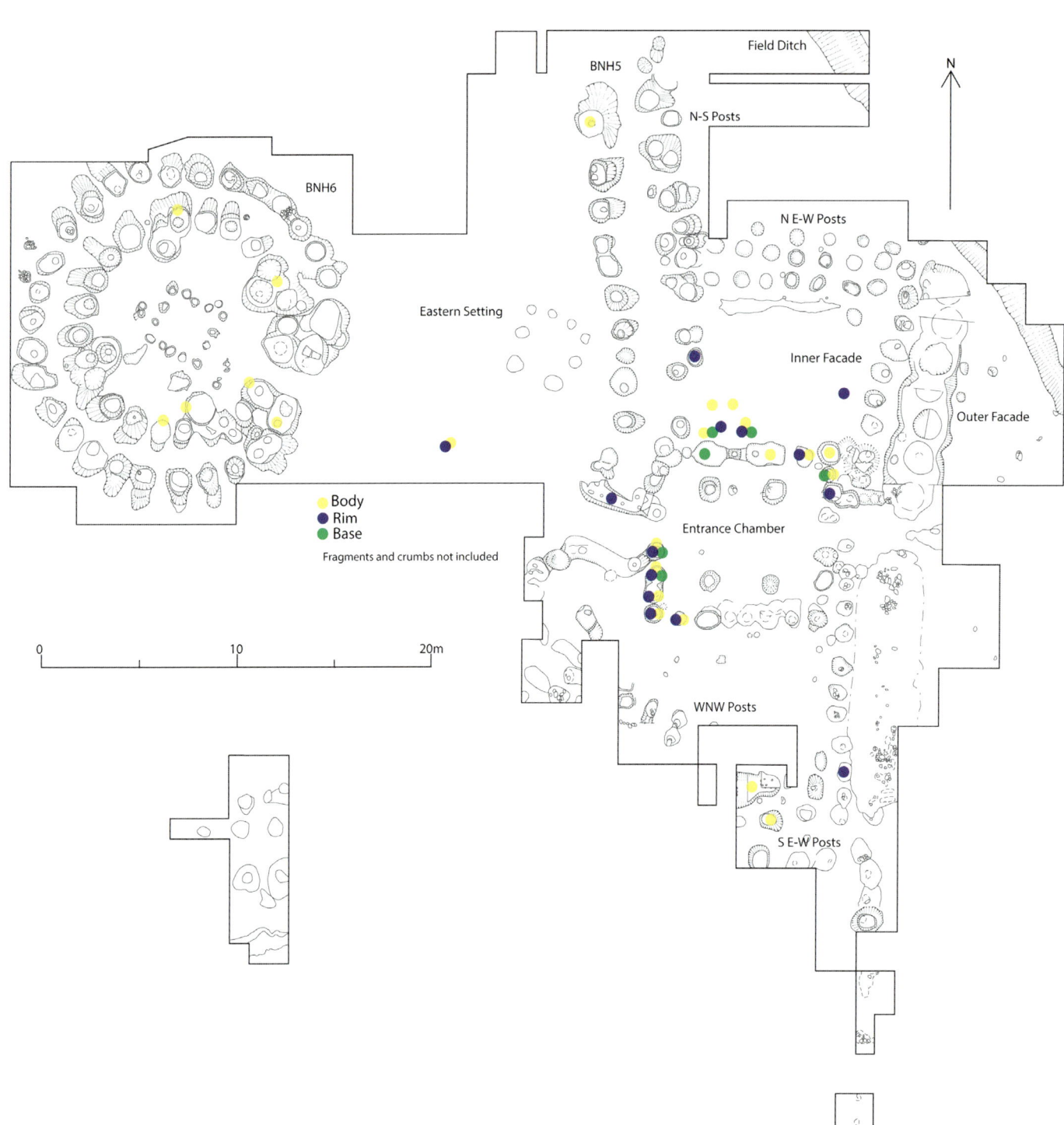

Plate 45 Distribution of the Grooved Ware pottery by sherd type.

Plate 46 Distribution of Coarse Ware pottery at Ballynahatty showing the three distinct concentrations: a) chamber (C591) area; b) outside the façade; c) west of BNH6.

Plate 47 Sherds from a Globular Bowl found in postholes EC15–17 in the Entrance Chamber.

Plate 48 Reconstruction of Vessel 1 (mounted on plaster cast of interior), recovered from chamber C591.

Plate 49 Reconstruction of Vessel 2 (mounted on plaster cast of interior), recovered from chamber C591.

Plate 50 Grooved Ware sherds from Ballynahatty with sooty accretions (burnt-on organic residue) on their interior surface.

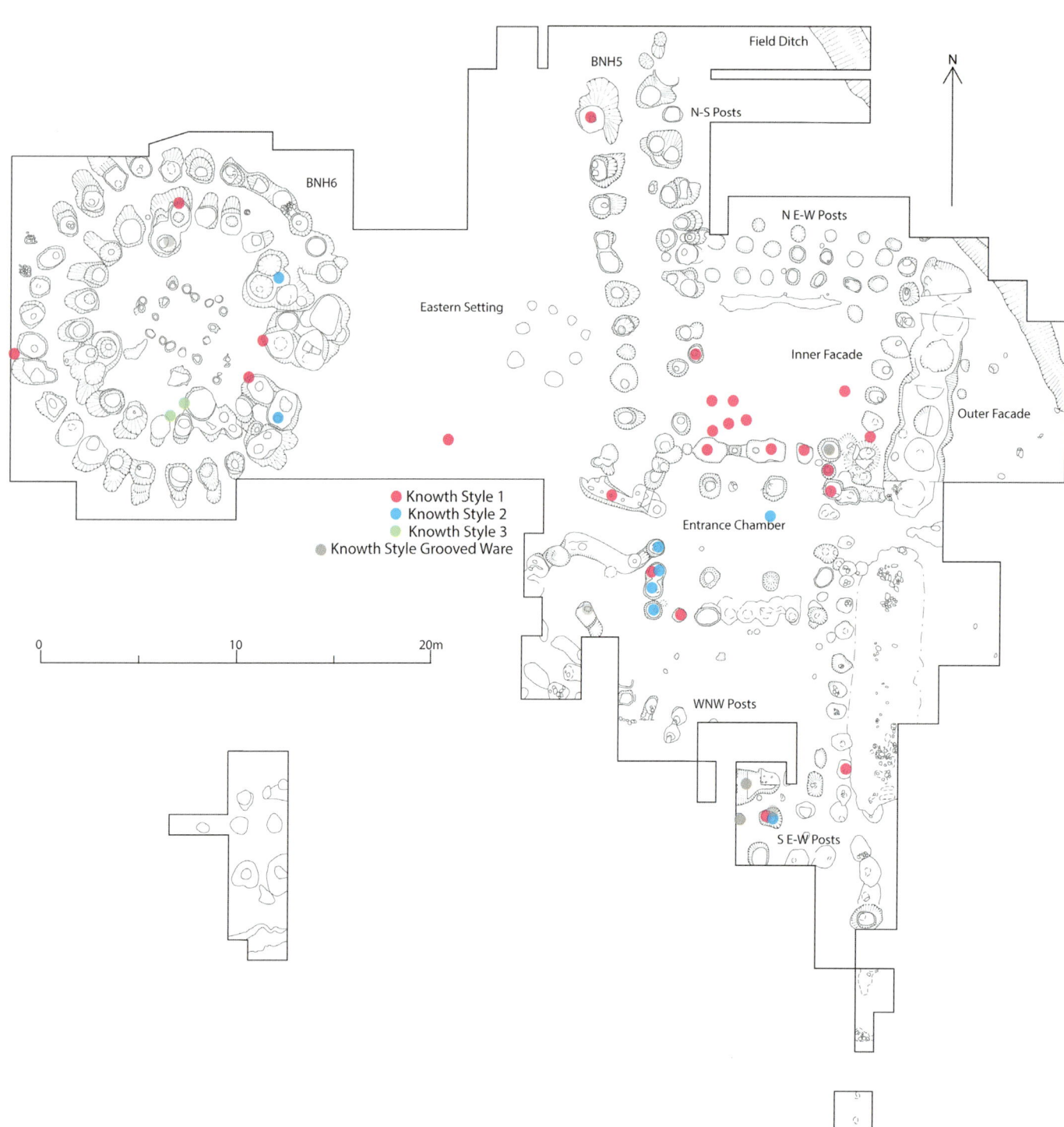

Plate 51 Distribution of the Grooved Ware pottery by type (Knowth Style 1, 2 and 3) across the various feature groups at Ballynahatty.

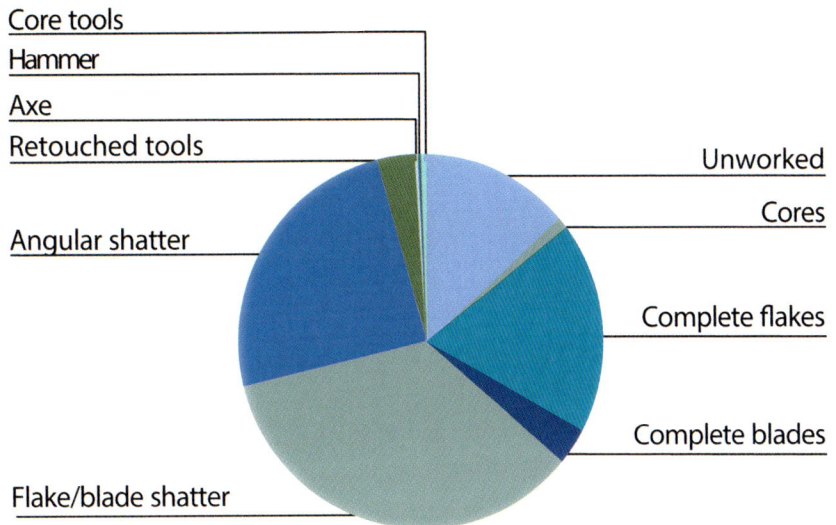

Plate 52 Composition of the primary technology assemblage.

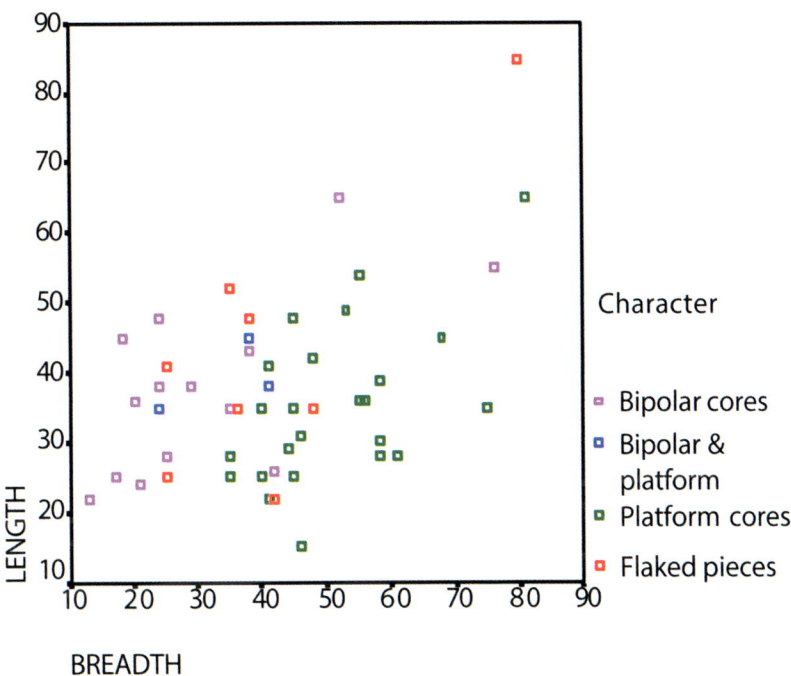

Plate 53 Length by breadth (mm) of cores.

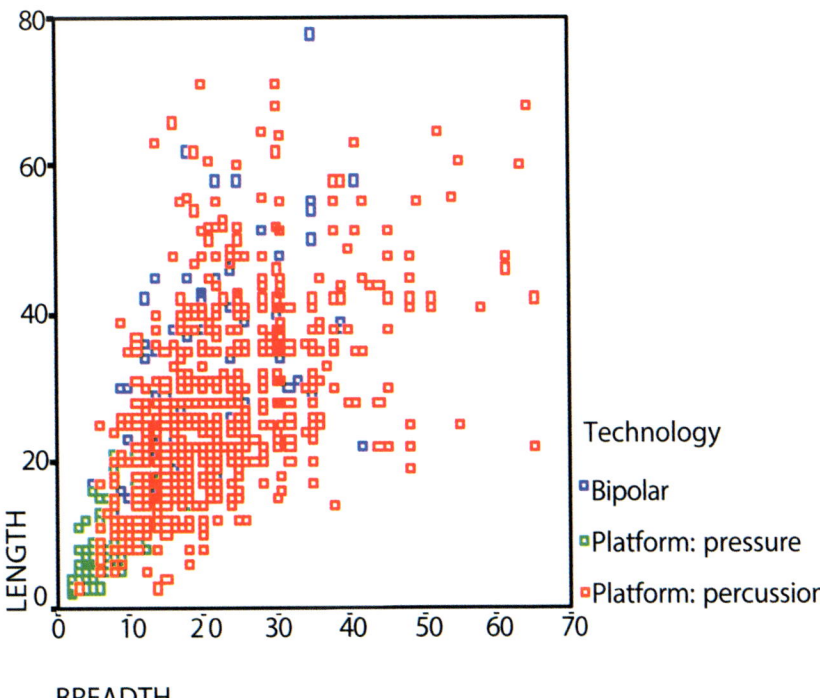

Plate 54 Length by breadth (mm) of flakes and blades produced by platform percussion flaking and bipolar techniques.

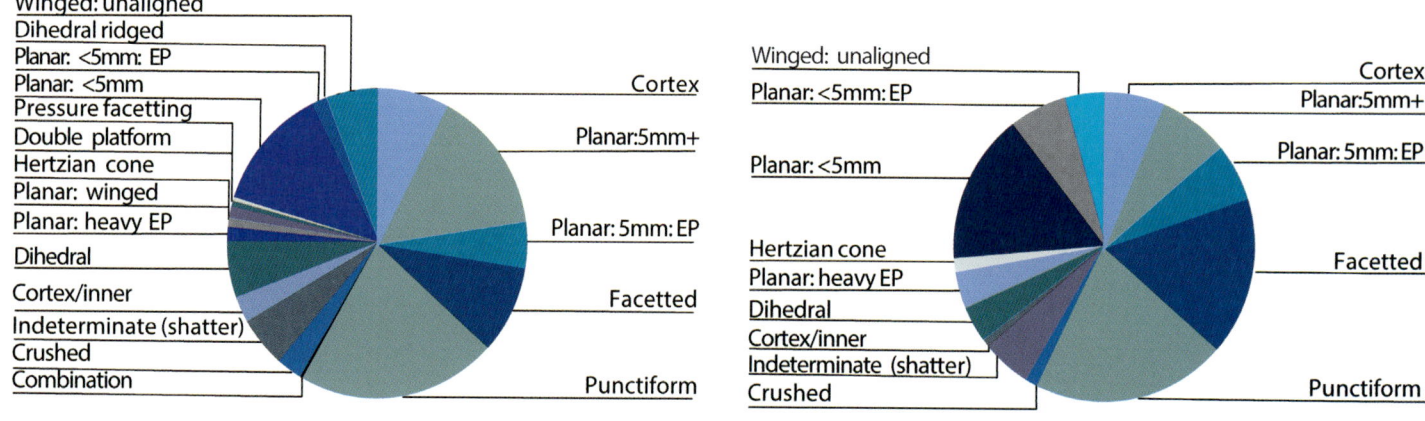

Plate 55 Platforms found on complete flakes (left) and blades (right).

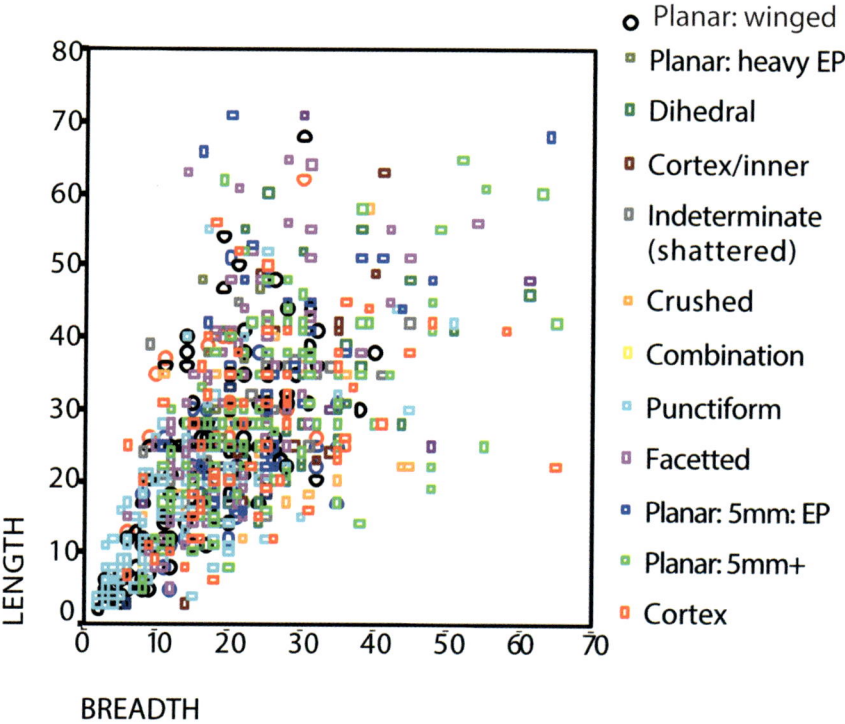

Plate 56 Length by breadth (mm) of flakes and blades, indicating type of platform present.

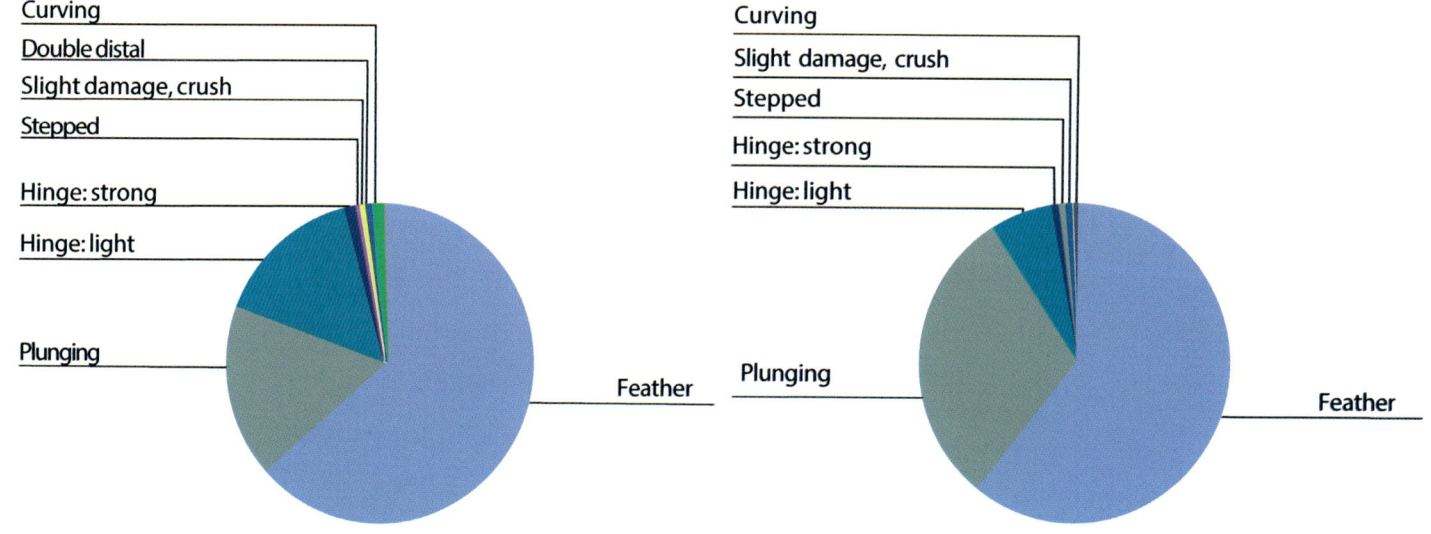

Plate 57 Types of terminations found on complete flakes (left) and blades (right).

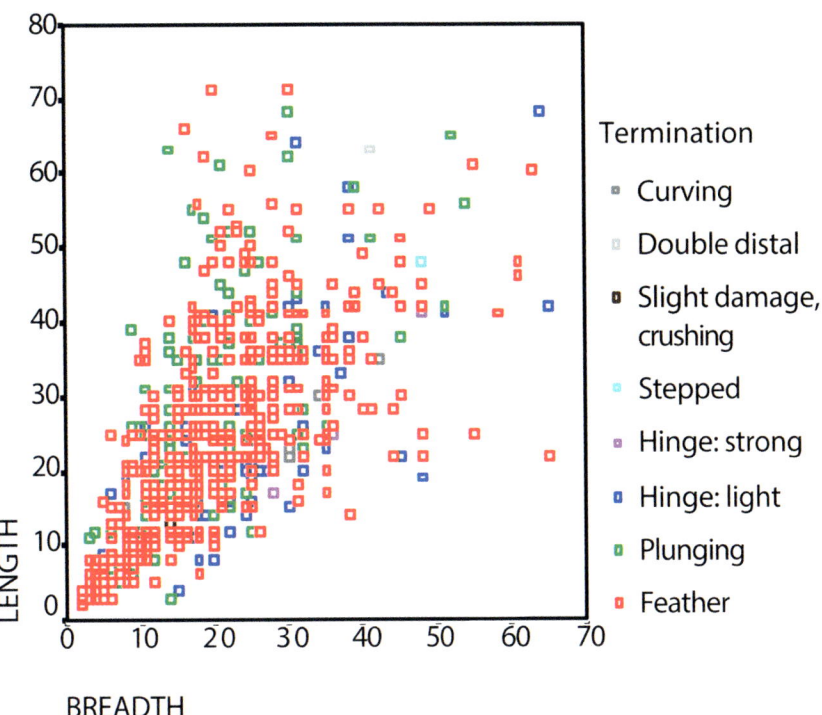

Plate 58 Length by breadth (mm) of flakes and blades, indicating types of terminations.

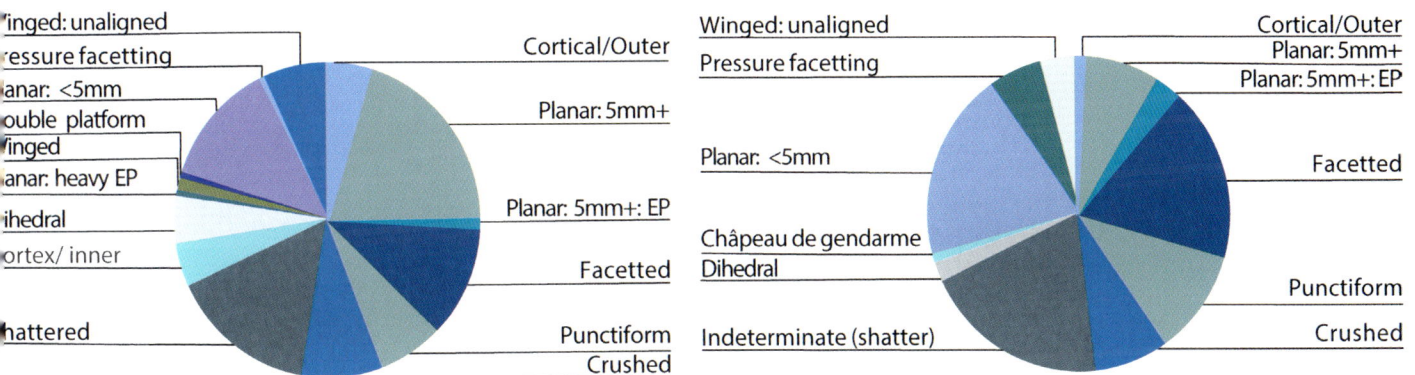

Plate 59 Types of platforms found on proximal flake (left) and blade shatter (right) much of the crushed/shattered proximal features constitute bipolar pieces.

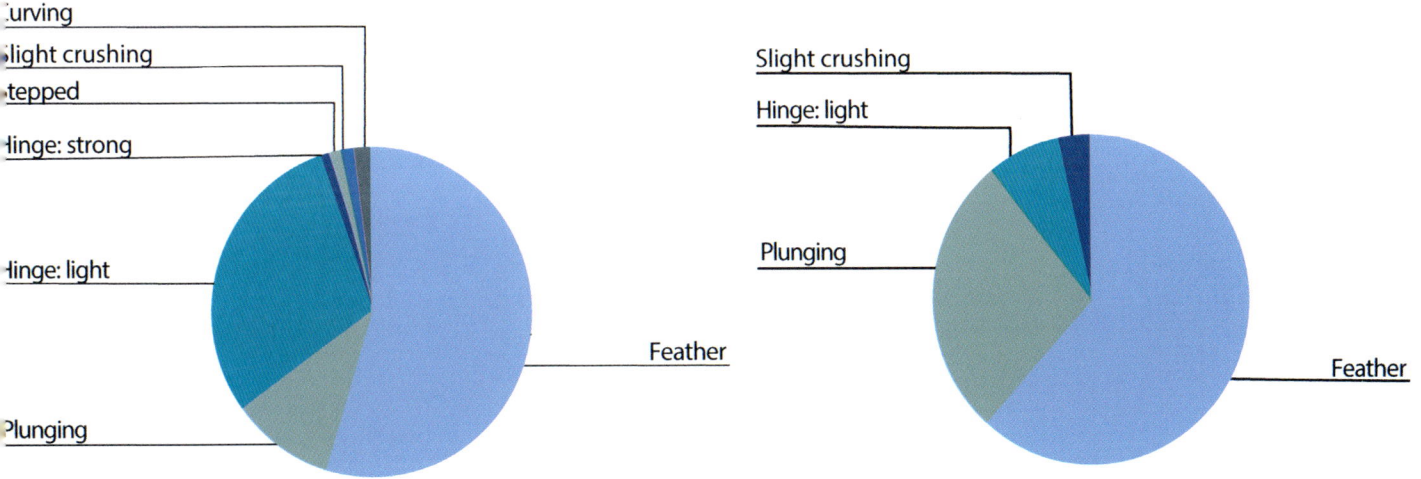

Plate 60 Types of terminations found on complete platform flakes (left) and blades (right).

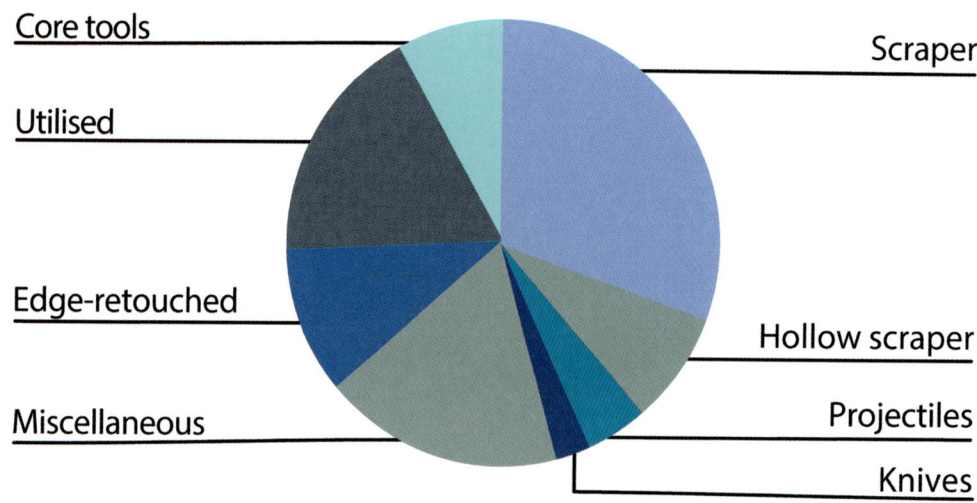

Plate 61 Basic composition of the tool assemblage.

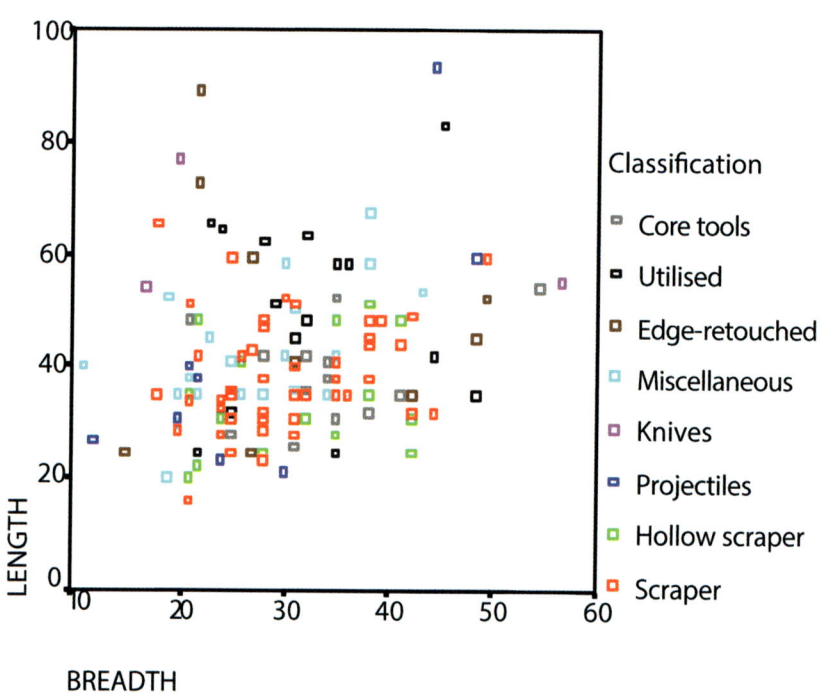

Plate 62 Length by breadth (mm) of modified tools.

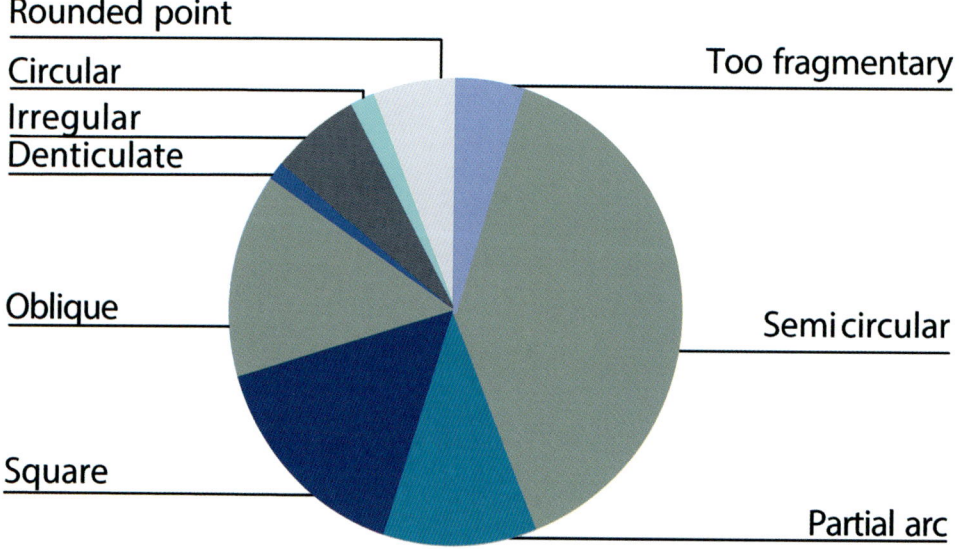

Plate 63 Composition of the scraper assemblage.

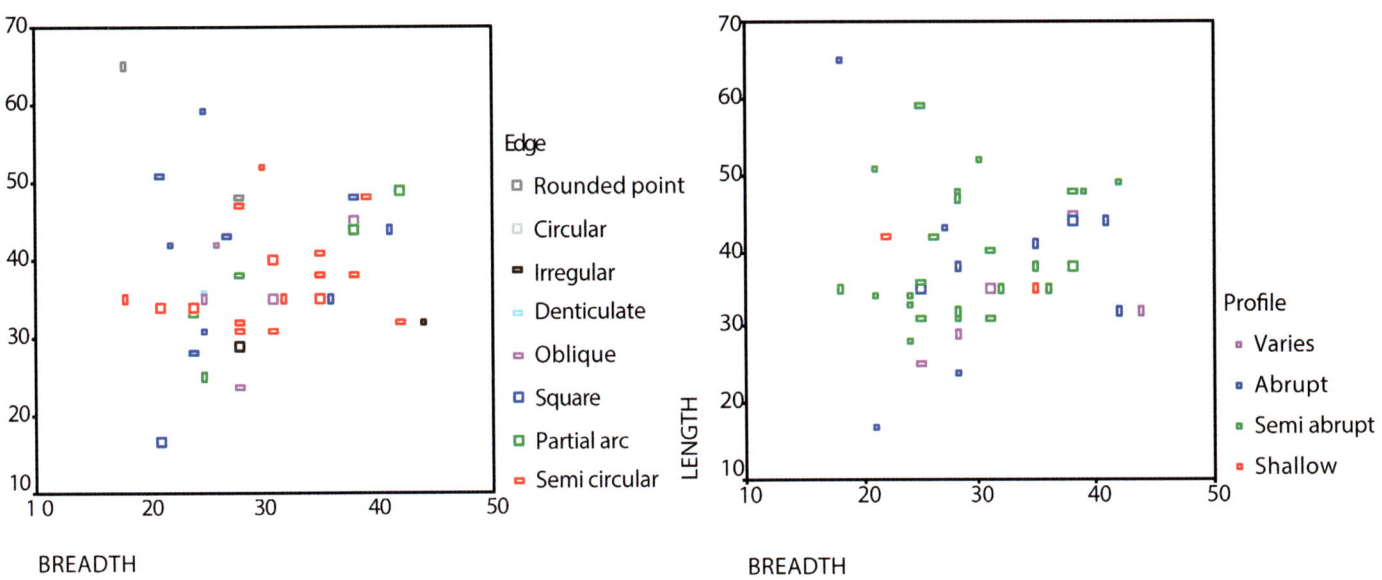

Plate 64 Length by breadth (mm) of complete scrapers (left) and angle (right) of scraping edges on scrapers.

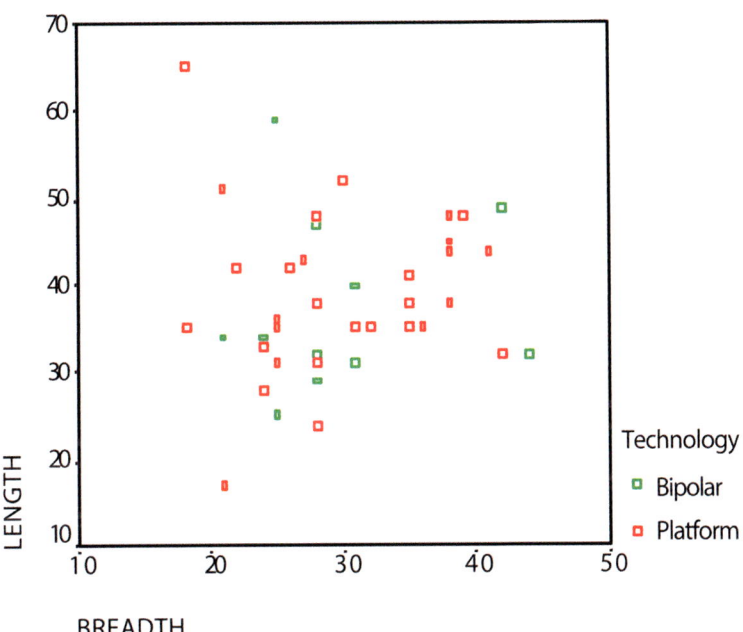

Plate 65 Length by breadth (mm) of complete scrapers, showing type of removal.

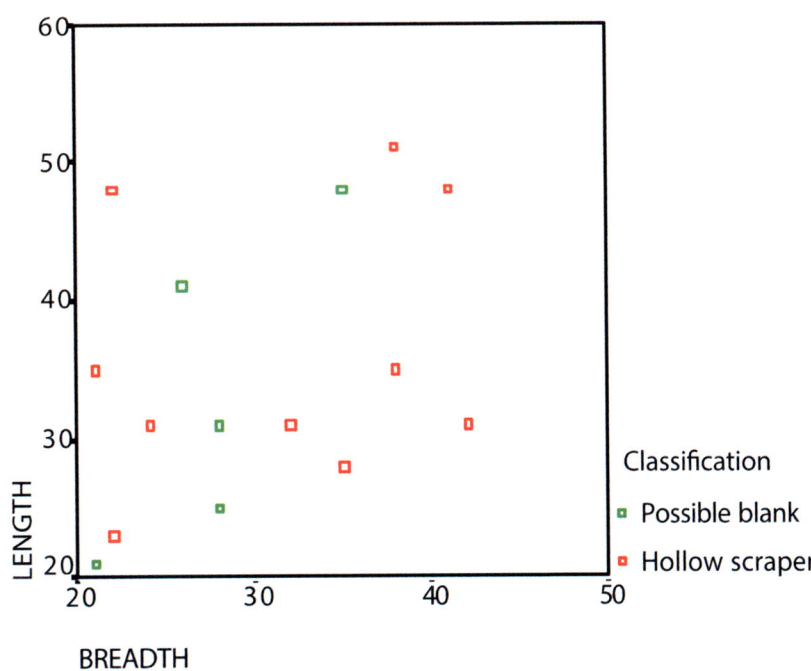

Plate 66 Length by breadth (mm) of hollow scrapers and possible blanks.

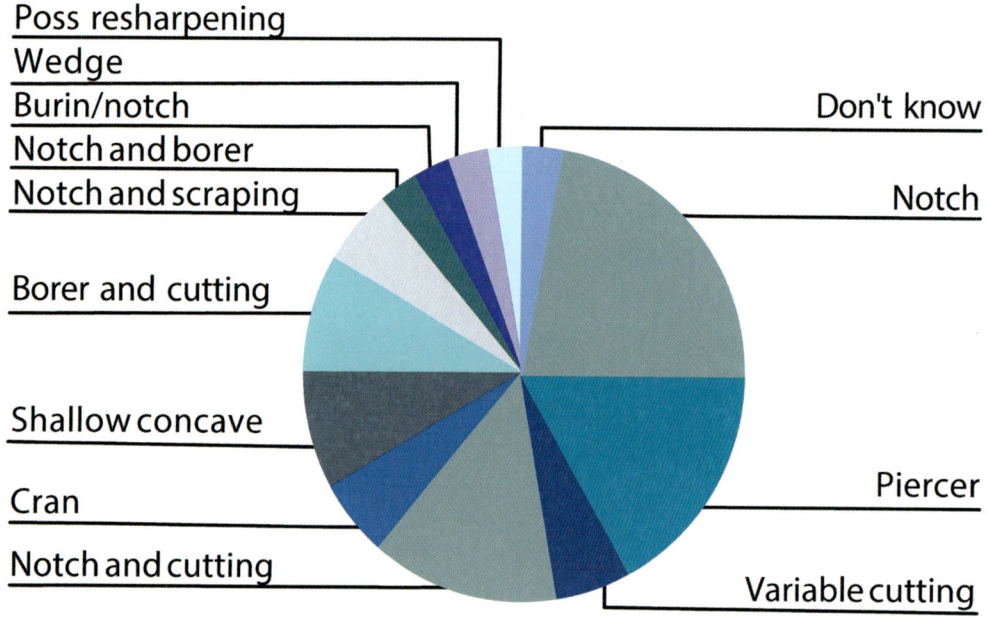

Plate 67 Types of 'miscellaneous' tools.

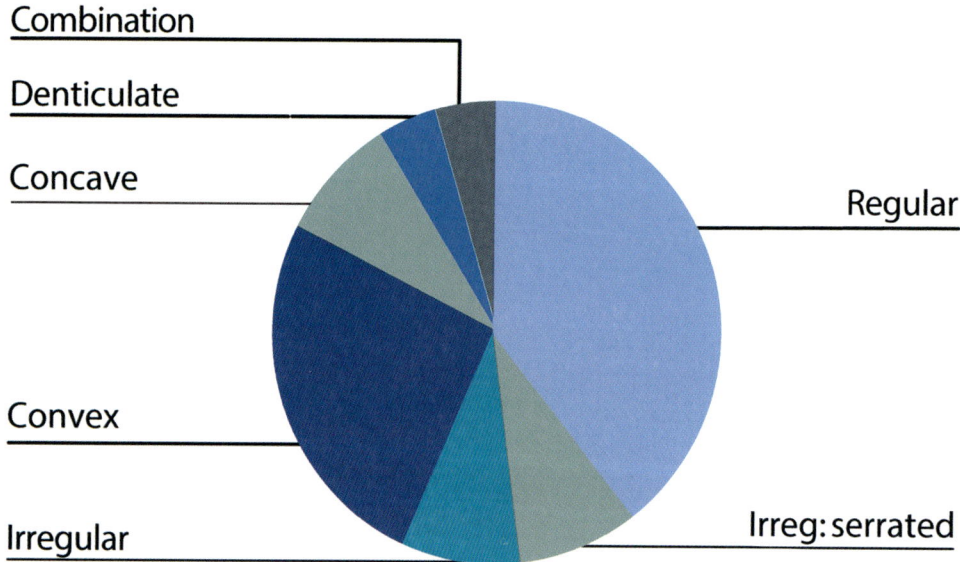

Plate 68 Types of functioning edges present on edge retouched tools.

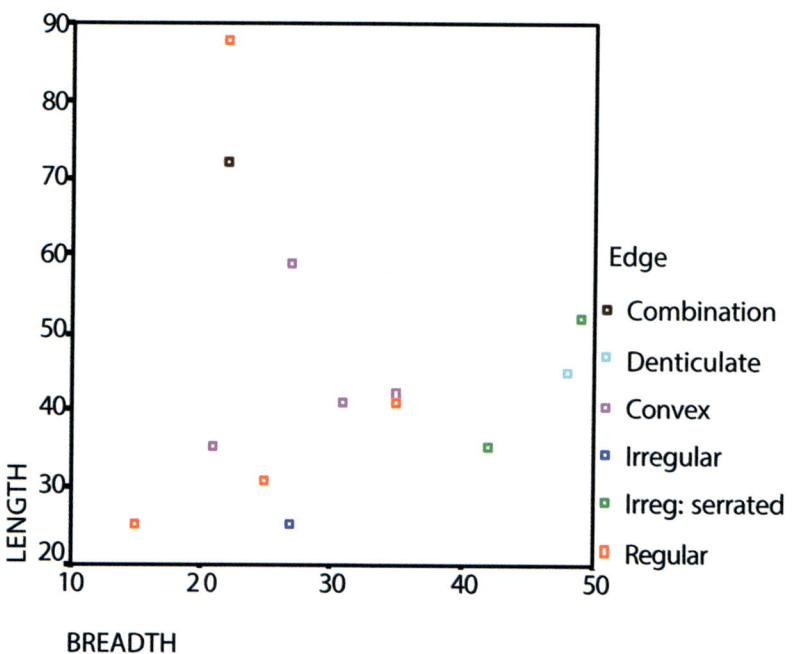

Plate 69 Length by breadth (mm) of edge retouched tools showing types of edges present.

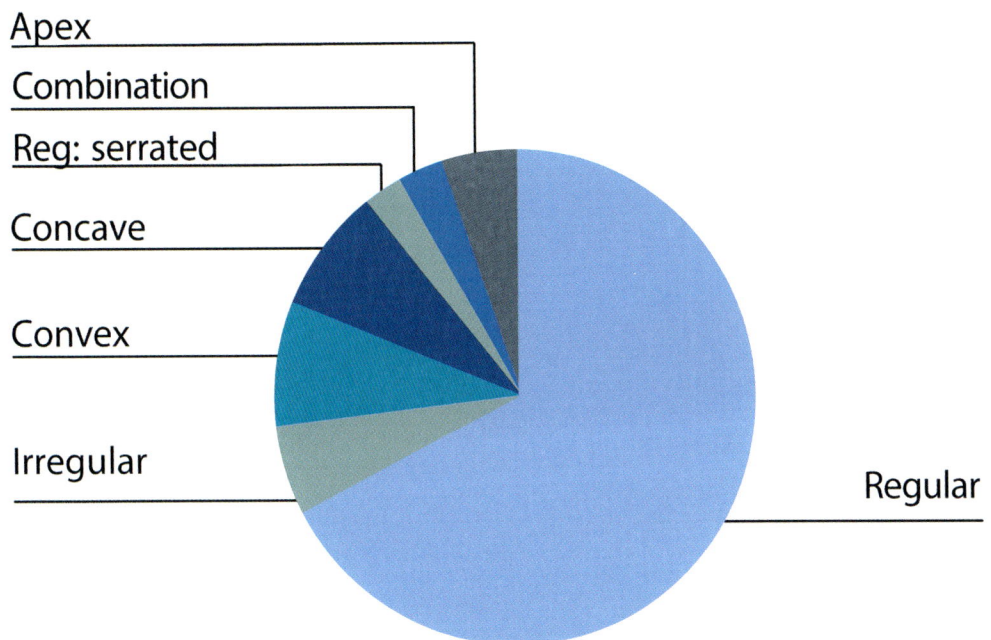

Plate 70 Edge morphology of utilised tools.

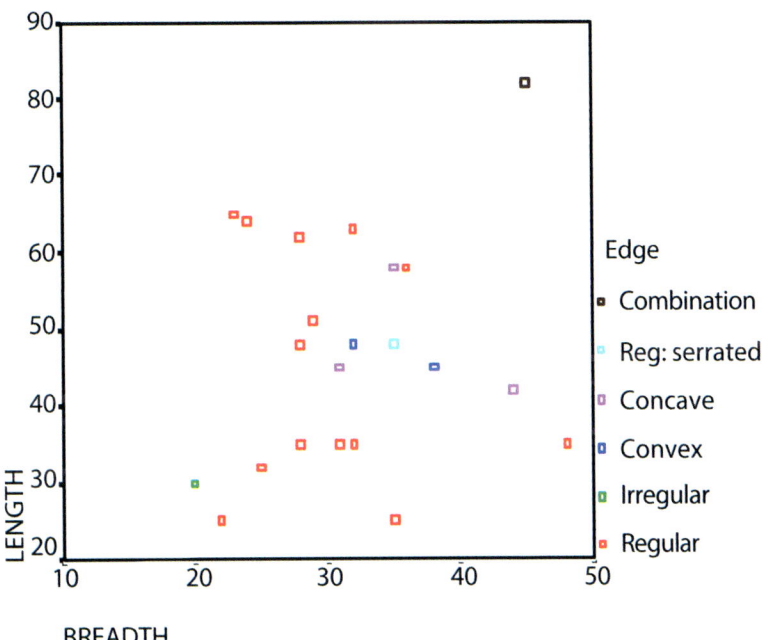

Plate 71 Length by breadth (mm) of utilised tools.

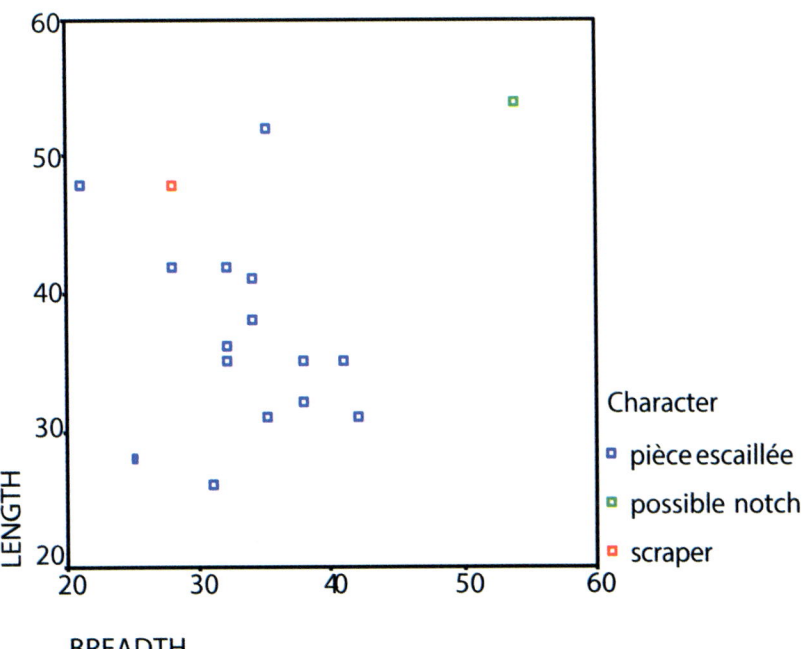

Plate 72 Length by breadth (mm) of core tools, showing possible function.

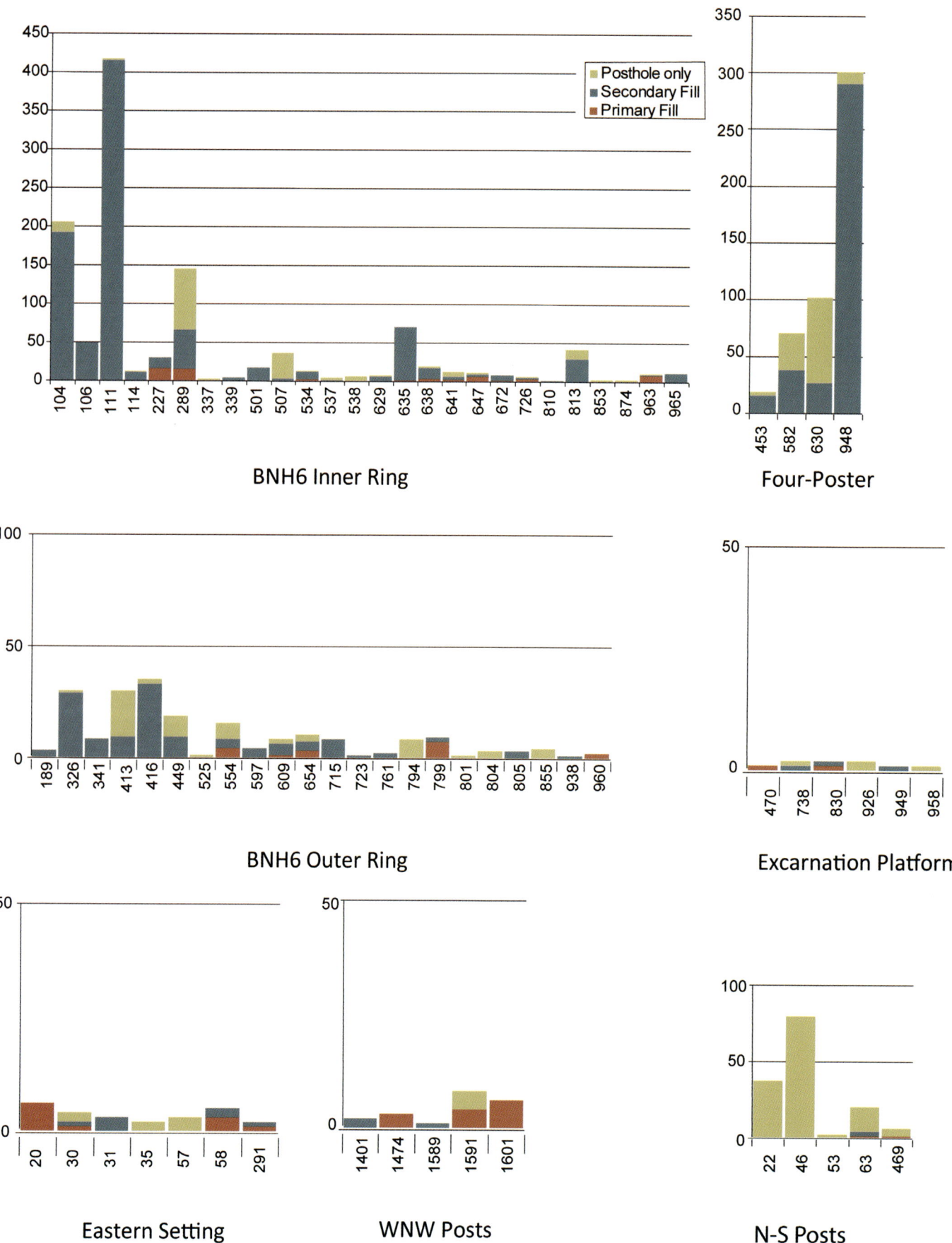

Plate 73 Distribution of lithics in the primary and secondary fills of postholes in each feature group. X-axis shows the posthole number and the Y-axis shows the number of lithic artefacts recovered.

7. The lithic assemblage: chipped stone

Table 7.14 Fragment size of flake shatter (Nelis 2003, 740).

Fragment size (mm)	Indeterminate	Indeterminate: burnt	Thermally damaged	Lateral fragment	Total	%
<5	4	30	–	1	35	3.4
5–9	26	192	22	11	251	24.3
10–19	74	264	72	23	433	41.8
20–29	31	98	37	25	191	18.5
30–39	23	28	13	10	74	7.1
40–49	6	14	3	4	27	2.6
50–59	2	6	10	2	20	1.9
60–69	–	–	1	–	1	0.1
70–79	1	–	–	1	2	0.2
80–89	–	1	–	–	1	0.1
Total	167	633	158	77	1035	100

Table 7.15 Flake and blade shatter, including whether created by percussion, pressure or bipolar flaking.

Types	Flake shatter	Blade shatter	Total
Percussion	151	85	236
Possible pressure flaking	8	6	14
Bipolar	15	11	26
Total	174	102	276

Table 7.16 Angular shatter recovered, showing types and condition (Nelis 2003, 742).

Types of angular shatter	No.
Burnt (beyond classification)	855
Thermally damaged	262
With indeterminate ventral features	64
Without ventral features	35
Possible bipolar debitage	2
Total	1218

Similar to complete flakes and blades, the platforms evident on proximal shatter are varied (Plate 59), (Nelis 2003, 741); the substantial number of shattered or indeterminate platforms might suggest that a proportion of shattering occurred during knapping. Both large and small planar platforms, with or without edge preparation, account for the bulk of identified platforms and faceting on complete flakes and blades is more commonly found on broken blades than broken flakes.

The terminations present on distal flake and blade shatter are similar to those found on complete examples: the majority have feathered terminations, and most of the remainder comprise hinged or plunging terminations; again, flakes shatter is more commonly found to have hinging terminations, while blades tend to have plunging terminations (Plate 60).

7.2.2.9 Angular shatter

Almost one-quarter of the entire flint assemblage is comprised of angular shatter (1218 pieces/24.3% of flint assemblage); in addition, three pieces of chert angular shatter were recovered (Table 7.16).

The main component of this assemblage is burnt material, which has been damaged beyond further identification (855 pieces / 70.2% of angular shatter; 17.1% of flint assemblage). The bulk of the remainder would appear to be thermally damaged indeterminate debitage, (262 pieces / 21.5% of angular shatter). It is probable that some of the angular shatter represents the undiagnostic remains of burnt flakes and blades, cores and retouched tools. The remainder of angular shatter is knapping debitage which has not been further damaged (101 pieces / 8.3% of angular shatter); many of these knapping chunks and spalls have, on one face, the partial remains of ventral curvature, and a further small number may be angular debitage resulting from bipolar reduction (Table 7.16). Only small numbers of angular shatter are fully corticated (i.e. one face is fully corticated in 13 pieces). Approximately, one-third of angular shatter is partially corticated (411 pieces / 33.7% of angular shatter), but the bulk of the material is uncorticated (794 pieces / 65.2%).

Angular shatter ranges in size from less than 5 mm to over 100 mm in maximum length with most between 5 and 29 mm (909 pieces / 74.6% of angular shatter). Burnt material constitutes almost three-quarters of the total angular shatter assemblage, and accounts for approximately four-fifths of material measuring less than 19 mm; therefore, the burnt condition of the material may be a contributory factor to the existence of most of the small-scale angular shatter (Table 7.17).

7.2.3 Secondary technology

The flint assemblage includes 211 retouched flint artefacts, (showing secondary shaping or edgework) which accounts for one in 25 artefacts, with scrapers forming almost

Table 7.17 Fragment size of angular shatter and micro-debitage, indicating extent of cortex and condition (Nelis 2003, 743).

Angular shatter fragment size (mm)	Primary	Secondary	Inner	Total flint	% of total no.	Burnt flint	Burnt flint % of fragment size
<5	–	1	99	100	8.2	86	86
5–9	1	38	234	272	22.3	223	81.9
10–19	8	120	272	400	32.8	313	78.3
20–29	3	107	127	237	19.5	146	61.6
30–39	–	74	45	119	9.8	62	52.1
40–49	3	36	8	47	3.9	15	31.9
50–59	–	22	4	26	2.1	4	15.4
60–69	–	77	3	10	0.8	4	40
70–79	–	1	1	2	0.2	1	50
80–89	–	3	1	4	0.3	1	25
90–99	–	–	–	–	–	–	–
100+	–	1	–	1	0.1	–	–
Total	13	411	794	1218	100	855	70.2

Table 7.18 Basic composition of retouched flint assemblage, showing extent of cortex and burning present (Nelis 2003, 746).

Retouched tools	Cortex			Total	%	Burnt no.	Burnt %
	Primary	Secondary	Tertiary				
Scrapers	2	31	31	64	30.3	14	21.9
Hollow scrapers	–	8	9	17	8	–	–
Projectiles	1	3	6	10	4.7	–	–
Knives	–	2	3	5	2.4	2	40
Miscellaneous	1	13	9	23	10.9	4	17.4
Edge retouch	1	13	9	23	10.9	4	17.4
Utilise	–	20	17	37	17.5	1	2.7
Core tools	1	13	3	17	8.1	1	5.9
Possible axe fragments	–	–	2	2	0.9	1	50
Total	5	112	94	211	100	24	11.4
%	2.4	53.1	44.5	100	–	–	–

one-third of these artefacts. The majority of scrapers from Ballynahatty have rounded scraping edges: most are semi-circular, and some have partial arcs; one scraper was almost fully circular, with the rounded edge incorporating most of the flake. Most of the remaining scrapers have scraping edges that are more squared or oblique than rounded, and a small number have narrow, almost pointed scraping ends (Plates 61 and 62; Table 7.18).

Ballynahatty scrapers are quite diminutive, ranging from 17 mm to 65 mm long (Figs 7.1 and 7.2); however, the vast majority of the assemblages measure 44 mm or less in length. The size of scrapers does not appear to correlate strongly with the type of scraping edge present; the same can be said of the angle of retouch, although there appears to be a very slight tendency for long, blade-like scrapers to have semi-abrupt scraping edges, and shorter, broader scrapers to have more abrupt, and sometimes variable, edges.

Most of the other artefacts are petit tranchet and derivative arrowheads, including the 'original' transverse type, as well as short and elongated pointed derivative forms. Some of the elongated pointed forms seem to have been utilised for cutting along their unretouched edge. The remaining assemblage includes a number of relatively undiagnostic bifacial fragments, as well as a single leaf-shaped arrowhead fragment.

The retouched assemblage includes 17 hollow scrapers and possible blanks, all of which are slightly irregular and abraded. Generally, the tool assemblage includes artefacts which are non-conformist and possibly quite expedient, with tools that appear to serve specific functions. Notched tools, often in combination with a piercing or cutting edge, are common, while edge-retouched pieces included forms with the working edge regular, slightly convex or concave, serrated or irregular, and similar tools which were utilised without retouch. In the main, the natural morphology of

7. The lithic assemblage: chipped stone 115

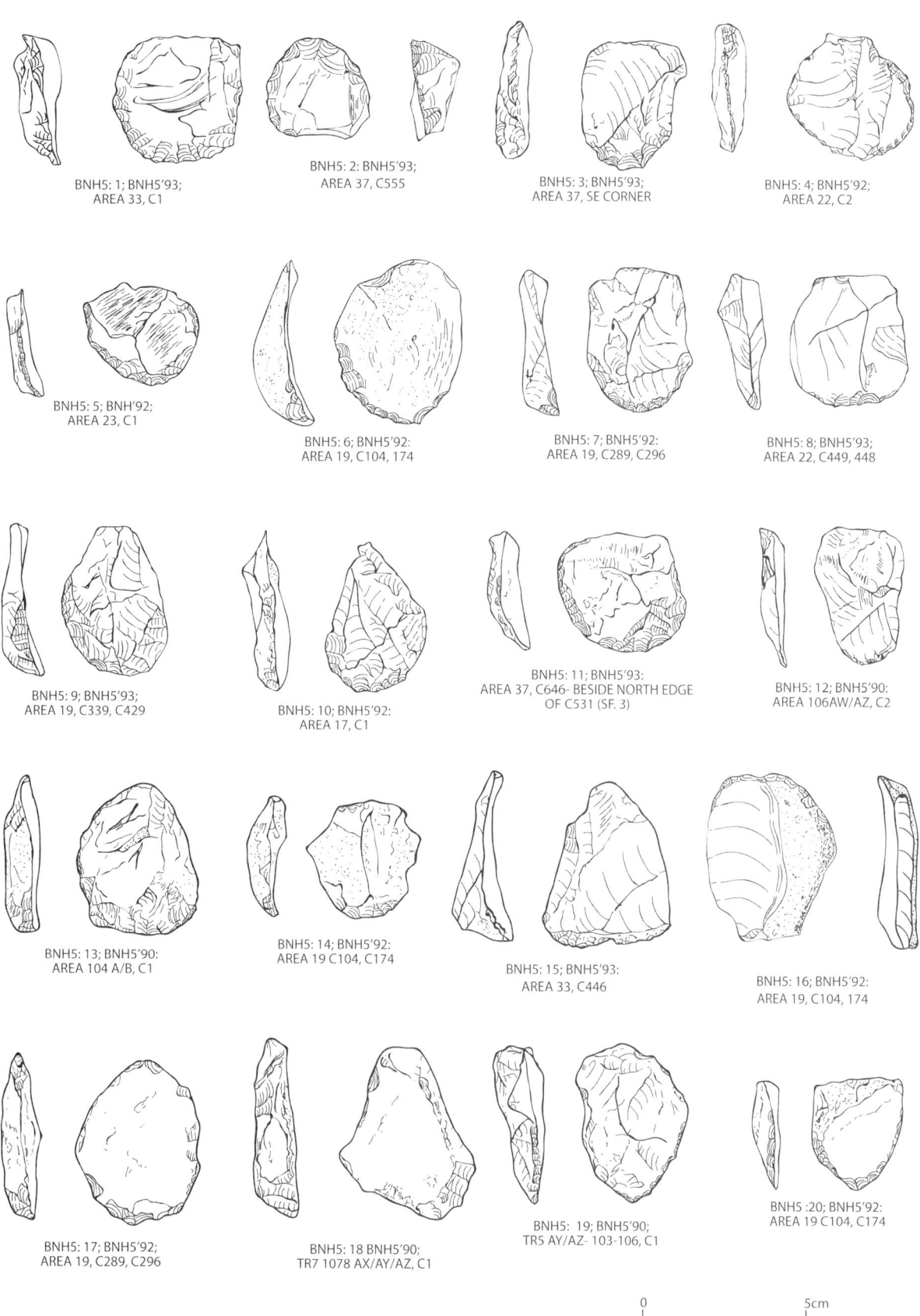

Figure 7.1 Lithics, showing a selection of scrapers.

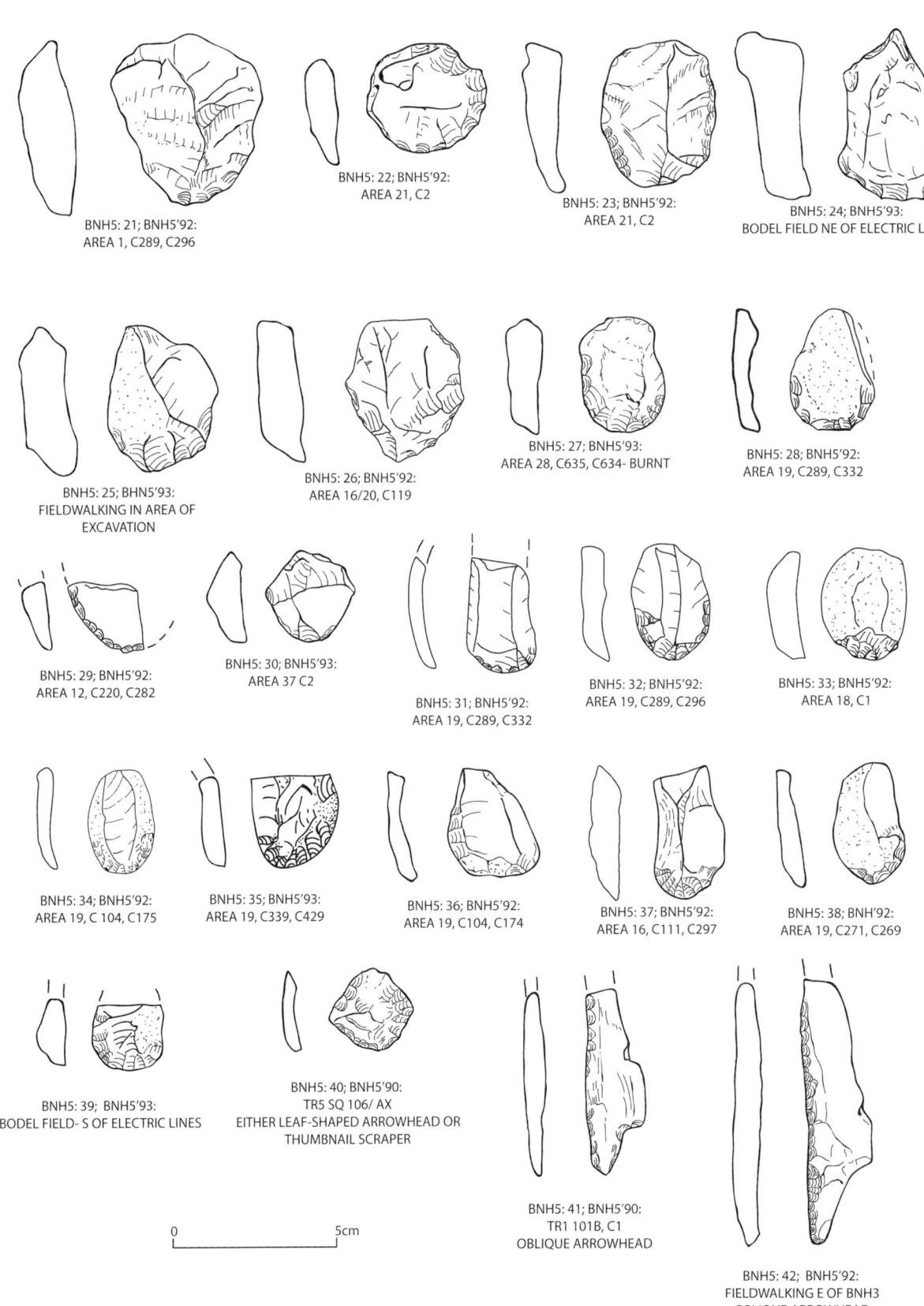

Figure 7.2 Lithics, showing a selection of scrapers and arrowheads.

the utilised edge was regular; therefore, it is probable that the artefacts were selected due to their suitability, with edge-retouched and utilised pieces likely to have been used as cutting tools.

Core tools are unusually numerous, comprising bipolar cores or scalar reduced flakes, and suitable for use as wedge tools. The quantity of these artefacts, and their dominance of the core tool assemblage, implies that they were created with function in mind, rather than indicating merely the opportunistic re-use of cores, which in general tends to contribute to a much more varied core tool assemblage. The length by breadth of core tools shows that most *pièces escaillées* or wedge tools, are roughly square and, in general, measured around 35–40 mm in length and breadth. Some artefacts are core tools (17 pieces/almost one in four cores), but the remainder involve the simple use or retouch of flake and blade removals (Plate 72; Nelis 2003, 745 & 770).

Scrapers constitute almost one-third of the retouched assemblage (64/30.3%), and account for four of the five chert tools found; the remaining chert tool is a retouched flake. The assemblage includes a number of hollow scrapers and associated possible blanks (17/8.5%). Core tools constitute almost one-tenth of retouched tools (17/8.1%). Expediently made and utilised tools account for most of the tool assemblage, totalling 93 artefacts, or just over two-fifths (44.1%). They include miscellaneous tools, piercers, notched tools, edge-retouched tools (*i.e.* variable retouched edges), unretouched utilised tools and two possible flaked axehead fragments (2/0.9%). Projectiles and knives are rare (9 projectiles, 4.3% of retouched assemblage; 5 knives, 2.4%).

Tools formed on fully corticated material are rare and include scrapers, a possible but irregular projectile, an edge-retouched piece and a core tool (based on a decorticated flake). The remaining scrapers are equally divided between partially corticated and uncorticated material (Table 7.18). Hollow scrapers, knives and projectiles are marginally more likely to use uncorticated material, and miscellaneous, edge-retouched and utilised pieces have a slight preference for partially corticated material; core tools tend to have some cortex present, and the possible axehead fragments are uncorticated.

Just over one-tenth of the tools are burnt. Despite the small number of knives, two out of five have been burnt; scrapers, in particular, are commonly burnt (14/21.9% of scrapers), as well as edge-retouched tools (4/17.4% of edge-retouched tools), and there are examples of burnt miscellaneous tools and utilised tools, as well as one core tool. Hollow scrapers and projectiles are the only tool types which do not have burnt examples.

7.2.3.1 Scrapers

A total of 64 flint scrapers were recovered from Ballynahatty; a further core tool may also have functioned as a scraper. Four scrapers are chert, three of which are formed on bipolar flakes (Plate 63).

Of the 64 flint scrapers, all have their functioning edge primarily located at the distal end, with most extending along at least part of the laterals. Thirteen pieces are broken, but only three are so fragmentary that the form of the scraping edge could not be determined. The majority of scrapers have rounded/convex scraping edges: most are semi-circular (25/40.9% of scrapers), and some have partial arc scraping edges (7/11.5% of scrapers); one scraper is almost fully circular, with the rounded edge incorporating most of the flake (Table 7.19).

Most of the remaining scrapers have scraping edges that are more squared (10/16.4%) or oblique (9/9.9%) than rounded, and a small number have a narrow, almost pointed scraping end (4/6.6%). The remaining scrapers have irregular edges (4/6.6%), and there is one denticulate piece (Nelis 2003, 770). The majority of scrapers were created on platform flakes (50/81.9%) but a substantial proportion utilised bipolar flakes (14/18.1%: *ibid.*, 770). All scraping edge types are represented by platform flakes and the bipolar scrapers echo the general trends, with most having semi- or partial arcs (Tables 7.15 and 7.19); however, two of the three irregular scrapers were formed on bipolar flakes.

Fully corticated scrapers are rare and, in the context of this assemblage, their scraping edge has largely been determined by the natural form of the removal, in this case, semi-circular (2), oblique (2) and pointed (1) (Table 7.20).

All scraper types utilise uncorticated material and, in general, the most commonly occurring types also frequently used partially corticated material (Table 7.21). Over one-fifth of scrapers, including the commonly found types, are burnt.

Most scrapers are edge-retouched only (20/32.8% of scrapers) or have combined edge and semi-invasive retouch (30/49.2% of scrapers) (Plate 64). The extent of retouch applied to scrapers is similar for platform and

Table 7.19 Types of scrapers, showing use of platform and bipolar removals (Nelis 2003, 748).

Types	Platform	Bipolar	Total
Too fragmentary	2	1	3
Semi circular	17	8	25
Partial arc	5	2	7
Square	9	1	10
Oblique	9	–	9
Denticulate	1	–	1
Irregular	2	2	4
Circular	1	–	1
Rounded point	4	–	4
Total	50	14	64

Table 7.20 Extent of cortex present on scrapers, indicating burnt examples (Nelis 2003, 749).

Scraping edge	Cortex			Total	Burnt	% of total
	Primary	Secondary	Tertiary			
Too fragmentary	–	–	3	3	–	–
Semi-circular	2	12	11	25	4	16
Partial arc	–	4	3	7	1	14.3
Square	–	7	3	10	2	20
Oblique	2	3	4	9	3	33.3
Denticulate	–	–	1	1	–	–
Irregular	–	1	3	4	2	50
Circular	–	–	1	1	–	–
Rounded point	1	1	2	4	2	50
Total	5	28	31	64	14	21.9
%	7.8	45.9	48	–	–	–

Table 7.21 Extent of retouch on scrapers, showing removal types (Nelis 2003, 749).

Location of retouch	Platform	Bipolar	No.
Too fragmentary	3	–	3
Edge retouch only	16	4	20
Semi-invasive & edge retouch	23	7	30
Semi- & fully invasive retouch	3	–	3
Semi-invasive only	3	2	5
Scaling edge retouch	2	1	3
Total	50	14	64

Table 7.22 Statistics for length, breadth and thickness of scrapers (Nelis 2003, 750).

Scraper statistics	Length (mm)	Breadth (mm)	Thickness (mm)
Mean	38.22	29.92	9.24
Median	35	28	9
Range	48	31	11
Minimum	17	18	4
Maximum	65	49	15
Percentiles (25)	31	25	7
Percentiles (50)	35	28	9
Percentiles (75)	44	35	11

bipolar removals, except that fully invasive pressure flaking, which is generally rare, is not found at all on bipolar scrapers (Table 7.21).

Some scrapers exhibit scaling retouch and these are among the smallest in the assemblage. One of the square-ended scrapers (and one of the chert pieces) exhibits scaling damage on the ventral side of the scraping edge which might indicate that they were used in a manner that pushed, rather than pulled, against the scraping edge. Such damage is relatively rare on scrapers of the Irish Neolithic.

Scrapers at Ballynahatty tend to be quite diminutive, ranging in maximum length from 17 mm to 65 mm, with most measuring 44 mm or less in length (Table 7.22). The size of the scraper (Plate 65) does not appear to relate significantly to the type of scraping edge present; similarly, the angle of retouch, although there appears to be a very slight tendency for long, blade-like scrapers to have semi-abrupt scraping edges, and shorter, broader scrapers to have more abrupt, and sometimes variable, edges. Furthermore, the size of the scraper does not appear to be determined by the knapping method used to produce the removal. Isolating 'thumbnail' scrapers within an assemblage can be problematic since the scrapers in the Ballynahatty assemblage tend to be quite small and their identification is often based on size attributes.

However, examples such as those found at Donegore Hill causewayed enclosure, Co. Antrim (Mallory et al. 2011, 151), which are clearly distinct from the remainder of the scraper assemblage, show that their diminutive size is only one aspect of these tools.

7.2.3.2 Knives and projectiles

Knives and projectiles are rarely found at Ballynahatty; only two burnt, minimally retouched knives were recovered, both of which are partially corticated but otherwise have little in common (Table 7.23). The assemblage includes petit tranchet and petit tranchet derivative arrowheads, types which are poorly understood compared with the more 'formal' tool types identified in the Irish Neolithic (but see Woodman et al. 2006). Such arrowheads have rarely been found in excavations and are mostly known as unprovenanced finds gathered by antiquaries. The assemblage at Ballynahatty includes two transverse arrowheads (Nelis 2003, 770, 7.4a); one is quite finely pressure flaked bifacially, the other is a finely pressure flaked, short, pointed example and both are fractured at the tip (e.g. ibid., 7.4b). In addition to these are two elongated examples (ibid., 7.4c), and a possible tip fragment was

Table 7.23 Types of knives, petit-tranchet and derivative types, extent of cortex and burning (Nelis 2003, 752).

Knives and projectiles	Cortex			Burnt	Total
	Primary	Secondary	Tertiary		
Knives	–	2	–	2	2
Monolateral: long: distal blade fragments	–	1	–	1	1
Bilateral: leaf-shaped	–	1	–	1	1
Petit tranchet & derivatives	–	2	5	–	7
Transverse	–	2	–	–	2
Short pointed	–	–	1	–	1
Elongated pointed	–	–	3	–	3
Complete	–	–	2	–	2
Possible fragment	–	–	1	–	1

Table 7.24 Types of projectiles, and extent of cortex and burning (Nelis 2003, 753).

Projectile	Cortex			Burnt	Total
	Primary	Secondary	Tertiary		
Projectiles	1	1	4	–	6
Thinned bifaces: arrowhead preform	–	–	1	–	1
Thinned bifaces: possible fragment	–	–	–	–	–
Possible arrowhead: uniface leaf-shaped	1	1	–	–	2
Possibly minimally retouched fragment	–	–	1	–	1
Leaf-shaped arrowhead: fragment	–	–	2	–	1
Base fragment	–	–	1	–	1
Possible fragment	–	–	1	–	1
Javelin: unpolished	–	–	1	–	1

identified. Both the complete examples may have been utilised for cutting along their unretouched edge.

The assemblage also includes an unfinished single leaf-shaped arrowhead fragment (Table 7.24). Two unifacially worked leaf-shaped artefacts may be projectiles together with a minimally worked fine flake fragment that may also have been intended as a projectile. A further two bifacial tools (one fragmentary, one complete) may be arrowhead preforms. The most striking artefact of this assemblage is the complete javelin head (*ibid.*, 7.4e). Morphologically, it is perfectly symmetrical, although it is relatively minimally worked, unpolished and marginally pressure flaked along most edges. These date to an earlier phase of the Neolithic than the timber structures.

7.2.3.3 Hollow scrapers

The assemblage of retouched tools includes 17 slightly irregular hollow scrapers and possible blanks (Nelis 2003, 770, fig. 7.4e; 11 retouched, 6 possible blanks: Table 7.25). Most complete tools are on slightly irregular flakes, even when trapezoidal, and all have slightly uneven hollows which tend to be quite worn; some combine piercing (No. 4955) or cutting (No. 4960, 5014) functions.

The types of platform appear to relate to the type of flake, and trapezoidal flakes (Table 7.25 and 7.26)

Table 7.25 Hollow scrapers and possible blanks, showing types of flakes utilised (Nelis 2003, 754).

Platform types	Hollow scrapers & associated tools	Possible blanks	Total
Regular flake	2	4	6
Trapezoidal flake	5	2	7
Core trimming	2	–	2
Laterally fractured	2	–	2
Total	11	6	17

exclusively carry ridged platforms, that is, one that rises to an apex where the force of percussion is applied (*e.g.* dihedral, *chapeau de gendarme* or a combination of both); those found on regular flakes have flat platforms (*i.e.* no apex: *e.g.* cortical, planar, flat faceting).

Core-trimmed flakes, when used, do not seem to have been prepared for use prior to removal and may represent the opportunistic use of an appropriate flake. The only very 'typical' example of a hollow scraper within the assemblage, on a fine trapezoidal flake with an extensively worked *chapeau de gendarme* platform, has a distal concave fracture, with the remaining straight distal edge apparently utilised for cutting. The length by

Table 7.26 Hollow scrapers, showing types of flake and platform (Nelis 2003, 754).

Platform types	Type of removal	Regular flake	Trapezoidal flake	Laterally split	Core trimming	Total
Not present: broken		1	1	–	–	2
Cortical/ outer		1	–	–	1	2
Planar: EP		2	–	–	1	3
Split/ shorn		–	–	2	–	2
Facetted		2	–	–	–	2
Dihedral		–	2	–	–	2
Châpeau de gendarme		–	3	–	–	3
Châpeau de gendarme/dihedral		–	1	–	–	1
Total		6	7	2	2	17

Table 7.27 Miscellaneous, edge-retouched and utilised flint tools, showing extent of cortex and burning (Nelis 2003, 755).

| Retouched tools | Cortex | | | Total | % of retouched tools | Burnt no. | % burnt |
	Primary	Secondary	Tertiary				
Possible axe fragments	–	–	2	2	1	1	50
Miscellaneous	–	20	13	33	15.6	1	3
Edge retouch	1	13	9	23	10.9	4	17.4
Utilised	–	20	17	37	17.5	1	2.7
Total	1	53	39	93	100	6	6.5
%	1.1	57	41.9	–	–	–	–

breadth scatter plot (Plate 66) reflects the variability of flakes selected for the production of these tools.

7.2.3.4 Miscellaneous, edge-retouched and utilised tools

The tool assemblage included a substantial component of artefacts which were non-conformist. These include two possible but unconvincing flaked axehead fragments. Most of the tools appear expedient as utilised or minimally edge-retouched flakes and blades (Table 7.27). A number appear to serve one or more specific functions and are here termed miscellaneous tools (Table 7.27; Plate 67), but most appear to have served a single function. Common notched tools (13 pieces) have variable notches, so it may be of questionable value to group these together, but they are included here nevertheless since they probably served as spoke-shaves. A number of tools combine features such as notch and/or retouched edge (inferred to have served as a cutting tool), borers and, less commonly, scraping edges. Several piercers were recovered, mostly as single function tools. The assemblage includes a probable wedge (Nelis 2003, 770, fig. 7.3 a) and two indeterminate pieces (one perhaps a resharpening flake, not derived from a scraper). None of the miscellaneous tools has been heavily burnt, although one combination tool (with cutting and notch functions) may have been heated.

Most miscellaneous tools have been formed on flakes rather than blades, both complete and shattered

Table 7.28 Types of miscellaneous tools (Nelis 2003, 757).

Miscellaneous tools	Flakes	Blades	Total
Notched tools	9	5	13
Small notch	7	1	8
Concave/ large notch	1	2	3
Burin/ notch	–	1	1
Cran	1	1	2
Piercer	5	1	6
Wedge	–	1	1
Combination tools	8	5	13
Notch and cutting	4	1	5
Notch and scraping	2	–	2
Notch and borer	–	1	1
Borer and cutting	2	1	3
Variable cutting	–	2	2
Possible resharpening flake	1	–	1
Indeterminate fragment	1	–	1
Total	24	12	36

(Table 7.28) (ibid., 757). Typically, they utilise leaf-shaped (13) or sub-rectangular (11) flakes but also make use of more amorphous flakes or blades (11); perhaps of note is that round flakes are rarely used (1), and instead were selected for use as scrapers. Only two bipolar pieces (1 notched tool, 1 combined cutting tool) are included in

7.2.3.5 Edge-retouched tools

Over one-tenth of the flint tool assemblage comprises edge-retouched artefacts (23/10.9%; Tables 7.29 and 7.30; Plate 68). In most cases, edge retouch is regular (9 pieces) or slightly convex (6); a few are slightly concave edge-retouched tools, and four are irregularly retouched, of which a couple are effectively serrated. One tool has a denticulate edge and another has a combination of retouch on its functioning edge (Table 7.29) (Nelis 2003, 758). Just over half of these tools functioned as cutting tools (13), whilst the function of the remainder is less obvious but may have been cutting.

Fully corticated tools are rare (1 convex piece) but reflects much of the retouched assemblage, where partially corticated flint (13) was utilised more commonly than uncorticated material (9). With the exception of three artefacts which utilise bipolar flakes (1 regular, 1 convex, 1 combination), platform removals are most common. Only four edge-retouched pieces are burnt; these include two regular pieces, as well as a convex and a concave tool.

Most edge-retouched tools have a shallow retouched edge; only the convex tools deviate significantly, having many partially or fully abrupt edges, as well as shallow edges (Table 7.30) (*ibid.*), and only the denticulate tool has a variable retouch angle.

The size of edge-retouched tools varies, with some of the larger examples being among the largest retouched tools within the entire assemblage (Plate 69). However, there is no clear pattern between the size of the artefacts and the type of modified retouched edge; in most cases, artefacts with at least one fine, shallow edge seem to have been selected, apparently regardless of the overall morphology or size of the removal.

Table 7.29 Types of retouched edges on edge-retouched tools, showing possible function (Nelis 2003, 758).

Edge type	Possible function			Total
	Cutting	Unclear	Possibly cutting	
Regular	7	2	–	9
Irregular	–	2	–	2
Convex	2	2	2	6
Concave	1	1	–	2
Irregular: serrated	1	1	–	2
Denticulate	1	–	–	1
Combination	1	–	–	1
Total	13	8	2	23

Table 7.30 Edge-retouched tools and angle of retouch (Nelis 2003, 758).

Angle of scraping edge / Edge type	Shallow	Semi–abrupt	Abrupt	Variable	Total
Regular	8	1	–	–	9
Irregular	2	–	–	–	2
Convex	3	2	1	–	6
Concave	2	–	–	–	2
Irregular: serrated	1	1	–	–	2
Denticulate	–	–	–	1	1
Combination	1	–	–	–	1
Total	17	4	1	1	23

7.2.3.6 Utilised tools

A total of 37 artefacts were recognised as having been utilised without retouch, accounting for almost one-fifth of the tool assemblage (17.4%); in addition, it is probable that at least some similarly utilised tools exist within the debitage assemblage but without the visual damage which distinguishes the artefacts discussed here (Plate 70; Table 7.28).

In some cases, the utilised edge is regular (25 pieces/67.6% of utilised tools) and the natural morphology of the edge was utilised. It is probable that the suitably edged artefacts were selected. One artefact with a regular edge now bears a heavily serrated edge, evidently the result of use. Most of the remainder have subtle convex (3) or concave (3) natural edges, as well as irregular (2) or even angular edges (apex: 2 pieces) (Table 7.31). Typical of most tools, decortical flakes were rarely used (1 regular), but partially corticated and uncorticated artefacts are present in similar quantities (19 secondary; 17 inner). Only one utilised piece, with a regular edge, was heated, and no utilised tools have been heavily burnt. In most cases, utilised pieces appear to have served as cutting tools (29/78.4%) but, for some, the function is unclear (5); individual tools seem to have been used for scraping, as a spoke shave and with a combination of uses in mind.

Commonly, the edge damage caused by use occurs either as discontinuous (16) or worn (10) edge damage; continuous edge damage and scaling edge damage occur less frequently (6 and 5 pieces respectively). All types of edge damage are evident on regular tools and some tools were evidently used more extensively than others, which might also suggest that they were used on a variety of materials. In one case, the utilised edge has been quite heavily serrated, apparently through use; such edge damage is usually inferred to be an indicator of a tool used for cutting hard and brittle material, such as bone. Significantly, all the concave tools bore worn and smoothed working edges, possibly suggesting how they were used: worn edges are often seen as indicative of use on soft, pliable materials.

Table 7.31 Edge-retouched tools, and angle of retouch (Nelis 2003, 760).

Type of retouch	Possible function	Unclear	Cutting	Possibly scraping	Spoke shave	Combination	Total
Regular		2	21	1	–	1	25
Reg: serrated		–	1	–	–	–	1
Irregular		2	–	–	–	–	2
Convex		–	3	–	–	–	3
Concave		1	1	–	1	–	3
Apex		–	2	–	–	–	2
Combination		–	1	–	–	–	1
Total		5	29	1	1	1	37

Table 7.32 Utilised tools, showing types of edge damage (Nelis 2003, 761).

Edge type	Type of edge damage	Discontinuous nibbling	Continuous nibbling	Worn/ nibbling	Scaling edge damage	Total
Regular		10	5	6	4	25
Irregular		2	–	–	–	2
Convex		2	–	1	–	3
Concave		–	–	3	–	3
Combination		1	–	–	–	1
Apex		1	1	–	–	2
Regular: serrated		–	–	–	1	1
Total		16	6	10	5	37

Table 7.33 Utilised tools, showing angle of retouch (Nelis 2003, 762).

Utilised tools	Angle of scraping edge	Shallow	Semi– abrupt	Abrupt	Variable	Total
Regular		21	4	–	–	25
Irregular		2	–	–	–	2
Convex		3	2	–	–	5
Concave		2	–	–	–	3
Irregular: serrated		–	1	–	–	1
Apex		2	–	–	–	2
Combination		–	1	–	–	1

As with edge-retouched tools, there is a strong preference for shallow, lateral edges in most types of tools (30/81.1% of utilised tools) (Table 7.33). Tools with semi-abrupt edges are significantly less common and include two of the three convex and a number of the regular tools; however, none of the tools makes use of naturally abrupt or variable edges. The size of utilised tools suggests that there is perhaps a subtle preference for long to broad flakes and this reflects the selection of removals with long lateral edges (Plate 71).

7.2.3.7 Core tools

A total of 17 cores appear to have been utilised as tools or created with the objective of tool use in mind (Table 7.34). Most are bipolar cores or scalar reduced flakes (*e.g.* Nelis 2003, 770, fig. 7.3e), which would have been suitable for use as wedges (12 bipolar wedges or *pièces escaillées*; 2 reduced flakes); one additional wedge utilised a platform core with a keeled edge. The quantity of these artefacts, and their dominance within the core tool assemblage, seems to indicate that they were created with function in mind rather than representing merely the opportunistic re-use of cores which, in general, would perhaps result in a much more variable core tool assemblage.

The remaining core tools include a bipolar core which may have been used as a scraper and a combined platform and bipolar core that may have served as a notched tool. Only one piece, a bipolar *pièce escaillée*, is burnt. With the exception of three *pièces escaillées* (2 bipolar cores, 1 reduced flake), which are uncorticated, all core tools retain partial cortication. In general, the source of raw material was indeterminate; however, at least one bipolar *pièce escaillée* made use of a small beach pebble, and the keeled wedge may have used a river pebble. The length:breadth of core tools show that most *pièces escaillées*/wedges are roughly square in plan and, in general, measure around

Table 7.34 Types of core tools, indicating types of cores utilised (Nelis 2003, 763).

Core type Tool type	Platform	Bipolar	Reduced flake/ scalar	Platform & bipolar	Total
Wedge/*pièce escaillée*	1	12	2	–	15
Possible scraper	–	1	–	–	1
Possible notch	–	–	–	1	1
Total	1	13	2	1	17

35–40 mm in length and breadth (Plate 72); the largest of the core wedges is the platform and keeled core.

7.3 The distribution of the lithics

Analysis of the distribution of the lithic material has revealed a number of interesting patterns, with certain areas acting as focal points. Within the Ballynahatty complex, lithic deposition appears to have been concentrated on BNH6, the Entrance Chamber and BNH5. Flake and angular debitage, much of which shows evidence of burning, dominate BNH6, while retouched tools appear more frequently in the BNH5 area, and cores appear to be scattered across the site but are most commonly associated with BNH6, the Entrance Chamber and the Facade of BNH5.

7.3.1 Topsoil and ploughsoil assemblage

A third of the lithic assemblage (1895 items, 32%) was excavated from the topsoil and ploughsoil layers (and the total also includes a small number of lithics for which there is no context information apart from the area). The importance of ploughsoil as a context for lithics and the occurrence of large portions of total lithic assemblages from topsoil at prehistoric sites has been noted elsewhere (Cross 1999, 183). Figure 7.3 demonstrates that the lithics are distributed in the topsoil over the majority of the excavated area. They are, however, concentrated in the topsoil overlying BNH5 and BNH6.

7.3.2 Feature and context assemblage

The remainder of the lithic artefacts were recovered from contexts and features uncovered during the excavation (Fig. 7.4). Figures 7.5 and 7.6 reveal high lithic deposition associated with some feature groups, as noted in BNH6 Inner Ring and the Entrance Chamber. Considerable quantities of lithics were recovered from the Four-Poster feature. The lithics were largely recovered from the secondary fill of the posthole features. Of the 2548 lithic artefacts which could be assigned to the primary or secondary fill of a posthole, a total of 2224 were found in the secondary fill (Fig. 7.7; Plates 73 and 74). The secondary fills at Ballynahatty packed the postholes following the destruction of the site when the burnt timbers left cavities.

7.3.3 Discussion of the distribution of the lithic assemblage

The large assemblage of lithic artefacts indicates not only the presence of a functional assemblage of Middle and Late Neolithic artefacts but also one which has been intimately involved in the ritual practices performed at the site. The deposition of burnt flint debitage, for example, appears to relate to the destruction of the site.

The lithic assemblage can be assigned to at least two periods. An earlier phase was suggested by the presence of hollow scrapers and the javelin head, which are likely to be directly associated with the earlier Coarse Ware phase. The second phase occurred when the lithics were incorporated into the primary fills of the postholes, particularly within the Entrance Chamber. The most prolific and later phase of lithic activity, however, was involved in the destruction of the site when large quantities of lithics were incorporated into the secondary fills of postholes. The assemblage from Ballynahatty adds to the growing body of evidence suggesting that there may be a Grooved Ware 'package' of artefact types. Petit tranchet and petit tranchet derivative arrowheads have been recovered in association with Grooved Ware pottery elsewhere (Gibson 2015), a pattern repeated at Ballynahatty.

Patterns in the artefact distribution are perhaps more apparent in the lithics than is the case with the pottery, simply due to the greater quantity of material. A number of patterns are discernible. The deposition of lithic artefacts in the postholes was evidently an important practice and particularly so with the destruction of the monument. The lithics are found in the postholes of all the feature groups (unlike the pottery); however, there appears to have been a deliberate selection of postholes where artefacts were deposited. It seems that the bulk of the assemblage relates to BNH6 (over 1100 pieces) and most of the remainder with the Entrance Chamber at BNH5 (just over 200 pieces; Nelis 2003, 769, fig. 7.2) compared to a total of 200 from the BNH6 Outer Ring.

Within the BNH6 Inner Ring, the secondary fills of the postholes on either side of the entrance and postholes IR22 (C111) and IR12 (C635) (northeast and northwest of the centre, respectively) contained the greatest quantities of lithic artefacts. In contrast, a substantial portion of the lithics recovered from the Entrance Chamber was found in primary fill. As also appears to be the case with the pottery, some lithics were included with the primary fills of this feature group. Lithics were also recovered from the ground surface of the area west of BNH6, and with the open area in the northern Annexe. Most material from BNH6 comprises

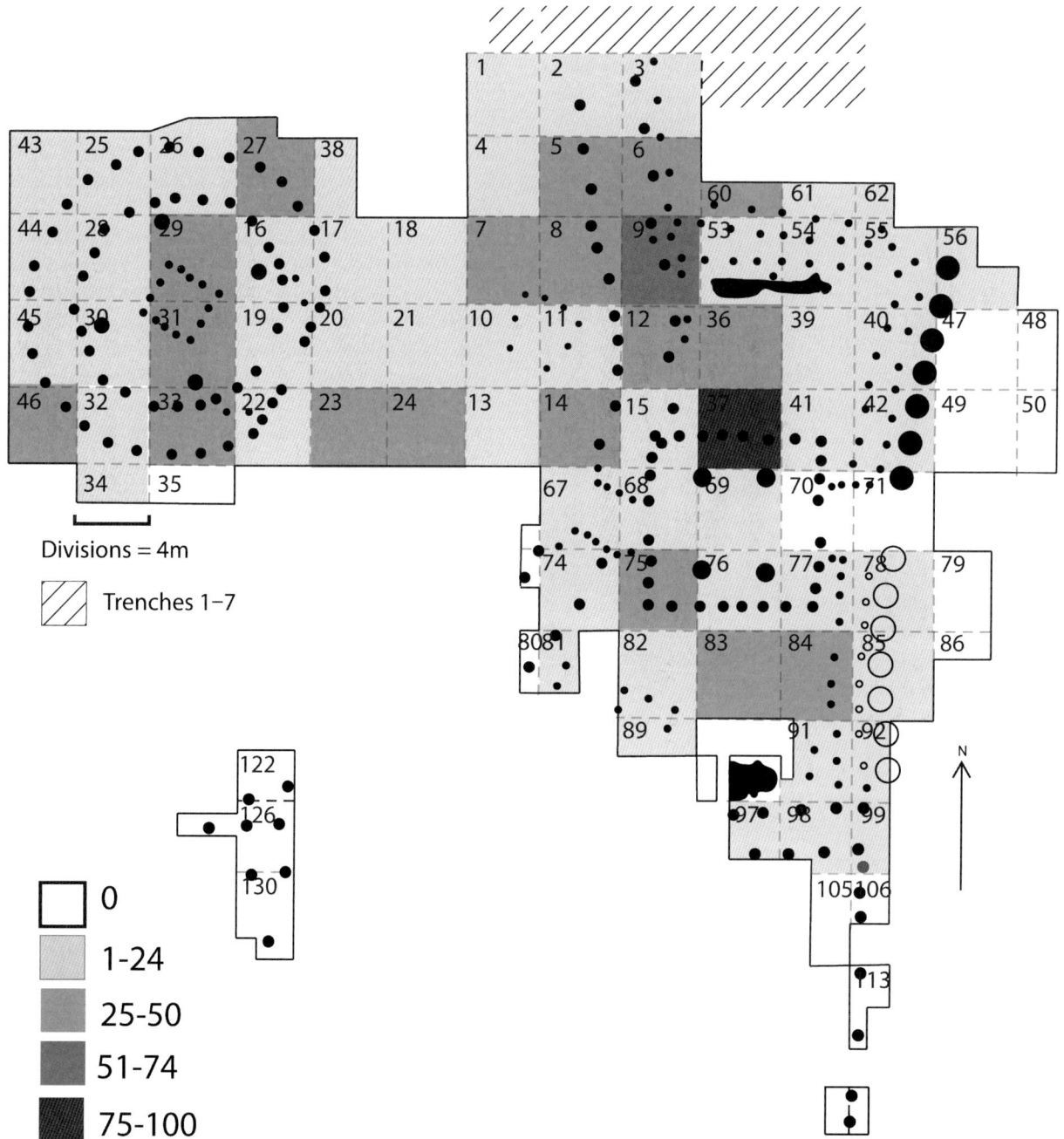

Figure 7.3 Distribution of the lithics recovered from topsoil and ploughsoil contexts.

flake and angular debitage and only a small proportion are tools, mainly scrapers and edge-retouched/utilised tools; similar proportions are found within the material from the Entrance Chamber.

The entrances are emphasised by the deposition of artefacts in large numbers in both the Entrance Chamber and the postholes flanking the entrance through BNH6 Inner Circle, perhaps marking these important transitional places between the outside of the monument and the interior of BNH5 and also between the space defined by BNH5 and the interior of BNH6. The constituent of the assemblage from BNH5 and associated areas seems to contain a more substantial proportion of tools than is evident throughout the rest of the site.

It is interesting to note that large numbers of lithics were recovered from the postholes of all the eight 'free standing' posts, both the Four-Poster and within the Entrance Chamber. In each case, however, the two southernmost postholes contained a greater proportion of artefacts. The lithics in the Entrance Chamber are concentrated in greater numbers towards the south, whilst the pottery distribution is more concentrated in the southwest. If the Entrance Chamber is split on an axis running through the entranceways, it is apparent that almost 80%

7. The lithic assemblage: chipped stone

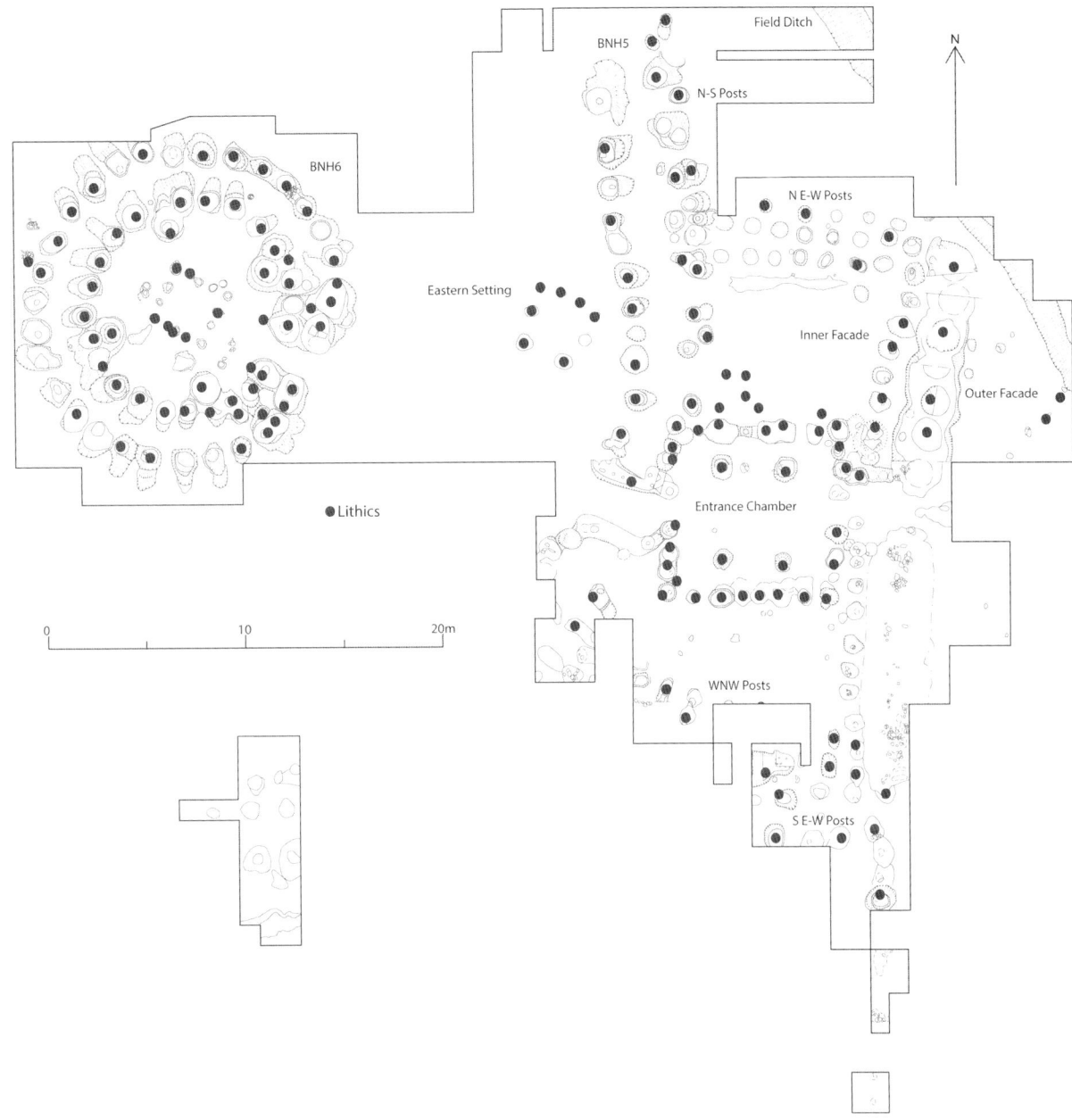

Figure 7.4 The distribution of the lithics by feature group.

of the lithics recovered from this feature were found to the south and just over 20% to the north. The pottery distribution in BNH6 Outer Ring is focused on postholes to the north and east.

The condition of the lithics in most areas the majority is in a fresh condition but in BNH6 the overwhelming majority is burnt (Plate 75). There are also subtle distributional trends with regard to artefact types: cores are scattered across the site but most are found in BNH6, the Entrance Chamber and at the facade of BNH5. While the provenance of some retouched tools remains unclear, many were retrieved from BNH5 (although they constitute a small element of the assemblage from that area); these include almost half the total number of scrapers and many of the remaining tool types; most of the remaining scrapers and other tools were recovered from the Entrance Chamber.

Parts of the site were severely damaged by ploughing, especially in the excavated area of BNH5, and much of the lithic assemblage was recovered from ploughsoil. In the case of most of the arrowheads and knives recovered from topsoil, distribution patterns offer little to clarify the relationship between the petit tranchet derivative arrowheads assemblage and the monumental complex. Most of the hollow scraper assemblage probably relates to the area around the Burning Pit and the particularly abraded condition of these tools might suggest that they pre-date most of the artefacts recovered. Petit tranchet derivatives

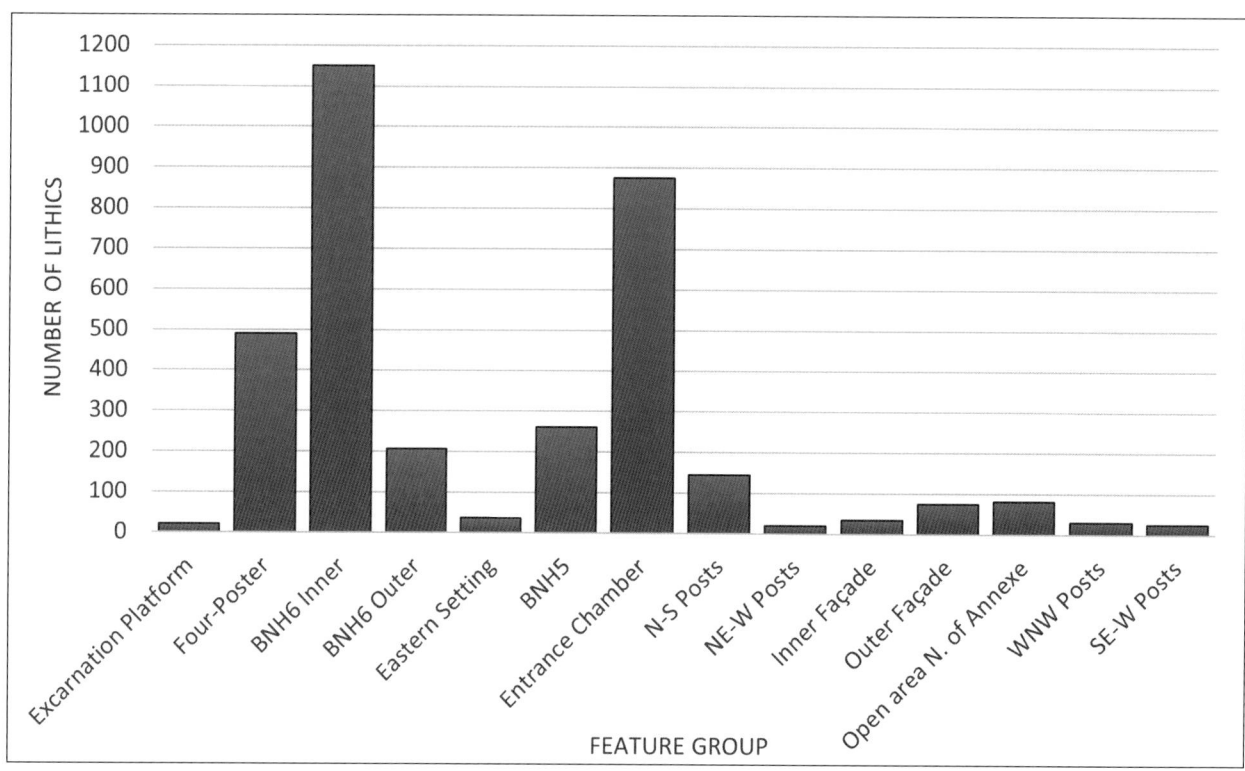

Figure 7.5 Number of lithic artefacts from each feature group.

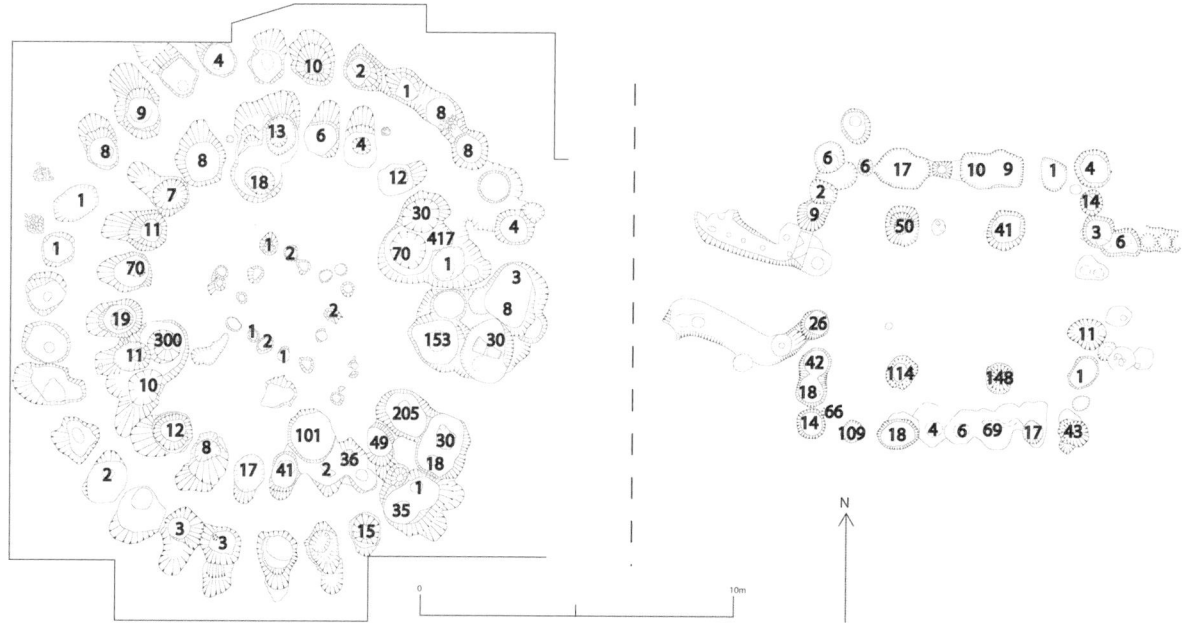

Figure 7.6 Number of lithics recovered from the postholes of BNH6 and the Entrance Chamber. The lithics from Ballynahatty were concentrated in these two areas, with some postholes containing substantial deposits of lithic material.

and hollow scrapers seem to be reasonably common in the area but their relationship with the Ballynahatty complex is far from clear: most are not derived from any of the deposits on the site and petit tranchet derivatives, in particular, seem to be more commonly found as stray finds from the area (in common with examples held by the Ulster Museum).

Analysis of lithic distributions from other Irish timber circles suggests that structured deposition was a common element at these sites. A detailed analysis of the distribution of the lithic assemblage at Knowth indicated a pattern of deposition within the pits which was related both to the number and mass of the artefacts deposited. The results indicate that (on an axis of symmetry

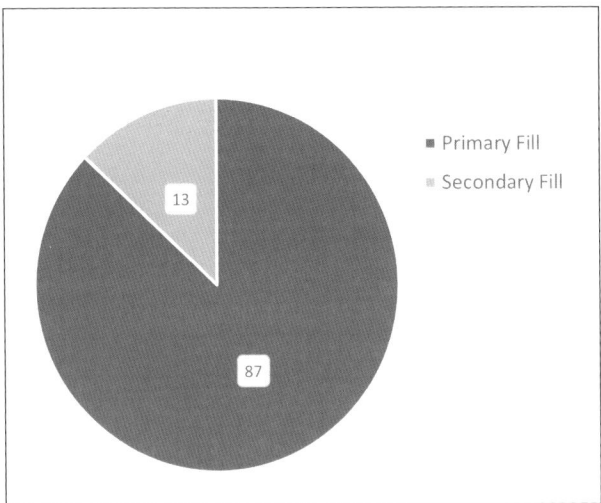

Figure 7.7 Percentage of lithic artefacts which were recovered from the primary and secondary fills of postholes.

through the entrance) similar numbers of artefacts were deposited in mirror-imaged pits, especially pits at the entrance and in the four large internal pits (Dillon 1997, 179–183). The deposited artefacts were mostly debitage, but the pattern also extended to scrapers. Such symmetry was not evident at Ballynahatty (Fig. 7.7), although it is apparent that particular postholes were selected for special treatment and received substantially greater quantities of lithics than others. These include the postholes on either side of the entrance within the BNH6 Inner Ring, the Four-Poster posthole 4P2 and also the two southernmost of the free-standing posts in the Entrance Chamber.

At many timber circle sites, the central four-post structure appears to have been the focus for lithic deposition. Over half of the lithic assemblage from Lagavooren (247 lithics: 57%) was recovered from the four central postholes (Sternke 2012, cxliii). A similar pattern of deposition is also evident from the three timber structures at Balgatheran, Co. Louth. Originally interpreted as timber circles, the excavator later argued that they were more domestic in nature (Ó Drisceoil 2004; 2009) but there is much to suggest that they could have had a ritual function, especially the largest timber structure, Building 1. Like BNH6 at Ballynahatty, most of the postholes from Building 1 had either been removed or left to rot *in situ* before being deliberately backfilled with deposits rich in pottery and lithics (Ó Drisceoil 2009, 80). Just over a third (36%) of the total lithic assemblage from Balgatheran was excavated from Building 1, the majority of which was located within the postholes of the central four-post structure, along with pottery sherds and the remains of a polished stone axehead. The number of lithics recovered from Buildings B and C was much smaller, but they were also focused on the central four-post feature. Despite the small number of lithic finds at Kilmainham 3, it is possible to discern a similar pattern. Only 14 lithics were recovered from the 40 postholes that formed the main exterior circle, while ten were found within the central four-poster (Nelis 2011).

At Glenshane Quarry, Claudy, Co. Derry, over half of the overall lithic assemblage (805 lithics: 54%) was excavated from the timber circle (Bailey 2020, xviii). However, unlike many of the other sites, Glenshane Quarry lacks a central four-post feature, but it does contain numerous internal pits. One of these pits (C468) appears to have received the bulk of the lithic material, with over 369 lithics coming from this pit alone (*ibid.*, xx).

This discussion highlights the fact that that the distribution of lithic evidence from timber circle sites across Ireland suggests the deliberate ritual deposition of lithic material. Special attention appears to have been focused on the entrance postholes and central four-post structures where lithic deposition was concentrated. The composition of lithic material from timber circle sites is primarily composed of flake and blade debitage rather than finished tools, suggesting that flint knapping was taking place at these sites. The fresh condition of the lithics at Knowth led Dillon (1997, 179) to propose that the flakes and blades may have been knapped specifically for deposition within the postholes. The preponderance of debitage at these sites suggests that the ritual behaviour with lithic artefacts at timber circles seems to differ quite fundamentally from the ritual behaviour evident at earlier monuments, where the deposition of grave goods focused on retouched tools. At those monuments, deposition was concerned with human burial; in contrast to Ballynahatty and several other timber circle sites (including Knowth, Balgatheran and Glenshane Quarry), lithic artefact deposition seems to relate to the laborious closure of a site which also had human burial. Such a change in focus, from ritual lithic activity primarily concerned with retouched flint artefacts to the less aesthetically impressive debitage, is a striking shift in ritual behaviour and offers an unusual insight in the creation of an assemblage beyond simply the production of tools for use and deposition.

7.4 Summary and discussion

The lithic assemblage recovered during the excavations at Ballynahatty constitutes a significant amount of material, numbering over 5000 artefacts, and with a combined weight of 34.3 kg. While this is not by any means the largest of the Irish assemblages relating to Neolithic activity, its size contrasts starkly with the small numbers retrieved from other sites of monumental ritual activity in the north of Ireland. Typically, earlier megalithic monuments, such as court tombs and passage tombs, tend to have very small quantities of lithic artefacts. Similarly, Late Neolithic, Chalcolithic and Early Bronze Age monument types (*e.g.* wedge tombs, stone circles and henges), some of which may be contemporary with Ballynahatty, have produced few and scattered lithic

Table 7.35 Location of the flint artefacts at Ballynahatty by feature group.

Area	No. artefacts	Context & no. lithics retrieved	Detail of deposit & posthole (PH)
Excarnation Platform	9	EP2 (C949), EP3 (C926), EP4 (C958), EP8 (C470), EP9 (C738), EP13 (C830)	6 postholes of platform in centre of BNH6 in primary & secondary fills
Four-Posters	489	4P1 (C630) = 101 4P2 (C948) = 300 4P3 (C453) = 18 4P4 (C582) = 70	Uneven distribution between postholes. Assemblage is 8% of site total
BNH6 Inner Ring	1146	Examples: IR1 (C104) = 205 IR25 (C289) = 153 IR23 (C810) = 1 IR 22 (C111) = 417	Secondary fills rich, 23 of 24 PH contained lithics (IR24 (C858) contained none). IR1 & IR5 located on either side of main entrance to Inner Ring
BNH6 Outer Ring	204	Examples: OR1 (C413) = 30 OR4 (C416) = 35 OR32 (C326) = 30	Lithics in 22 of 33 PH in secondary fills, & dense in PH by entrance
Eastern Setting	25	7 of 8 PH yielded lithics	Primary & secondary fills
BNH5 Inner & Outer Ring	242		From 19 PH, no distinctions in distributions, 33 lithics in primary fills
North–South line of posts	144	5 postholes	116 lithics found in 1990 season
Northern East–West Posts	10	NEW2 (C1261) = 1 NEW3 (C1244) = 7 NEW 14 (C1127) = 1 NEW21 (C1128) = 1	Primary & secondary fills
Open Area in northern Annexe	64	13 contexts	Shallow scoops/features & floor area associated with Grooved Ware pottery & cist structure
Entrance Chamber	845	secondary fills = 697 primary fills = 149 EC14 (C1276) = 109 EC37 (C1461) = 148 EC38 (C1255) = 114	Postholes with high density in 2 southernmost PH of freestanding structure in entrance
Inner Façade	223	IF4 (C764), IF5 (C765), IF7 (C892), IF8 (C895), IF17 (C1314), IF18 (C1319)	Even distribution in primary & secondary fills
Middle Façade		MF6 (C1360), MF7 (1669)	
Outer Façade	80		–
West–North–West posts	20	C1401, C1589, C1474, C1591, C1601	5 postholes, 13 lithics from primary fills
Southern East–West Posts	18	SEW5 (C1544), SEW10 (C1576), SEW12 (C1450) C1669).	Even distribution between PH & primary–secondary fills
Area west of BNH6	13		Shallow features in areas 45 & 46 with lithics associated with 2 zones of split stones & Coarse Ware sherds
Deposits of cremated remains	5	C943 = 5 C1012 = 1 C1014 = 1 C1630 = 5	Association with deposits of cremated remains: C943 & in area 76 (C1630)

finds. Only a small number of lithics were found during Collins's excavations at the Giant's Ring, lying southeast of Ballynahatty, and may not relate to the construction of use of the henge.

Ballynahatty's lithic assemblage has parallels with assemblages recovered from Late Neolithic sites elsewhere, such as those recovered from the Late Neolithic/ Beaker phases in the vicinity of the Newgrange passage

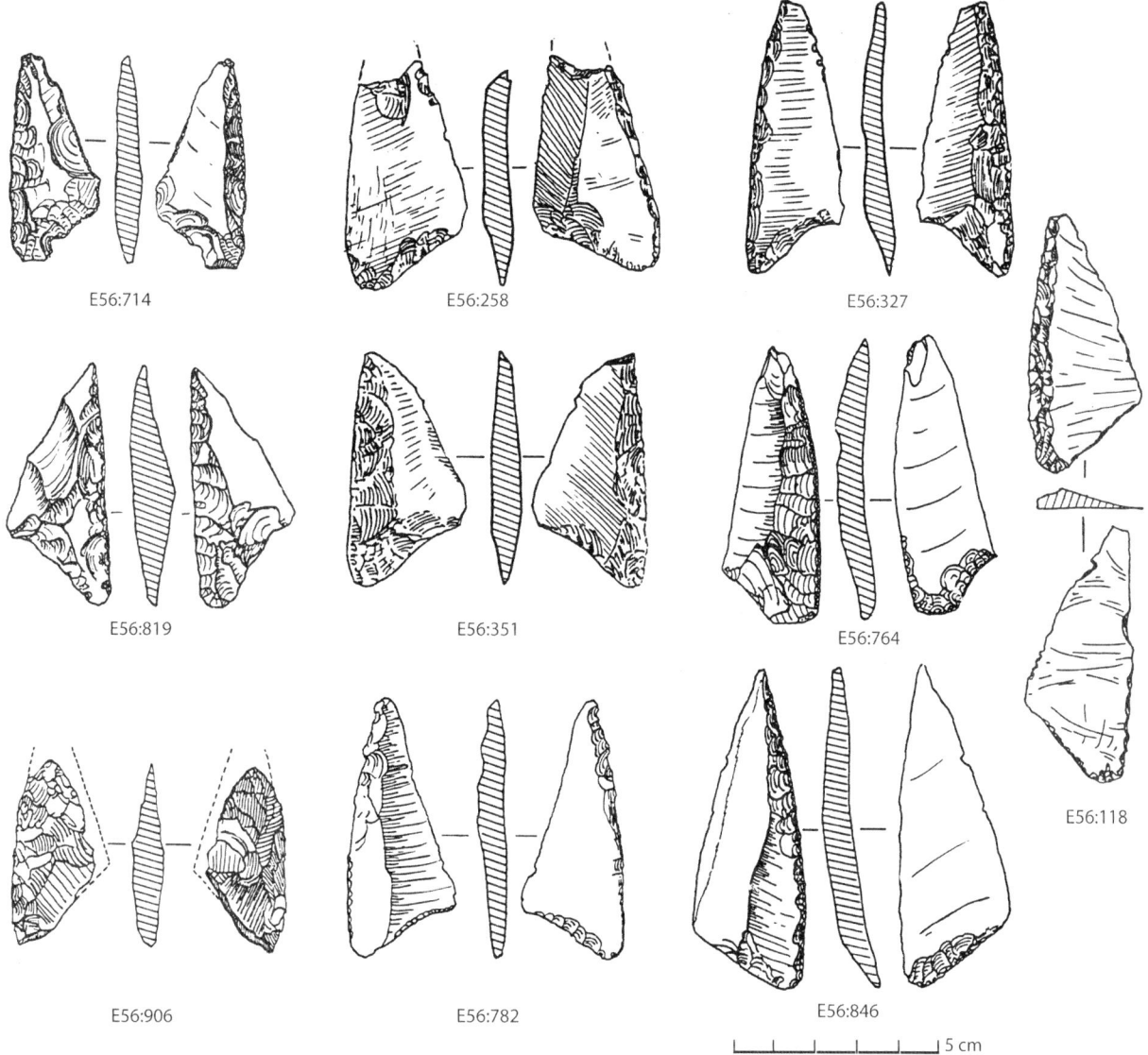

Figure 7.8 The petit tranchet derivatives from Newgrange, Co. Meath (after Lehane 1983, 148–149, figs 62–63).

tomb. Some 11,000 lithic artefacts, the majority from nodular flint, were recovered from contexts which included multiple arcs of 'great pits' (Lehane 1983, 118). As with the Ballynahatty assemblage (Figs 7.1 and 7.2) the retouched and utilised tools of this assemblage were dominated by scrapers, and petit tranchet derivatives were also present (*ibid.*, 131, 146; Fig 7.8).

At Balgatheran, a timber circle in Co. Louth, preliminary results indicate that, as at Ballynahatty, bipolar reduction is a major factor of the reduction strategy; however, unlike at Ballynahatty, and possibly as a consequence of its distance from flint resources, greater variability in raw material usage is evident, with chert and quartz being commonly reduced, and the flint industry almost exclusively depends on diminutive beach pebbles.

The quantity of lithic material evident at Ballynahatty is not reflected in the Knowth timber circle assemblage, where only 801 lithics were recovered. Flint was the dominant raw material used and a source on the northeast coast of Ireland was deemed likely for most of the material (Dillon 1997, 162). As at Ballynahatty, most of the assemblage recovered from Knowth consisted of flakes and blades (57% at Ballynahatty; 62% at Knowth (Dillon 1997, 179). Amongst the tool assemblage, scrapers, including broken ones, dominated, with over 80 examples. Petit tranchet or 'transverse' arrowheads (the term used by Dillon), were also recovered from Knowth (*ibid.*, 170, 195). The lithic technology at Knowth appears to be dominated by platform reduction, with no apparent evidence of bipolar techniques. Although much smaller, the lithic assemblage from the nearby timber circle at Kilmainham, Co. Meath, was similar in composition to that found at Knowth, While the majority of the material was flake debitage, a small number of retouched tools were found. Over half of all the retouched material from the site was recovered from the timber circle, with scrapers being the most common form. As at Knowth, platform reduction appears to have been the dominant technique (Nelis 2009, lxv, lxx).

At some timber circles, such as Lowpark, Co. Mayo and Rathmullan, Co. Meath, both bipolar and platform reduction techniques appear to have been utilised. The recovered lithics from Lowpark indicate that they were practising local reduction of imported materials (Gillespie n.d., 50). As found at other timber circle sites, the majority of the lithic material from these two sites consisted of flake debitage, while scrapers dominated the modified retouched assemblage.

The lithic material from the recently excavated timber circle at Glenshane Quarry, Claudy, Co. Derry (Nichol and Donaghy 2020) displays variability in the raw material usage with flint, quartz and other stone material present. As is typical of other sites, flake debitage accounts for the majority of the assemblage, while the formal tools are dominated by scrapers. The nature of the lithic material suggests that tool maintenance rather than manufacturing was the main activity at this site (Bailey 2020, xxvii).

The lithic assemblage from Ballynahatty is substantial, and its composition and context suggest that it is a functional assemblage, much of which is integrally linked to activities within the complex. As described above, the assemblage is dominated by complete and shattered flakes and blades and a few cores. The reduction strategy shows variation and flexibility, with many cores depending on bipolar reduction techniques. Some cores are based on platform reduction (some without supporting the core base), and yet others combine flaking methods on the same core. The reduction of some flakes and small pebbles indicates a certain degree of pressure on raw material, but nonetheless, local glacially-derived flint was largely overlooked, implying that other flint sources must have been preferred and accessible.

The retouched tool assemblage mainly consists of scrapers, with most of the remainder being minimally retouched or utilised as cutting tools (some of which combine two or more functions), together with piercers and notched tools. Several wedge tools were found, based either on bipolar cores (*pièces esquillées*) or scalar reduced flakes, indicating that in addition to cutting, scraping and piercing, splitting was a function that the flint assemblage was required to fulfil. Apart from scrapers, more 'formal' tool types are rare: knives are virtually unknown at the site and expedient cutting tools sufficed. Projectiles are rare, and most are petit tranchet or derivative types, representing all of Collins's three defined types: transverse (petit tranchet), and both short and long pointed (petit tranchet derivatives) (Collins 1981). The long, pointed examples within the assemblage, however, seem to have functioned as knives, with their unretouched sharp edges exhibiting wear. The most impressive member of the projectile assemblage is perhaps the javelin head, which is only semi-invasively pressure flaked. The rest of the projectiles are few in number and include fragmentary arrowheads and unifacially and bifacially worked pieces.

In general, the composition of the lithic assemblage from Ballynahatty is similar to that of assemblages recovered from other Irish timber circle sites, although the overall number of lithics is much larger. The evidence from the excavated Irish timber circle sites illustrates that there is a degree of variability in the raw material, although flint always accounts for the vast majority of the assemblages, with chert and quartz found in small quantities at many sites. The bulk of the lithic material at most of the timber circles appears to be dominated by flake debitage, which suggests lithics were worked on site. As at Ballynahatty, most timber circles have also produced a range of retouched tools, with scrapers being the most common.

8

Other artefacts from the excavation

8.1 Introduction

This Chapter deals with the 66 miscellaneous stone objects, including coarse stone items, which were not included in Eiméar Nelis's study of the Ballynahatty lithics (Chapter 7), and also with the six beads (and a possible bead) that were found during the excavation, of which one is definitely of glass, three are probably glass, one resembles faience and one – the large, possible bead – is probably of lignite.

The 66 miscellaneous stone objects are listed in Table 8.1, and items of note, including five axeheads (Plates 76 and 77) and two stone balls/hammerstones (Plate 78), are described below. The range of stone types includes quartz, chert, rock crystal and porcellanite, the last coming from Tievebulliagh or Brockley on Rathlin Island, Co. Antrim. A summary of the distribution of these miscellaneous stone items is offered below.

8.2 Distribution

The stone appears to follow the distribution of the pottery, flaked flint and (to some degree) bone. Of the 27 objects recovered from the primary and secondary fills of the postholes, the majority were recovered from the secondary fills (as with the flaked flint and bone) (Fig 8.1). Stone also appears to have been placed within features which saw concentrations of other artefact types: for example, posthole 4P2 (C948 of the Four-Poster arrangement) and IR1 (C104), IR22 (C111) and IR12 (C635) of BNH6 Inner Circle, and C1257 of the Entrance Chamber. The distribution of these stone finds mirrors the patterns observed at other timber circle sites, where certain postholes acted as a focus for deposition. The timber circles at Glenshane, Co. Derry, Scart, Co. Kilkenny, Balgatheran, Co. Louth and at Bettystown, Knowth and Lagavooren, Co. Meath, all had polished stone axeheads placed within the

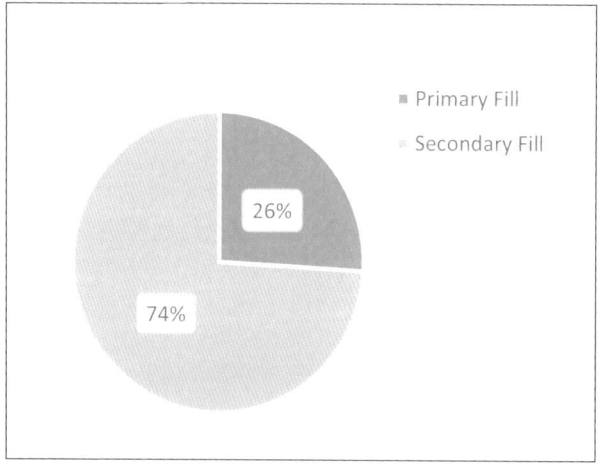

Figure 8.1 Percentage of the coarse stone artefacts recovered from the primary and secondary fills of the postholes.

postholes of the central four-poster structure, generally in the southeastern posthole (Eogan and Roche 1997; Eogan 2000; Monteith 2008; Ó Drisceoil 2009; Stafford 2012; Nicol and Donaghy 2020).

8.3 Selected artefacts

8.3.1 Stone axeheads

Considering the enormous amount of woodworking undertaken at Ballynahatty, it is remarkable that only two complete axeheads and fragments of three more were found (Plates 76 and 77).

a: AX23.6111 BNH92. Length: 89 mm; width (at blade): 53 mm; max depth: 26 mm; weight: 162 g (Plate 76a). Found during topsoil clearance (C1) in area 3 above posthole OR1 of the BNH5 outer enclosure.
This axehead has been ground to shape, although flake scars are still visible along one edge. The blade is sharp,

Table 8.1 Miscellaneous stone objects from Ballynahatty, listed according to their feature groups.

Feature Area	Posthole/context (C)	Primary/secondary	Finds no.	Classification	No.
Area west of BNH6	C934	–	6148	Stone flakes	6
Excarnation Platform	P3 (C926)	–	–	Pebble	1
Four-Poster	4P2 (C948)	Secondary	6132	Pebble	1
		–	6143	Coarse stone flakes	1
		–	6151	Burnt stone	1
		–	6154	Flaked stone	1
		–	6155, 6156	Burnt stone: large & rounded	2
	4P4 (C582)	–	6130	Quartz fragment	12
BNH6 Inner Ring	IR1 (C104)	Primary	6152	Burnt pebble	1
	IR1 (C174)	Primary	6109	Stone ball/hammer stone	1
	IR22 (C111)	Secondary	6110	Burnt quartz pebbles	2
			6124	Small quartz pebble	1
	IR19 (C537)	Secondary	6118	Rock crystal	1
	IR18 (C538)	Secondary	6125	Quartz fragment	1
	IR12 (C635)	Secondary	6120, 6121	Flaked stone	2
	IR 9 (C963: 957)	Secondary	6157	Porcellanite chip	2
BNH6 Outer Ring	OR25 (C723)	Secondary	6127	Rock crystal	1
	OR18 (C794)	Secondary	6137	Flat pebble	1
	OR9 (C805)	Secondary	6133	Stone flake	2
			6141	Pebble	1
	OR20 (C856)	Primary	6160, 6161	Pebbles (2)	1
BNH5 Inner & Outer Ring	IR8 (C735)	–	6410	Quartz fragment	1
	OR6 (C220)	–	6117	Burnt flint ball	1
Entrance Chamber	EC26 (C532)	–	6126	Rock crystal	1
	EC24 (C720)	–	6145, 6150	Rock crystal	2
	EC17 (C1257)	–	6177	Chert	1
Inner Façade	IF5 (C765)	Secondary	6131	Small quartz pebble	1
North–South Posts	NS9 (C63)	–	6113	Quartz pebble (worked?)	1
	NS7 (C469)	–	6116	Half a stone bead	1
Open area in northern Annexe	C771	–	6138	Rock crystal	1
	C904	–	6159	Rock crystal	1
	C865	Secondary	6153	Hammer stone	1
Outer Façade	763/865: 1161	Secondary	6164	Dolerite axe	1
West–North–West Posts	WNW13 (C1589)	Secondary	6175	Pitchstone	1
Burning Pit	C1452	–	6176	Burnt Quartz	1
Backfill	–	–	6111	Basalt stone axe	1
Topsoil/ploughsoil	C1/ 2	–	6108	Worked quartz	1
		–	6115	Butt of a porcellanite axehead	1
		–	6119	Worked quartz	1
		–	6128	Polished porcellanite fragment	1

whereas the butt has been shaped to a blunt point and has two flake scars, one of which may have been sustained post polishing. The shape is a product of the original flake, although the determining factor in its manufacture may not have been regularity or utility but the stone itself. This is a very fine-grained basalt with snowflake texture and the only known source in Ireland is from the Tardree rhyolites in mid-Antrim. Much effort has been expended on the finish. Apart from a few grinding scratches, the surface has been finely polished to show the crystal structure, and the sharp blade is intact and undamaged.

b: AX23.6164 BNH96. Length: 132 mm; width (at blade): 68 mm; max depth: 37 mm; weight: 519 g (Plate 76b). Found in outer façade, C763/865: C1161.
Axehead (b) is a regularly shaped dolerite axehead with a rounded butt and abraded, heavily used blade. There is no abrasion on the butt. This is more than three times the weight of axehead (a) and is an unpolished working axehead, though care has been taken in grinding it to shape as there are no traces of flake scars. The axehead retains its body depth close to the blade, giving a cutting angle of 80° (axe (a) is 50°). If used on wood, it would have relied on its weight to splinter rather than on the blade to cut.

c: AX23.6115. Length: 32 mm; width (max): 24 mm; depth (max): 13 mm; weight: 15 g (Plate 77c). Found C1, south of BNH6 entrance.
Butt fragment of a small polished porcellanite axehead. Its angled butt is a characteristic feature of many porcellanite axeheads (Sheridan 1986). Plate 77c shows the dorsal side with remaining flake scars. Flake scar on ventral side.

d: AX23.6157. Length: 44 mm; width (max): 22 mm; weight: 11 g (Plate 77d). Found in BNH6 IR9 (C963)
Body fragment of polished porcellanite axehead.

e: AX23.6128. Length: 45 mm; width (max): 29 mm; weight: 13 g (Plate 77e). Found in C1/2 of north Annexe.
Body fragment of a polished porcellanite axehead.

As noted above, the source of the porcellanite will have been Tievebulliagh mountain on the Antrim Plateau, or else Brockley on Rathlin Island, both sources known to have been intensively exploited during the Neolithic period (Sheridan 1986). Around 7500 axeheads made of porcellanite are known from Ireland and Britain (Cooney and Mandal 1998) and it is the dominant rock type used to make stone axeheads in Ireland.

It is impossible to tell whether these complete and fragmentary axeheads from Ballynahatty were associated with the Late Neolithic period of activity or to the preceding episodes during the Early or Middle Neolithic; the findspot contexts are mostly either topsoil or the secondary fill of postholes.

8.3.2 Stone 'balls', possibly hammerstones, plus quartz pebble hammerstone

a: Flint 'ball' (AX23.6117). Diameter (max): 65 mm, (min): 57 mm; weight: 312 g (Plate 78a). From BNH5 OR6 C220.
This may be a natural geological oddity, a spherical nodule of flint which was collected and possibly used.

b: Stone 'ball' (AX23.6109). Diameter (max): 72 mm, (min): 65 mm; weight: 440 g (Plate 78b). From BNH6 IR1 C174.
The dolerite 'ball' was recovered from the primary deposit of the first posthole to the left of the entrance on the BNH6 Inner Ring. The placement of the stone here may be considered a deliberate act of deposition as this posthole marks the entrance into the interior of the timber circle. This trait is replicated at other timber circle sites, where the entrance postholes are marked with artefacts. It is a well-finished, near-round, ground stone ball. An area of pitting, *c.* 20 mm in diameter (seen in the upper part of Plate 78b), suggests that it was used as a hammerstone at some point. The size and shape of the ball would have made it an unwieldy knapping tool for anything other than the initial reduction of a flint nodule into a few cores. It is unlikely that a percussion hammerstone was the primary function of such a (relatively) finely shaped stone. Alternatively, if the pitting resulted from the primary function, it may have nothing to do with knapping flint to make tools. Spherical stones are known to have been used as 'rubbers' or 'mullers' for the grinding of grain on a saddle quern, but while the 'ball' is slightly irregular in shape, it does not seem to have any faceting that could arise when used for such a purpose.

Decorated stone balls of this approximate size have been found elsewhere, almost all in Scotland and mostly from Aberdeenshire, with a few from England and Ireland, dating to the end of the Neolithic period. Many are elaborately carved, usually with knobs, though a few are smooth and polished. Dorothy Marshall (Marshall 1977, 63–64; 1984, 628) identified 411 decorated balls of this size and discussed their possible uses in ball games and as oracles, weapons and symbols of prestige, concluding that a ritual use was most likely. More recent interpretations, of varying degrees of plausibility, have included: symbols of power, bearings to move large stones, weights and trial pieces for carving stone (Garrow and Wilkin 2022, 114; Stewart-Moffitt 2022; note that the total is now around 525). In Ireland, there are several smooth versions of stone balls from earlier passage tomb contexts which may have inspired the Scottish decorated versions (Sheridan and Brophy 2012; see section 5.2.4). Three stone balls are reported from Northern Ireland, including one from Ballyknock, Co Armagh (Arm 014:020), reportedly found under an enclosure in a chamber with many human bones (Welsh and Welch 2021, 136).

The Ballynahatty 'balls' are clearly not comparable to the Scottish carved stone balls, nor are they as fine or regular in shape as the Irish examples of smooth stone balls found in passage tombs. It may be that there is a prosaic reason for their presence at the site; it is not necessary to invoke a connection with passage tomb practices here.

c: Smooth quartz pebble with granular inclusions and one flat side (AX23.6110). Dimensions: 78 × 65 × 61 mm,

weight: 445 g. From C254, a deposit of dark, charcoal-rich sandy loam with burnt cobbles in BNH6 IR22.
It appears to have been affected by heat, resulting in a number of fracture lines, one of which has produced a flat platform relatively cleaner than the stained brown/black exterior of the pebble. The edge of the platform is abraded and there are two other areas of coarse, heavy pitting suggesting its use as a hammerstone.

8.3.3 Fragment of Arran pitchstone

AX23.6175. Dimensions: c. 6 × 3 × 2 mm (Plate 79). Found in the upper fill of one of the West-North-West postholes.
This is either a fragment of a microblade (which is possible) or else a chip from a larger object (less likely; Torben Ballin, pers. comm.) of black aphyric pitchstone (an obsidian-like volcanic glass).

This is an important find as it will have originated on the Isle of Arran in the Firth of Clyde, most probably in one of the sources on the eastern side of the island, possibly at Corriegills, where this type of pitchstone outcrops. Finds of pitchstone outside Scotland are very rare and the Ballynahatty find joins the small set of examples from Ireland (Ballin 2009, fig. 24). All the Irish finds are from near the east coast, including a core found at Nappan on the Antrim Plateau.

While pitchstone is known to have been exploited from as early as the Mesolithic period, the distribution of pitchstone artefacts expanded significantly during the Early Neolithic, when it tended to be used to make microblades – possibly used as special-purpose tools, given the great sharpness of the raw material. If the Ballynahatty object is part of a microblade, that indicates that this belongs to the Early Neolithic phase of activities at the site, and its presence in the upper fill of a Late Neolithic posthole will be as a residual item. (The fact that it is not heat-damaged is consistent with such an interpretation.) While pitchstone use is known to have extended as late as the Late Neolithic (*e.g.* at Ness of Brodgar on Orkney) and indeed the Chalcolithic or Early Bronze Age period, the Ballynahatty piece is not diagnostic of the kind of relatively large pitchstone artefacts that were in use during the Late Neolithic (unless, of course, it is a chip detached from one such object). The balance of probability is that it is of Early Neolithic date, within the first half of the 4th millennium BC; as such, it is one of many pieces of evidence for contacts between farming communities in northeast Ireland and southwest Scotland at the time.

8.3.4 Possible chisel

AX23.6174. Length: 17 mm; diameter: 3.2–4.6 mm (Fig. 8.2). C1589: 1595, WNW posthole
A small cylindrical stone object of greywacke (the country rock of the area), which has been irregularly shaped along

Figure 8.2 Broken tip of a possible carved stone chisel.

the body with several flattened areas and may originally have been longer. The tip has been evenly ground from either side to a chisel point to produce a flat 'blade'. Its function is uncertain, but it was probably used to work softer materials such as wood and animal hide or as a scriber.

8.3.5 Beads

Six beads were found during wet-sieving of samples in the laboratory (Plate 80). Three appear to be of glass, and are therefore post-Neolithic, but were found in secure Neolithic contexts (a–c). Each of these small beads come from different contexts in the northwest sector of BNH 6 over a 7.00 m area.

a: Dark, globular, glassy bead (?). Diameter: 3.6mm (Plate 80a). OR19.
It has the same colour and surface as beads (b) and (c). However, although there is a pit on one side suggesting a perforation, it is blocked. If it ever had a full perforation, it may have been damaged by burning, as there is a slight distortion on one side. It was found in OR19, at a depth of 0.20 m, in a dark brown sandy layer with charcoal and stones. Although this was initially interpreted as primary fill, it was probably the upper level of an *in situ* post, which had been pulled down into the post mould as it decayed.

b: Dark, doughnut-shaped, glassy bead. Diameter: 2.8 mm, depth: 2 mm (Plate 80b). From a sample taken at 1.00 m depth in the Neolithic secondary deposit of IR11.

c: Dark, doughnut-shaped, glassy bead. Diameter: 2.9 mm, depth: 2.2 mm (Plate 80c). Found in C552, a wedge of grey, charcoal-rich soil between 0.40–0.85 m deep, which was a part of the secondary fill of 4P3.
The only explanation for the location of these very small beads is the process of bioturbation. They are similar in colour, size and shape and could have been part of a single necklace lost in this area. It is unlikely that more beads of this small size would have been identified in the intermediate areas during regular trowelling.

d: Granular pale blue-green bead, possibly of faience. Diameter: 2.4 mm, depth: 1.6 mm (Plate 80d). From the early medieval 'burning pit' C1453.

e: Part of a translucent perforated blue glass bead. Diameter: 2.8 mm, depth: 2 mm (Plate 80e). From the early medieval 'burning pit' C1453.
Compositional analysis using portable X-ray fluorescence spectrometry confirmed that this is a soda-lime-silicate bead, its composition consistent with an Iron Age or early medieval date (Mary Davis, pers. comm.).

f: Stone bead (AX23 6116). Length: 21 mm; width: 8 mm (Plate 80f). From NS7, C469.
This is a fragment of a possible Neolithic bead with a smooth outer surface and pecked perforation. Its original diameter would have been *c.* 24 mm with a 5 mm perforation.

8.3.6 Possible bead of lignite

(AX23 6172) Diameter: 55 mm; weight: 69 g (Plate 81). Found in the passage of the Entrance Chamber between the four central posts close to three finely made flint artefacts (an end scraper and blades).
Incomplete, oblate artefact, possibly a very large bead, of lignite with a central perforation, measuring *c.* 55 mm in diameter. An irregular, vertical break through the perforation has produced two halves, and in the break can be seen the laminar structure, which is causing it to split horizontally. Overall, the lower third has been lost as well as the upper part of one-half. However, most of the other half is intact and shows a finely crafted circular shape and smooth surface finish, which, in its original state, may have presented as glossy black. The central, opposed, hourglass perforations have been made from the upper and lower surface. The upper one is complete and is 22 mm deep, while the lower perforation survives to 9 mm. This suggests that the original total depth was 44 mm, giving the artefact an oblate shape. The perforation shows that it was part of a composite artefact, meant to be strung on a rope or mounted on a rod, although at 6 mm at its narrowest point, such a rod would have been insubstantial. The function is uncertain, and it would appear too light for a spindle whorl, net sinker or loom weight, and the hole is too small to be for the shaft of a macehead. The quality of the finish seems to exclude a more utilitarian function and suggests some form of personal adornment. Despite being large and fairly heavy, it could have been worn as an outsized bead. Outside of Orkney, Late Neolithic beads are very rare. This would have been a precious possession.

9

Human remains from excavations at Ballynahatty

Eileen Murphy

9.1 Cremated bone from excavations at Ballynahatty 5 and 6

Unburnt bone does not survive in the soils of Ballynahatty unless protected in a cist context. However, over 200 cremated bone samples were collected during the excavation. Twenty-eight samples of cremated bone were recovered from the topsoil and ploughsoil contexts. The bone was spread over most of the excavated area, with usually only one or two fragments from each 4 m square area (Fig. 9.1). In a total of 71 cases, the bone was recovered from either the primary or secondary fill of the postholes. The majority (n=60) were recovered from the secondary fills (Fig. 9.2 and Table 9.1). A number relate to individual deposits of cremated remains, including that found within the stone chamber, C591 (5.2), but a substantial quantity of bone was recovered from the postholes and other features. For radiocarbon dates see Table 10.1.

Although most of the bone appears to be human, the presence of pig bones within these deposits has previously been noted (Hartwell 1994, 13), as well as the remains of cattle and sheep/goat. Small quantities of human remains occur in the postholes of various 'feature groups', providing a new insight into the rituals and ceremonial behaviour attested at Ballynahatty. The relatively few identifications should not be taken as representative of the total remains recovered, as it was not possible to identify many of the samples due to their small and undiagnostic nature. All material from each 'feature group' was assessed to establish which animal species were present and to identify further human remains. This information is tabulated after the separate description of individual deposits of cremated remains (Table 9.1).

9.1.1 Deposits of cremated remains
Details of the quantity, size and degree of fragmentation of the bones were provided to enable an assessment of the completeness of the deposit (in accordance with procedures recommended in McKinley 1994). In addition, the degree of fragmentation of the bones was recorded using a series of three stacking sieves with 2.00 mm, 5.00 mm and 10.00 mm grids. This approach allows assessment of the nature of the potential information that could be obtained from the bone deposit and also enables aspects of the pyre technology to be reconstructed. The colour of the cremated bone was examined so that the efficiency of the cremation process can be ascertained. Where possible, details of the age, sex and pathology of each individual were provided. The exception to this is the analysis of C588, which was undertaken in the early years of the excavation by Dr J.L. Wilkinson (formerly of the Anatomy Department of University College, Cardiff), and, although reformatted, the report is included in full.

Numbers relate to deposits of cremated remains that are associated with Coarse Ware or are from contexts roughly contemporary with the 1855 monument:

9.1.1.1 C591–C596: BNH5 Chamber
The human remains within the chamber were associated with two pots which appeared to have been disturbed, causing commingling of the calcined bone in antiquity. The only exception was a small quantity of bone which was recovered from within Vessel **2** (26.60 g).

The total weight of cremated bone from within the chamber, including the Vessel **2** material, was 844.20 g, of which 64.7% was unidentifiable. The vast majority of the fragments were white in colour, indicating that the bone was burned at a high temperature and completely oxidised in the process. Some 29.8% of the bone fragments were too large to pass through a 10.00 mm sieve, 42.1% were 5.00–10.00 mm in size and 28.1% measured 2.00–5.00 mm. As such, the vast majority (70.2%) of remains were smaller than 10.00 mm in size, indicating that the bone was highly fragmentary in nature.

Figure 9.1 Distribution of bone fragments and cremated deposits from the excavations.

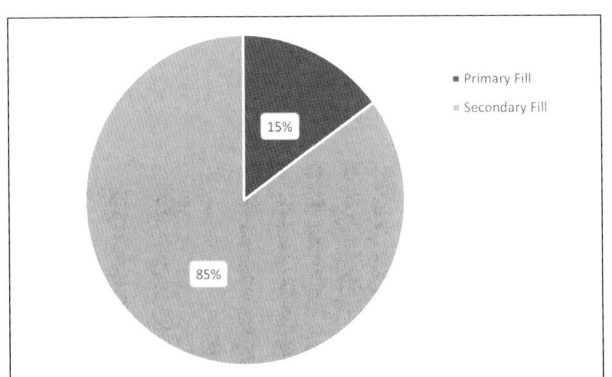

Figure 9.2 Proportion of bone fragments recovered from the primary and secondary fills of the postholes (120 fragments).

SKULL

The total weight of the skull fragments was 108.80 g, which represented 36.3% of the total weight of bone. A total of 160 mature (*i.e.* adult) skull fragments were identified, comprising 95 unidentifiable vault fragments, 17 parietal, five frontal, five occipital, ten temporal (including a near complete left petrous bone), one zygomatic, three unidentifiable irregular cranial bones, three mandible fragments (including a left mandibular condyle) and 13 teeth, mostly roots. The largest skull fragment had a greatest length of 41.00 mm, and the average vault thickness was 4.00 mm. Cranial sutures present were open in appearance suggesting the individual was a young adult (18–35 years). In addition, a further 17 juvenile skull

Table 9.1 *Distribution of bone fragments recovered from the primary and secondary fills of the postholes and from the early medieval 'Burning Pit'.*

Feature	Posthole	Context	No. bone samples
Excarnation Platform	EP3	926	1
	EP7	595	1
Four-Poster	4P3	453	8
	4P1	630	4
	4P2	948	2
Area west of BNH6	–	930	1
	–	934	1
	–	868	1
BNH5 Inner & Outer Ring	–	221	1
	–	1268	2
BNH6 Inner Ring	IR1	104	4
	IR3	507	2
	IR4	853	1
	IR12	635	4
	IR13	647	1
	IR14	629	1
	IR18	538	2
	IR20	114	1
	IR22	111	6
	IR25	289	8
BNH6 Outer Ring	OR20	856	1
	OR26	609	1
	OR28	415	1
	OR29	597	1
	OR30	189	1
	OR31	341	1
	OR32	326	3
Burning Pit	–	1453	2
Cremations	–	588	20
	–	895	7
	–	943	2
	–	1012	1
	–	1014	1
	–	1016	1
	–	1628	1
	–	1630	2
Entrance Chamber	EC6	1260	1
	EC15	1493	2
	EC18	1616	1
	EC24	720	1
	EC27	531	5
	EC29	898	1
	EC30	783	1
	EC38	1255	1
	EC37	1461	2
	EC40	1546	1
Inner & Outer Façade	MF10	1586	4
	IF5	765	1
	IF7	892	1
Open area in northern Annexe	–	500	1
	–	522	2
	–	531	1
	–	646	1
	–	720	1
	–	740	1
	–	771	4
	–	768	1
Southern East–West Posts	SEW	1635	1
	SEW	1544	5
Total			139

fragments were present, including three frontal, three parietal and 11 unidentifiable vault. The largest had a greatest length of 43.00 mm and the average thickness was 1.00 mm, suggesting the individual was a young child less than 6 years of age. Only mature skull fragments were recovered within Vessel **2**.

AXIAL SKELETON

The total weight of the fragments from the axial region of the skeleton was 22.90 g, which represented 7.6% of the total weight of bone. The identifiable mature fragments comprised 11 rib, 21 vertebral neural arch and two vertebral body fragments and included the posterior aspect of a cervical or thoracic vertebral body, the dens of the axis and the facet for the dens on the first cervical vertebra. The largest fragment measured 22.70 mm × 17.40 mm. No evidence of osteoarthritis was visible on any articular facets, perhaps suggesting they had derived from a young adult. In addition, two unfused rib heads and small rib shafts were identified and were likely to come from the young child.

UPPER LIMBS

The total weight of fragments from the upper limbs was 43.40g, which represents 14.5% of the total weight of bone. The fragments of mature humerus comprised 16 shaft and two head and other identifiable adult fragments comprised four ulna shaft and six radius shaft fragments. The largest fragment was of ulna shaft and measured 50.00 × 10.90 mm. Juvenile bones identified consisted of an unfused distal radius and the hand phalanges, the proximal ends of which were both unfused. The morphology of the unfused distal radius was in keeping with that of a child less than 7 years (Schaefer *et al.* 2009, 186). The size of the hand phalanges was also indicative of a younger child.

LOWER LIMBS

The total weight of fragments from the lower limbs was 124.60 g, which represents 41.6% of the total weight of the bone. Mature fragments comprised five tibia shaft fragments, three fibula shaft fragments, one patella, a fused metatarsal head (possibly 1st), and five fragments of pelvis, including a portion of ischium. Mature femur fragments consisted of 39 shaft fragments, three head fragments, two distal epiphysis and one portion of distal metaphysis, which was large but possibly unfused. The latter bone may indicate that the older individual in the cist was an older adolescent or a very young adult since the distal femur fuses at 14–19 years in females and 16–20 years in males (Schaefer *et al.* 2009, 276). The largest fragment was a piece of tibial shaft which measured 75.60 × 17.00 mm. Juvenile fragments comprised eight pieces of long bone shaft and one of unfused ischium, which would have derived from an individual aged younger than 13–16 years (*ibid.*, 253).

ADDITIONAL MATERIAL

Eleven fragments of bone had a notably chalky and rounded appearance and may have derived from an earlier deposit of cremated remains. These included three adult cranial vault fragments and a possible piece of femoral shaft that measured 24.50 × 17.00 mm. One of these fragments was recovered from within Vessel **2**.

CONCLUSION

The remains within the chamber had been disturbed in antiquity, causing the bone elements originally associated with Vessels **1** and **2** to become commingled. The side of the chamber had suffered damage as posthole IF8 of the Annexe façade had been driven through it and at least half of each pot and their contents were subsequently redeposited in the primary fill of the posthole and the cavity of the chamber. The calcined bone represented at least two individuals – an older adolescent or very young adult, probably in the later teens, and a younger child definitely aged less than 6 years. The small amount of bone recovered from the interior of Vessel **2** was adult and it is possible that the remains of the older adolescent/young adult had been interred within this pot, suggesting that Vessel **1** was originally associated with the young child. A relatively small amount of the child's remains were identified but this is probably due to methodological difficulties and their remains may be well represented in the substantial body of unidentifiable material. Since the deposit comprised at least two individuals, its weight of 844.20 g was clearly below that which might be expected, given that the average weight of a modern adult human's cremated remains ranges between 2500 g and 3000 g (McKinley 1994, 75). It is, therefore, possible that the remains originally deposited within the pots did not represent those of the entire bodies and were token in nature. The presence of adult cremated bone with a chalky, rounded appearance may represent the accidental inclusion of older material from the pyre site. Alternatively, the fragments may have been deliberately included in the burial, perhaps due to the practice of curation of fragments of ancestral bone.

9.1.1.2 C779 disturbed material from above cist

This layer covered the chamber and the posthole of the Inner Façade, which had been driven through it. The bone seems almost certainly to have come from the disturbed deposits of cremated remains in the cist (1). The total weight of bone was 77.60 g, of which only 7.0% was identifiable. The calcined bone was white in colour, suggestive of an efficient cremation process. The remains were those of an adult, and identifiable fragments comprised one from a femoral shaft, one from a tibia shaft, one from a skull vault, one from a petrous temporal, and a premolar tooth crown. The femoral shaft fragment was the largest and measured 19.40 × 18.00 mm.

9.1.1.3 C1018/C1019: BNH5 Annexe, entrance

Contexts 1018, 1016, 1014 and 1012 represent a line of four deposits of cremated remains which have suffered varying amounts of plough damage. Although found at the entrance to the Annexe, they were close to the C591 chamber and are similarly associated with the same funerary tradition as seen in the 1855 chamber. The cremated bone from this feature consisted of two unidentifiable fragments with a greatest length of 8.00 mm. Both fragments were white in colour, suggestive of an efficient cremation process. This feature had been heavily truncated but, because of its association with cremation deposits C1016, C1012 and C1014, it is likely that these fragments represent the remnants of a larger deposit of cremated remains.

9.1.1.4 C1016/1017: East of BNH5 Annexe, entrance

The total weight of cremated bone from this context was 173.30 g, 69.0% consisting of unidentifiable fragments. All the fragments were white in colour; 18.8% were too large to pass through a 10.00 mm sieve, 68.5% passed through a 10.00 mm sieve, 12.6% through a 5.00 mm sieve, and 0.1% were small enough to pass through a 2.00 mm sieve. These values indicate that the majority of fragments were smaller than 10.00 mm in size and, therefore, of a highly fragmentary nature. A fragment of bone was dated to 3360–3100 cal BC (UBA-42747; AMS 4513±28 BP). See Table 10.1.

SKULL

The total weight of the skull fragments was 37.70 g, which represents 21.4% of the total weight of bone. There were 80 skull fragments in total, consisting of 21 frontal, 37 parietal, four occipital, two sphenoid and 17 unidentifiable vault fragments. There was also a fragment of a zygomatic arch. A large proportion of the skull fragments were smaller than 10.00 mm in size, although there was a parietal fragment with a length of 30.00 mm, which was 3.50 mm thick. This parietal fragment contained a suture which was nearly completely obliterated. All the skull fragments were adult.

Axial skeleton

The fragments of the axial part of the skeleton weighed 5.20 g, which represented 3.0% of the total weight of bone. There were three midshaft rib fragments, seven vertebral neural arch fragments and one vertebral body fragment. All fragments were adult.

Upper limbs

There were no identifiable fragments from the upper limbs.

Lower limbs

The fragments from the lower limbs weighed 11.70 g, which represented 6.6% of the total weight of bone. There were five midshaft and two distal condyle fragments of the femora, which measured up to 28.00 mm in length and had a thickness of 5.00 mm. Four midshaft fragments and one proximal condyle fragment of the tibiae were recovered. The tibia fragments had a greater length of 32.00 mm and a thickness of 4.00 mm. All fragments were adult.

Conclusion

The cremated bone from this context appears to be from a mature adult of unknown sex. Since only 173.30 g of bone was found – much less than the 2500–3000 g for an average adult, as noted above – the remains could be considered as a token deposit. However, the upper part of this deposit had suffered plough damage and, although there is a substantial amount of bone (175.30 g), it is not possible to estimate what proportion has been lost. It is quite probable that the bone was deliberately removed from the original cremated skeleton rather than representing debris left behind when the cremated body was gathered up for burial.

9.1.1.5 C1012/1013: BNH5 Annexe, entrance

The cremated bone from this context weighed 19.40 g, and 76.0% of it consisted of unidentifiable fragments. All the cremated fragments were white in colour. Some 3.6% were too large to pass through a 10.00 mm sieve, 54.6% passed through a 10.00 mm sieve, 41.3% through a 5.00 mm sieve, and 0.5% were small enough to pass through a 2.00 mm sieve. These values reveal that the great majority of fragments were smaller than 10.00 mm in size and that the cremated bone was highly fragmented.

Skull

The total weight of the skull and teeth fragments was 3.20 g, which represents 16.0% of the total bone weight. There were 16 skull vault fragments; it was not possible, however, to identify any of these to specific bones. The largest skull fragment measured approximately 11.00 mm in length and was 3.00 mm thick. A fragment of the head of a mandibular condyle was also present. The skull fragments appeared thin, suggesting that they might have come from a juvenile individual. It is possible, however, that they are from the thinner areas of the skull vault, such as the squamous part of the temporal bone. The size of the fragments and the lack of any sutures makes it impossible to know which explanation is correct. Eleven fragments of tooth root were present. It was possible to identify two mandibular premolars. All tooth roots were from permanent teeth and were completely developed.

Axial skeleton

The only fragment from the axial region of the skeleton were three vertebral body fragments. These weighed 0.30 g, which represents 1.5% of the total weight of bone.

Upper limbs

The upper limbs were represented by a fragment of the distal end of the radius from the region where it articulates with the scaphoid and the lunate and two fragments of intermediate hand phalanges. The distal end of the radius was fused, and the hand phalanges were adult-sized. The weight of the upper limb fragments was 0.90 g, which represents 4.6% of the total weight of bone.

Lower limbs

The only fragment from the lower limbs was from the surface of the proximal tibial condylar area. The fragment was fused and weighed 0.30 g, which represents 1.5% of the total weight of bone.

Conclusion

The bones present in this context appear to be from a single adult individual. There is no duplication of any of the bone fragments, and they all appear to be adult and of the same build. It is not possible to obtain information on sex or more accurate ageing information due to the small quantity and highly fragmentary nature of the bone.

Since only some bones of the body are represented, and the total weight of the bone is only 19.40 g, it is probable that the cremated bone represents an incomplete survival or a token deposit. It is possible that certain fragments of the original corpus of cremated material were separated and buried in this context. Alternatively, it is possible that the upper levels of the deposit were destroyed by modern cultivation, a suggestion that seems more feasible when the small size of the fragments is considered.

9.1.1.6 C1014/1015: BNH5 Annexe, Outside entrance

The total weight of cremated bone from this context was 154.70 g of which 81% was unidentifiable. All the fragments were white in colour; 15% were too large to pass through a 10.00 mm sieve, 63% passed through a 10.00 mm sieve, 21% through a 5.00 mm sieve, and 1% were small enough to pass through a 2.00 mm sieve. The majority of fragments were smaller than 10.00 mm in size, indicating that the bone was highly fragmentary in nature.

SKULL

The total weight of the skull fragments was 5.00 g, which represented 3.2% of the total weight of bone. There were 16 fragments, which consisted of 11 unidentifiable vault fragments, four parietal fragments and one fragment of the petrous part of the temporal bone. The largest skull fragment had a greatest length of 18.50 mm and a thickness of 2.00 mm. In all the vault fragments, the diploe was immature, indicating that the fragments were from a juvenile individual.

AXIAL SKELETON

The total weight of the fragments from the axial region of the skeleton was 1.20 g, which represented 0.8% of the total weight of bone. There were four fragments, all of which were from vertebrae. Three of the fragments were of neural arches, and one was from a vertebral body. One of the neural arch fragments appeared to be immature, but it was not possible to ascertain this for the other fragments.

UPPER LIMBS

The total weight of fragments from the upper limbs was 3.20 g, which represents 2.0% of the total weight of bone. The fragments of humerus consisted of two midshaft fragments, one fragment of the head and one fragment of the trochlea. The trochlea fragment was an unfused epiphysis. The state of fusion of the distal humerus indicates that the individual had an age-at-death of less than 13–19 years (after Brothwell 1981, 66). The humerus fragments measured up to 25.50 mm in size and had a thickness of 1.50 mm. Radius fragments present were a fragment of the head and neck and the distal epiphyses. Both fragments were unfused, indicating the individual had an age-at-death of less than 13–23 years (*ibid.*).

LOWER LIMBS

The total weight of fragments from the lower limbs was 20.50 g, which represents 13.0% of the total weight of the bone. Of the femur, there were 14 midshaft fragments, two of which were from the posterior surface and contained portions of the linea aspera and the gluteal tuberosity, and a fragment from the posterior surface of the distal end from the region immediately superior to the distal condyles. The femora fragments measured up to 37.00 mm and had a thickness of 4.00 mm. There was also a midshaft fragment and one fragment from the intercondylar area of the proximal tibia. The tibiae fragments measured up to 23.00 mm and 3.00 mm thick.

CONCLUSION

The remains appeared to be from a juvenile individual aged less than 13 years. The size of the proximal radius and the distal humerus fragments indicate that it is likely that the individual was younger than this and probably died around 9–10 years of age. The weight of bone, in conjunction with the fact that many bones are not represented in the assemblage, indicates that the bone may be representative of a token deposit, or it may be incomplete due to subsequent damage to the feature. The fairly substantial amount of bone (154.70 g) is suggestive that it was deliberately removed from the original cremated body rather than representing debris left behind when the cremated remains were gathered up for burial.

9.1.1.7 C588: BNH6 Inner Ring (J.L. Wilkinson)

GENERAL

The total weight of 1062.00 g represents the major part of one individual, though many fragments are now damaged beyond recognition. A fragment of bone was dated to 3310–2910 cal BC (UBA-42749; AMS 4400±31 BP; see Table 10.1).

SKULL (28.00 G)

18 fragments of vault up to 22.00 mm long and 3.60 mm thick, generally of slender build. Some endosteal fusion. One fragment of ethmoid bone. Four pieces of shattered petrous temporal bones. Maxillae: four fragments of alveolar region, including tooth sockets, but only the edges of the hard palate. Mandible: two condyles, five pieces of body, one with tooth sockets. Teeth: nine root fragments with adult-type closed apices; no crowns survived cremation.

AXIAL SKELETON (TOTAL 56.00 G)

Vertebrae (26.00 g): One almost complete upper lumbar vertebral body; its intervertebral surfaces are mature, and there is no arthritic lipping. Body of the axis vertebra, with its odontoid process broken off. Eleven small pieces of bodies, 26 vertebral arch fragments. Pelvis (18.00 g): 16 fragments, including one of a sacro-iliac joint surface; no arthritis. The 1st coccygeal vertebra is present: it was partly fused to the 2nd coccygeal vertebra (>30 years), but not to the sacrum. Ribs (12.00 g): 31 fragments up to 38.00 mm long, of slender build.

UPPER LIMBS (40.00 G)
Scapulae: one fragment of body. Humeri: five pieces of head, the maximum diameter 34.00 mm (incomplete); two small pieces of lower articular surface; 15 shaft fragments up to 50.00 mm long and 4.30 mm thick. Forearm bones: 13 fragments up to 46.00mm long. Hands: two carpal fragments, including one scaphoid. seven metacarpal fragments, including three heads and four shafts. Two heads of proximal phalanges.

LOWER LIMBS (112.00 G)
Femora: 54 shaft fragments up to 52.00 mm long, mostly <4.90 mm thick two fragments are 5.50 mm thick; five articular fragments of femoral condyles. Tibiae: seven shaft fragments up to 38.00 mm long. Feet: five metatarsal fragments, including a medium-sized head of a 1st metatarsal; two very slender proximal phalanges, two very small distal phalanges, including the terminal phalanx of the right great toe.

CONCLUSIONS
Age: probably 30–40 years (mature intervertebral surfaces without any arthritic lipping, partial fusion of coccyx, some endosteal fusion of skull sutures). Sex: female (slender to medium build of long bones, slender skull vault, small foot bones).

9.1.1.8 C943: BNH6 Outer Ring
The total weight of the bone fragments from this context was 812.20 g, of which unidentifiable fragments comprised 56.0% of the total. All fragments white in colour suggesting that the cremation process was efficient. Some 31.4% of the fragments were greater than 10.00 mm in size, 12.4% were 2.00–5.00 mm, and 1.6% were smaller than 2.00 mm. These values indicate that the remains were not very highly fragmented and that a large proportion of the bone was greater than 10.00 mm in size. The majority of fragments were from a juvenile individual but several fragments of adult bone were also present. It is likely that more fragments of adult long bone are present in the unidentified bone category. A fragment of bone was dated to 3340–3025 cal BC (UBA-42748; AMS 4467±29 BP; see Table 10.1).

(A) JUVENILE INDIVIDUAL
SKULL
The total weight of the skull and tooth fragments was 120.90 g, which represents 14.9% of the total weight of bone. There were 438 skull fragments in total, and of these 180 vault fragments could not be identified to the specific bone. One hundred and three parietal fragments were present with a size up to 28.60 mm in length and a thickness of 1.00–2.00 mm. The frontal bone was represented by 75 fragments, all of which were less than 10.00 mm in length. Eight fragments of temporal bone were present, including the right and left petrous parts of the bone. The occipital was represented by 30 fragments, which included a fragment of the basilar part of the bone from the area immediately adjacent to the foramen magnum and a fragment of the external occipital protuberance. The greatest length of the occipital fragments was 33.00 mm, and they had a thickness of 3.00 mm. All skull vault fragments had immature diploe, indicating that the skull was from a juvenile individual. Fifteen fragments of sphenoid were recovered, all of which were less than 10.00 mm in length. Nineteen maxilla and eight mandible fragments were present. All the tooth sockets appear to be those of deciduous teeth.

Sixty-five teeth fragments were recovered. There were 22 fragments of tooth root, which were not identifiable to specific teeth. Those which were specifically identifiable were from both maxillary and mandibular teeth. The mandibular teeth represented were both central incisors and canines. There were also four detached roots from the first molars and two detached roots from the second molars. The maxillary tooth roots were those of both central incisors, both lateral incisors and both canines. There was also one detached single root from a first molar and four from the second molars. All tooth roots were complete, and there was no definite evidence of root reabsorption. There were 24 fragments of tooth crown. The majority of fragments were only 1.00–2.00 mm in size and were unidentifiable to specific tooth. There were also four larger fragments which were grey in colour, indicating that they had not been as completely oxidised as the rest of the bone. These fragments were from molar teeth, and they were likely to be from the unerupted permanent first molars, which would have been protected by the bone of the mandible during the burning process. The appearance of the teeth corresponds to an age-at-death of approximately 4–5 years ± 16 months (after Ubelaker 1989, 64).

AXIAL SKELETON
The total weight of the bone fragments from the axial part of the skeleton was 68.90 g which represented 8.5% of the total weight. There was one midshaft fragment of the left clavicle, which measured 42.10 mm in length, and 43 midshaft rib fragments, which measured up to 36.00 mm in length. The vertebrae fragments consisted of 99 neural arch and seven body fragments. The neutral arches appear to be fused to the bodies, indicating that the individual was older than approximately 4 years of age (after Ferembach *et al.* 1980, 531). An immature-sized odontoid process of the axis was also present.

The innominate fragments included 17 miscellaneous fragments, 17 ilium, one acetabulum and three ischium fragments. The ilium fragments had a length of up to 35.00 mm and the ischium fragments measured up to 27.60 mm. The latter included the right and left ischial tuberosities which were both small and unfused. The ilium fragments included three fragments of the iliac crest; all of which were unfused. A fragment of the auricular surface was also recovered, which had an immature appearance. The state of fusion of the innominates can only indicate that the individual had an age-at-death of less than 17–25 years (after Brothwell 1981, 66).

UPPER LIMBS

The total weight of the fragments of the upper limbs was 53.20 g, which represented 6.6% of the total weight of bone. Fragments of humerus were 15 midshaft, nine head and three distal. The head fragments were unfused epiphyses and the distal fragments were from the metaphyses and were also unfused. The state of fusion of the humerus only indicates that the individual had an age-at-death of less than 13–25 years (*ibid.*). The humerus fragments measured up to 42.70 mm and had a thickness of 2.00 mm. Twelve fragments of midshaft ulna and one proximal ulna fragment were present. The proximal fragment was unfused. The ulna fragments were up to 45.80 mm in length and had a thickness of approximately 1.00 mm. Seven midshaft and two proximal radius fragments were present. The proximal fragments were unfused. The fragments of radius were up to 37.40 mm in length and had a thickness of 1.00 mm. The state of fusion of the ulna and radius indicates that the individual had an age-at-death of less than 13–19 years (*ibid.*). Eight metacarpal midshaft fragments, one complete proximal hand phalanx, two complete intermediate hand phalanges and two unidentified phalanx fragments were present. The phalanges were all from a juvenile individual.

LOWER LIMBS

The fragments of bone from the lower limbs had a weight of 111.80 g, which represented 13.8% of the total weight of bone. There were 20 midshaft, 11 distal epiphyses and nine proximal metaphyses fragments of the femur. The distal and proximal fragments were from both the right and left femora and were all unfused. The greatest length of the femur fragments was 48.40 mm, with a thickness of *c.* 3.00–4.00 mm. The state of fusion of the femora indicates that the individual had an age-at-death of less than 15–23 years (Brothwell 1981, 66). The tibiae were represented by 15 midshaft and seven proximal fragments. The presence of fragments which contained the posterior foramina indicated that both right and left tibiae were present. The greatest length of the tibia fragments was 57.70 mm, and they had a thickness of 3.00 mm. The proximal tibia fragments were unfused, indicating that the juvenile died at less than 16–23 years (*ibid.*).

(B) ADULT FRAGMENTS

The adult fragments recovered from this context were commingled with the remains of the juvenile and comprised a fragment of the petrous part of the temporal, a root from a maxillary canine tooth, a vertebral articular process with possible osteophytes and porosity, a proximal first metacarpal, a fragment from a metacarpal and eight femur midshaft fragments. The petrous part of the temporal bone was grey in colour which indicates that it was not completely oxidised.

CONCLUSION

The juvenile present in this burial was aged approximately 4–5 years old when he/she died. It is not possible to determine sex since they died before reaching puberty. The juvenile individual represents the bulk of the bone finds in the context and their remains seem to be relatively complete. The majority of the unidentifiable fragments are midshaft fragments of long bones. Since there is a reasonably large proportion of fragments greater than 10.00 mm in length, it is unlikely that the body was subjected to any further processing following cremation.

The adult fragments are in the great minority in this context. They probably represent contamination of the juvenile burial with fragments from another individual or individuals. The adult remains could represent the residue of an earlier deposit, which had been cleared out of the burial site. Alternatively, the adult remains could have become commingled with those of the juvenile at the pyre site. Although it is probable that there are other fragments of adult bone in the unidentifiable category, the quantity of adult fragments present in this context is much smaller than the amount of juvenile bone.

9.1.2 Identifiable burnt bone (including burnt animal bone) from other contexts

The burnt bone from other contexts is quantified in Table 9.2. Fragments of calcined animal bone were retrieved from C4 (above the Entrance Chamber); C289 (BNH6 Inner, secondary fill); C453 (Four-Poster, secondary fill); C507 (BNH6 Inner); C647 (BNH6 Inner) and C1453 (Burning Pit). Cattle, pig and sheep/goat were all represented, with cattle fragments predominating (n=11), followed by pig (n=7) and sheep/goat (n=4). All four of the sheep/goat fragments were from C1453 (the early medieval Burning Pit), and it is possible they had derived from a single animal aged over 6–10 months. This context also included calcined human and pig bones.

The age-at-death values obtained from two cattle bones indicated the presence of an animal aged less than 24–36 months and another aged over 34–36 months. The former may not have reached full size, while the latter probably would have (McCormick 1997, 822). It was possible to gain age-at-death determinations from five pig bones, which indicated slaughter at between 12 and 42 months, with one animal probably killed around 24–30 months. This may indicate a policy of slaughtering the animals when they reached full size. Two sheep/goat bones had been derived from animals older than 6–10 months.

Table 9.2 Burnt bone from other contexts.

UN	Identification	Context no.	Feature group
6189	Animal: cattle L phalanx 2 (fused > 34–36 mths), cattle R mandibular M3 tooth (unburnt or slightly burnt), Pig L phalanx 2 (fused)	4	Plough damaged. BNH5 OR posthole?
6193	Animal: possible cattle tooth fragment	119	No attribution
6202	Animal: Pig L scapula (fused > 12 mths)	289	Posthole, BNH6 IR
6265	Human: 5 cranial vault, 1 petrous, 4 femoral shaft (adult)	635	BNH6 IR (primary fill)
6201	Animal: fragment of large vertebra, probably cattle	289	Posthole, BNH6 IR
6211	Animal: cattle L astragalus (fused), 2 x tarsals	289	BNH6 IR (secondary fill)
6266	Animal: pig distal metatarsal (unfused < 24–30 mths)	647	BNH6 IR
6268	Animal: cattle R & L distal tibial epiphyses (unfused < 24–36 mths), 2 metapodials	507	BNH6 IR
6229	Animal: pig 2 frags L dist femur (unfused < 42 mths)	453	Four-Poster (secondary fill)
6231	Animal: pig R prox radius (fused > 12 mths)	453	Four-Poster (secondary fill)
6226	Animal	453	Four-Poster (secondary fill)
6230	Animal: pig L ulna	453	Four-Poster (secondary fill)
6225	Possibly human	531	North Entrance Chamber
6258	Human: 1 maxilla, 4 tooth root	531	North Entrance Cchamber
6372	Human: 2 vertebrae (adult)	1418	East Entrance Chamber
6327	Human	1544	Southern E–W postholes
6326	Human: possible mandible fragments	1453	Burning Pit
6356	Human: skull fragments (adult)	1453	Burning Pit
6362	Animal: pig dist metatarsal (fusion line visible, around 24–30 mths)	1453	Burning Pit (depth 0.15 m)
6371	Animal: sheep/goat L dist humerus (fused > 10 mths), scapula (fused > 6–10 mths) and possible sheep/goat pelvis and femur	1453	Burning Pit

Note that animal age-at-death estimation is based on the method of Silver (1969, 285–286).

The cattle and pig bones represented a mixture of good (cattle: mandible, vertebra, distal tibia; pig: scapula, femur, radius, ulna) and poor quality meat joints (cattle: phalange, astragalus, metapodials; pig: distal metatarsals). This may indicate that complete animals were butchered in the vicinity of Ballynahatty and that the quality of the meat cut was not a major consideration when depositing the bones. The sheep/goat bones derived from the early medieval Burning Pit C1453 were all of good quality meat joints (scapula, humerus, pelvis, femur).

9.1.3 Overall conclusions

Several periods of funerary activity are represented among these cremated remains: Middle Neolithic, from the bones dated to *c.*3300–3000 cal BC; Bronze Age burial activity from C1544 in south Annexe (UBA-42745; AMS 3751±39 BP, 2290–2035 cal BC); and the early medieval Burning Pit. The deposits of cremated bone at Ballynahatty that constituted the complete or near-complete remains of people comprised those of a middle-aged adult (30–40 years), possibly female, from C588 (BNH6 Inner Ring) and those of a 4–5-year-old child from C943 (BNH6 Outer Ring). The latter assemblage also contained several fragments of an adult human, which probably either represents contamination from an earlier act of burial or, it is possible that if there were earlier remains of an adult at the pyre site, these may have been accidently included when they gathered the child's bones. C951–C956 (BNH5 Chamber) contained two Coarse Ware pots which had been disturbed in antiquity. Fragments of bone recovered from within Vessel **2** indicated that it had contained the remains of an older adolescent/young adult who was also well represented in the commingled remains in the cist. The individual that may originally have been associated with Vessel **1** was a young child less than 6 years of age. Some fragments of bone recovered from within the chamber also appeared to have belonged to earlier burials.

Two features from the entrance of the BNH5 Annexe contained fairly substantial quantities of bone (C1016/1017: 175.30 g; C1014/1015: 154.70 g), and these two assemblages may have represented token deposits or may have been damaged as a result of ploughing activity. One of the contexts contained fragments of a juvenile aged *c.* 9–10 years (C1014/1015), while the other contained fragments of a mature adult of unknown sex (C1016/1017). Context C1012/1013, also from the entrance of the BNH5 Annexe, contained only 19.40 g of bone and it is, therefore, feasible that this represented the residue of a cremated adult body, while C1013/1019 produced only two fragments of undiagnostic cremated bone. Some 77.60 g of calcined bone, representing the remains of an adult, were also recovered from Area 41, fill C779, a layer that covered the burial chamber and the posthole of the Inner Façade and likely contained disturbed deposits of cremated remains from the chamber (BNH5 C591–C596).

Small quantities of definite calcined human bone were recovered from five further contexts – C635 (BNH6 Inner, primary fill); C531 (Entrance Chamber); C1418 (Entrance Chamber); C1544 (Southern East–West Posts); and the early medieval C1453 (Burning Pit).

The burnt animal bone found at Ballynahatty belongs to two discrete periods: the Late Neolithic, when the timber structures were in use, and the early medieval period when the Burning Pit was in use.

9.2 Human remains from the 1855 tomb: a re-evaluation

The remains of both calcined and unburnt bone were recovered from the interior of the tomb (Plate 82). The extant material can be interpreted in the context of the original report produced by MacAdam and Getty in 1855 and in the light of more recent research, which included radiocarbon dating (Schulting *et al.* 2012 and see Chapter 10, Table 10.2) and DNA analysis of one of the individuals, an adult woman (Cassidy *et al.* 2016).

9.2.1 Cremated bone

Four bags of cremated bones were present, and it is possible, but by no means certain, that Bags 1 and 2 correspond to the material retrieved from 'Urns' 1 and 2 within Chamber A of the tomb, while Bag 3 was recovered from 'Urn 3' within Chamber B and Bag 4 from the area suspected of containing a further urn within Chamber B. Urns 1–3 were described by MacAdam and Getty (1855, 359) as being in a very 'decayed' condition, but the fact that they were still *in situ* may indicate that the cremated remains were originally recovered in a complete condition. The material in Bag 4 is certainly more mixed in nature which seems compatible with the description associated with the possible location of a fourth pot, 'Urn 4'. Calcined bone was also recovered from within Chambers D, E and F, however, so it is also possible that the surviving material may have originated from these compartments. Indeed, it may be significant that Bags 2 and 4 contained small fragments of pottery, which link them to the 'Urns' since no pottery fragments were retrieved from Chambers D, E or F (MacAdam and Getty 1855, 360). Unfortunately, we cannot be certain that calcined bone from throughout the tomb has not been mixed together since the time of its original discovery as we have no original labels nor can we be sure of curatorial decisions that may have been made since their arrival in the Ulster Museum.

9.2.1.1 Bag 1

A total of 18 fragments were present, weighing 66.80 g, of which unidentifiable pieces comprised 33.3% of the total. The remains were all adult. The largest fragment was a piece of parietal which measured 50.90 mm by 47.60 mm. Skull fragments predominated and comprised four parietal fragments, one frontal fragment, one unidentifiable vault fragment and two mandible fragments. Post-cranial remains consisted of a fragment of axis, which included the dens and the superior articular surfaces, one piece of vertebral body, an ulna shaft fragment and a fragment of a metacarpal. Osteophytes were evident at the tip of the dens, perhaps suggesting a middle-aged or older adult. The fragments were largely white in colour, suggesting an efficient cremation process, although one parietal fragment and the fragment of metacarpal also displayed light grey discolouration, possibly suggestive of burning at a slightly lower temperature.

9.2.1.2 Bag 2

Some 27 fragments were present, weighing 59.30 g, of which unidentifiable pieces comprised 59.3% of the total. The remains were all adult. The largest fragment was a piece of tibia, which measured 65.60 mm supero-inferior by 13.50 mm. The remains were all post-cranial and comprised two proximal ulna fragments, two midshaft humerus fragments, four tibia fragments (two proximal and two midshaft) and one midshaft femur fragment. The pelvis was represented by pieces of iliac crest and acetabulum. The fragments were predominantly white in colour, suggesting an efficient cremation process, although some fragments also displayed grey discolouration (one proximal ulna, one midshaft humerus and the fragment of acetabulum). The interior of one proximal tibia fragment was dark grey, while the piece of midshaft femur was completely dark grey. The presence of the dark grey fragments is suggestive of burning at a lower temperature. Two small fragments of pottery were also present.

9.2.1.3 Bag 3

A total of 37 fragments were present, weighing 121.50 g, of which unidentifiable pieces comprised 13.5% of the total. The remains were all adult. The largest fragment was a piece of occipital which measured 68.50 mm by 37.80 mm. Skull fragments predominated and comprised ten occipital, ten frontal, eight parietal, one sphenoid and one inferior mandible fragments. Post-cranial remains were of pelvis and consisted of one fragment of ilium and one fragment of acetabulum. A piece of occipital displayed a large external occipital protuberance which is suggestive of a male individual. The fragments were predominantly white in colour, suggesting an efficient cremation process, although the portion of acetabulum also displayed grey discolouration. The endocranial surface of one occipital fragment was dark grey, while one fragment of parietal also displayed this colour. The presence of the dark grey fragments is suggestive of burning at a lower temperature.

9.2.1.4 Bag 4

The possible presence of a fourth pot was considered feasible based on the presence of 'dark half-burnt earth, intermixed with the bones marked at the spot marked 4', but the lack of pottery fragments of sufficient size caused MacAdam and Getty to be uncertain that this was definitely the case (MacAdam and Getty 1855, 360). Regardless of whether or not a pot was present, the nature of the remains retrieved from Bag 4 seems compatible with this description since the deposit was rather mixed. In addition to the cremated bone, four small fragments of pottery, one twig, one sliver of wood, two pieces of straw and four small stones were also present. The bone material also contained a high quantity of small fragments, most diversity of colours, and some unburnt remains were present.

A total of 127 fragments were present, which weighed 64.90 g, of which unidentifiable pieces comprised 70.9% of the total. The remains were all adult. The largest fragment was a piece of cranial vault, which measured 29.90 × 22.40 mm. Skull fragments comprised five parietal fragments, one sphenoid fragment and eight pieces of unidentifiable vault. Post-cranial fragments consisted of three of vertebrae, two ulna fragments (one proximal, one midshaft), one radial head, one femoral head, five femoral or humeral head and eight fragments from the small bones of the hand or foot. In addition, pelvic fragments comprised one ilium fragment and two acetabulum fragments. Most fragments were white or pale grey in colour, although two cranial vault fragments were dark grey. The ilium fragment and one piece of acetabulum displayed a combination of white and dark grey colouration. The presence of the dark grey fragments is suggestive of burning at a lower temperature. Some eight unburnt fragments were also present, but six of these were unidentifiable. The two identifiable pieces comprised a fragment of molar tooth crown and a possible portion of manubrium. A calcined fragment of the distal end of a sheep/goat metacarpal or metatarsal was also recovered.

9.2.2 Unburnt bone: Chamber D

MacAdam and Getty (1855, 360) reported that Chamber D contained a few burnt bones, upon which lay 'two tolerably perfect skulls, and sufficient fragments to prove that five skulls, or at least portions of five, had been there deposited'. Unfortunately, this area of the tomb was described by MacAdam as damaged, presumably by farming activities, before his visit to the site. His estimation of five individuals was based on the presence of three 'undisturbed' mandibles sitting on the base of the recess in addition to the presence of two crania, one of which was not associated with a lower jaw and the other with one that he described as 'very much broken' (*ibid.*). He also recorded that a small quantity of post-cranial human remains and some animal bone were recovered from the chamber. All of the surviving unburnt bone from the tomb displayed a varnished appearance and had been chemically treated, perhaps with the consolidant Bedacryl, a synthetic butyl-methacrylate resin (see Schulting *et al.* 2012, 14). Unfortunately, this makes it difficult to assess whether breakage of the remains occurred in antiquity or more recently. It is only possible to reliably identify breakage that took place after the bones were coated.

9.2.2.1 Skull 1

MacAdam's drawing for metrical analysis and description of Skull 1 (MacAdam and Getty 1855, 361) indicates that the cranium was complete while the mandible was incomplete, with the drawing indicating that the right side had been preserved. He was very clear about the nature of the additional unburnt cranial remains recovered from the chamber, with these comprising one maxilla and three mandibles. Unfortunately, this suggests that the surviving fragmentary cranium is all that remains of Skull 1 today.

The incomplete right and left parietals have been glued to each other and to the occipital, of which only the superior aspect is preserved. Seven additional fragments of parietal and a possible portion of sphenoid are also present. The left temporal is represented by an incomplete petrous portion, the mastoid process and part of the zygomatic arch.

MacAdam (*ibid.*, 360) concluded that the cranium was that of a female aged 20–25 years. Morphological characteristics of the incomplete cranium suggest this assertion is reasonable in terms of sex determination but, since the individual's dentition is no longer preserved, it is not possible to comment on the age of the individual. The surviving incomplete mandible in the assemblage does not display a breakage pattern that is compatible with the depiction of the mandible present in MacAdam and Getty's publication (*ibid.*), and it therefore seems likely that this is the incomplete mandible which he described as not being associated with a cranium (see below). Similarly, the preserved dentition of the isolated maxilla present in the extant collection does not match the image of the right side of Skull 1's maxillary dentition, which MacAdam appears to have been recorded in an accurate manner since there are clear differences in the dentitions of the two skulls in this image.

9.2.2.2 Skull 2 (A.64)

Skull 2 is extant in the condition in which it was recovered by MacAdam from the tomb (Plate 83); the breakage patterns at its posterior aspect correspond to the image in his metrical analysis drawing (MacAdam and Getty 1855, 361). Radiocarbon dating has indicated that the cranium dates to 3343–3020 cal BC (UB-7059; 4465±38 BP) at 95% probability, which places the individual into

Table 9.3 Dentition: Skull A64.

Maxilla					/	/	AM	/	B	B					
	18	17	16	15	14	13	12	11	21	22	23	24	25	26	27	28

AM = lost ante-mortem; / = lost post-mortem; B = broken post-mortem.

Table 9.4 Dentition 1.

Mandible	48	47	46	45	44	43	42	41	31	32	33	34	35	36	37	38
					B	/	/	AM	/	/	/	/	/	PU & A		B

AM = lost ante-mortem; / = lost post-mortem; B = broken post-mortem; PU = pulp exposed; A = abscess.

the later Middle Neolithic or possibly Late Neolithic periods (see Table 10.2). Analysis of carbon and nitrogen isotopes were suggestive of a terrestrial diet (Schulting *et al.* 2012, 10).

MacAdam (*ibid.*, 360) concluded that the cranium was that of a female aged 20–25 years. Morphological characteristics of the incomplete cranium (see Ferembach *et al.* 1980; Schwartz 1995) were compatible with this assertion and this has also been corroborated through aDNA analysis (Cassidy *et al.* 2016, 369). The level of tooth attrition was suggestive of a younger adult (18–35 years) (Brothwell 1981). Ancient DNA analysis also revealed that the woman probably had black hair and brown eyes. In common with Neolithic individuals elsewhere in Ireland, she was descended from people who had migrated from the Continent as pioneering farming groups. Her haplotype was strongly correlated with people of the southern Mediterranean, particularly in Sardinia and Spain, although there is also evidence of admixture with hunter-gatherer populations (Cassidy *et al.* 2016, 372); this admixture is likely to have occurred before her ancestors left the Continent (Cassidy *et al.* 2020). Accessory ossicles were present at the lambda and in the lambdoid suture. Details of the dentition are presented below. No evidence of dental disease was observed; the mandible was not preserved (Table 9.3). The right maxillary central incisor had been lost ante-mortem.

9.2.2.3 Dental remains

In addition to the two aforementioned skulls, MacAdam (MacAdam and Getty 1855, 362) reported on the presence of a mandible that he considered to derive from a middle-aged male (Dentition 1); the maxillary and mandibular dentition of a younger adult male (Dentition 2); and an incomplete mandible from a middle-aged female (Dentition 3). These remains are all extant and are described below.

DENTITION 1

The mandible is near complete, although the right mandibular condyle appears to have been broken in recent times since the exposed bone is very pale in colour. MacAdam was of the opinion that the mandible was derived from a middle-aged male. The morphology of the mandible is compatible with a male, while the state of tooth attrition is suggestive of a middle-aged individual (35–50 years). Details of the dentition are presented in Table 9.4.

The right first incisor had been lost ante-mortem but the socket had not completely infilled. The first molars were notably worn, with pulp exposure evident in the left tooth. The latter tooth was associated with a medium-sized externally draining abscess that may have been secondary to the extensive attrition. While the enamel on some teeth was damaged, slight deposits of calculus were visible on the lingual aspects of the right first premolar and first molar and the left third molar. Periodontal disease of medium severity (as defined in Brothwell 1981) was visible.

DENTITION 2

The remains comprised a complete maxillary dental arch and palate and a near-complete mandible in which the mandibular ramus was partially complete on the right side and missing altogether on the left side. Pale bone was evident at the breakage points suggesting some damage may have occurred in recent times. MacAdam described this dentition as the maxillary and mandibular dentition of a younger adult male. It is now difficult to be certain that this maxilla and mandible genuinely belong to the same individual, especially since the dental wear pattern differs on the upper and lower jaws. MacAdam was meticulous in his recording of the human remains, however, and it is likely that the jaws were associated *in situ*, causing him to have confidence in their association. The morphology of the mandible is compatible with a male, while the state of tooth attrition is suggestive of a middle-aged individual (35–50 years). Details of the dentition are presented in Table 9.5.

A lower incisor had been incorrectly glued into the socket for the left lateral incisor in modern times. The left maxillary and mandibular third molars were genetically absent. The right mandibular third molar had been lost ante-mortem and the socket was in the mid-stage of remodelling. While the enamel on many of the teeth was damaged, slight deposits of calculus were visible on the lingual aspect of the mandibular right second molar and the buccal and lingual aspects of the left canine. Periodontal disease of medium severity (Brothwell 1981) was visible.

Table 9.5 Dentition 2.

Maxilla	–						/	/	/	/						NP
	18	17	16	15	14	13	12	11	21	22	23	24	25	26	27	28
Mandible	48	47	46	45	44	43	42	41	31	32	33	34	35	36	37	38
	AM				/	/	/	/	/	/		/	/			NP

AM = lost ante-mortem; / = lost post-mortem; - = jaw not present; NP = genetically absent.

Table 9.6 Dentition 3.

Mandible	48	47	46	45	44	43	42	41	31	32	33	34	35	36	37	38
		/	/	/	/	/	/	/	/	B	–	–	–	–	–	–

/ = lost post-mortem; – = jaw not present; B = broken post-mortem.

DENTITION 3

The remains comprised the body and complete right dental arch and anterior left dental arch of a mandible. MacAdam described this dentition as the incomplete mandible of a middle-aged female. The morphology of the mandible is compatible with a female, while the state of tooth attrition is suggestive of a middle-aged individual (35–50 years). Details of the dentition are presented in Table 9.6.

The enamel on both surviving teeth was extensively damaged, meaning that it was not possible to assess for calculus formation and other dental diseases. Periodontal disease of medium severity (Brothwell 1981) was visible.

9.2.2.4 Post-cranial bones

MacAdam (MacAdam and Getty 1855, 360) identified the human post-cranial remains as three ribs and a humerus. Re-examination of the remains has identified four ribs; one left first rib, two left middle thoracic ribs and one right rib, which may have derived from a juvenile. The humeral fragment was an adult-sized left humerus shaft with prominent muscle attachments.

9.2.2.5 Animal bones

The animal bones were described as 'part of the pelvis of a small cow, and a few fragments of the bones of a sheep or goat' (*ibid.*). Re-examination has determined that the largest extant fragment (labelled as pelvis of a small cow, left side) is a right cattle pelvic bone which includes the acetabulum and extends to the ischial tuberosity. The dorsal surface displayed four indentations which appear to have been due to gnawing by a large carnivore (Plate 84). The three other animal bone fragments are all labelled as sheep or goat but, in fact, derive from cattle pelvis. They comprise one right and two left fragments of acetabulum.

10

Dating and chronology

Rowan McLaughlin

10.1 Introduction

The sequence of events at Ballynahatty have been fixed onto calendar time using radiocarbon (^{14}C) dating. This technique depends upon the small proportion of carbon in living tissue that consists of the natural ^{14}C radioisotope. This slowly decays after the death of the organism and remains detectable for many millennia. Given a sample of once-living tissue, it is possible to determine how many years have elapsed since death by measuring what remains of its natural radioactivity. This sample could be, for example, a piece of charred wood or seeds or a fragment of human bone. The Ballynahatty excavation and post-excavation period since 1990 have spanned a number of technical developments. As a result, radiocarbon dating for this project was done both by the traditional method of measuring the level of radioactivity of the sample (via liquid scintillation counting, or LSC) and, more recently, by directly measuring the proportion of ^{14}C atoms that remain in the sample (via acceleration mass spectrometry, or AMS).

Irrespective of how they are measured, radiocarbon laboratory data must be compared against an age-calibrated record of the radiocarbon history of the environment, known as a calibration curve. These are continuously updated, so the results of radiocarbon measurements obtained some decades ago can be updated using the latest calibration curves. A simple mathematical process is used to combine laboratory measurement and calibration data. The resulting chronological model, usually termed a 'radiocarbon date', is a probability distribution estimating of the calendar date of when the sampled organism died. Twenty such radiocarbon dates have been obtained from the excavations described in this volume.

Radiocarbon dates rarely directly date human activities or archaeological features, only the samples found within deposits. For relatively simple features, such as a hearth, for example, the date of a sample of charcoal found within it can be seen as a good proxy for the date of the feature. For more complex features or groups of features, like timber circles, estimating their date relies on a critical assessment and logical comparison of multiple dates and an assessment of their archaeological context.

The optimum way to achieve this is to use Bayesian chronological models. These provide a syntax to define formally the information we know about each sample and also computational tools that employ all this information together and generate estimates of the site's chronology that are as accurate as possible. They are also conservative in the sense that they fairly incorporate the uncertainty about the date of the sample that stems from measurement error and environmental history. Other statistical tools termed density models can be used to understand the tempo of the site where such contextual information is either lacking is not secure enough in terms of the information it conveys about the relationship of the date of the sample to the date of the feature. The aim of this chapter is to use these tools to investigate such issues, recalibrate the dates, and determine a chronology for the site.

10.2 Potential problems with radiocarbon dates

Radiocarbon dating has a number of complexities that befuddle the results and their interpretation. First, ^{14}C can derive from a number of different natural carbon reservoirs, which demands the use of alternative calibration curves. In the case of the Ballynahatty samples, however, the ultimate source of the carbon was the atmosphere, so this pitfall does not apply. The laboratory measurement error is variable, and some samples have been measured with greater precision than others. This is compounded by the shape of the calibration curve, which results in samples from certain centuries being imprecisely dated compared to others. At Ballynahatty, activity in the late 4th millennium BC will inevitably be less precisely dated than that of the 3rd millennium, as the calibration curve contains a plateau between approximately 3300 and 3000 BC.

Samples can be contaminated with both older and newer carbon, skewing the result in either direction. Laboratories have protocols for dealing with some of this, although it is sometimes impossible to know if a sample is contaminated, especially if there are only one or two dates from a feature; guesswork is often employed to flag an outlying date as suffering contamination. As discussed below, this issue has affected at least one date from the human remains from the adjoining 1855 tomb obtained by Schulting *et al.* (2012). Samples of bone, in particular, are susceptible to numerous channels of contamination and must be carefully pre-processed by the laboratory. One date from Ballynahatty (GrA-14812) was obtained during the development of methods for dating calcinated bone (Lanting and Brindley 1999), which could be grounds for suspecting the result, although in this case, it fits well with the rest of the sequence of dates relating to mortuary activity at the site.

Certain material, especially charcoal but also samples of human bone, have a built-in age, where the organism has locked-in to its tissues ^{14}C decades or centuries before it died. As discussed by McLaughlin *et al.* (2016), this is a major problem for certain episodes of Irish prehistory. For most charcoal samples, it is impossible to determine what kind of 'old wood effect' to expect, so all that can be done is to model the data and see whether certain samples are too old compared to others and not over-interpret the results if this proves to be the case.

Beyond these technical limitations, there are numerous issues with sample selection in the field. Residuality is always a problem in archaeological dating, where already ancient material from earlier phases of occupation at a site is incorporated into the features we are trying to understand. This problem can affect any radiocarbon date. In the LSC era, bulky samples of charcoal and other deposits were required, which often involved the combination of fragments of several different entities, each with their own built-in ages, potentially compounding the old-wood effect and residuality problems mentioned above. AMS dating is better because it can be done on single-entity samples. Although these too may be residual, the problem is easier to detect when multiple determinations are obtained for a feature or groups of features. This was the approach employed for the Ballynahatty excavations once all the radiocarbon determinations were known.

10.3 Methods and modelling approach

Contextual information can sometimes simplify the occasionally problematic relationship between sample, context and human activity. The date of human bones found in the cremation burials at Ballynahatty are reasonable proxies for the date of the digging and depositing of the remains and any associated artefacts. On the other hand, charcoal finds may date from settlement activity contemporary or more ancient than the construction of the posthole complex or even be later material introduced to the context through bioturbation. The modelling approach must therefore accept these limitations and critically assess whether the spread of dates from a group of features we presume to be related are consistent with this prior belief. The key observation made from studying the site in plan is that BNH6 was constructed before BNH5 and related features (Hartwell 2002; see Chapter 11). Whilst the fact that the timber posts have long ago rotted away means we cannot hope to date them directly, we can make progress by working under the assumption that material found in the postholes is a reasonable proxy for activity related to the structure's construction and use. A key aim of modelling the radiocarbon data is to determine whether they are consistent with the sequence of events inferred from the archaeology.

Modelling is a cyclical process. When the dates from the site were first assembled and calibrated, it became obvious that the material on human bone from the timber circle complex was consistently earlier than the charred plant material and, indeed, overlapped with the sequence of dates from the nearby 1855 tomb. Here Kernel Density Estimate (KDE) models can be employed to investigate the overall tempo of this activity. The KDE models are also useful in they can incorporate hundreds or thousands of individual radiocarbon dates from other sites into a comparative framework to test whether patterns of human activity across prehistoric Ireland are also reflected in the goings on at Ballynahatty (*cf.* McLaughlin 2019).

10.4 Ballynahatty radiocarbon dates

Radiocarbon dates from the site are listed in Table 10.1. The probability distributions of the dates are displayed in Figure 10.1. The spread of dates illustrates how archaeological materials often preserve signals of occupation before and after the periods where we expect to find activity based on the morphology of the extant features. In the case of Ballynahatty, the obvious concentration of activity at the site in the Neolithic and Early Bronze Age periods (*c.* 4000–2000 cal BC) is well attested by the dates, but there is outlying activity occurring millennia earlier and later, in the Mesolithic and early medieval periods. The model presented in Figure 10.1 treats these early and late phases as individual instances and not part of a defined phase of activity.

Mortuary activity at the site presents some problems in defining a bounded phase of activity. Whilst cremation deposits suggest formal funerary activity, some of the dated material is fragments of human bone found in features that are part of the timber circle complex. It is not clear whether the distribution of these dates in

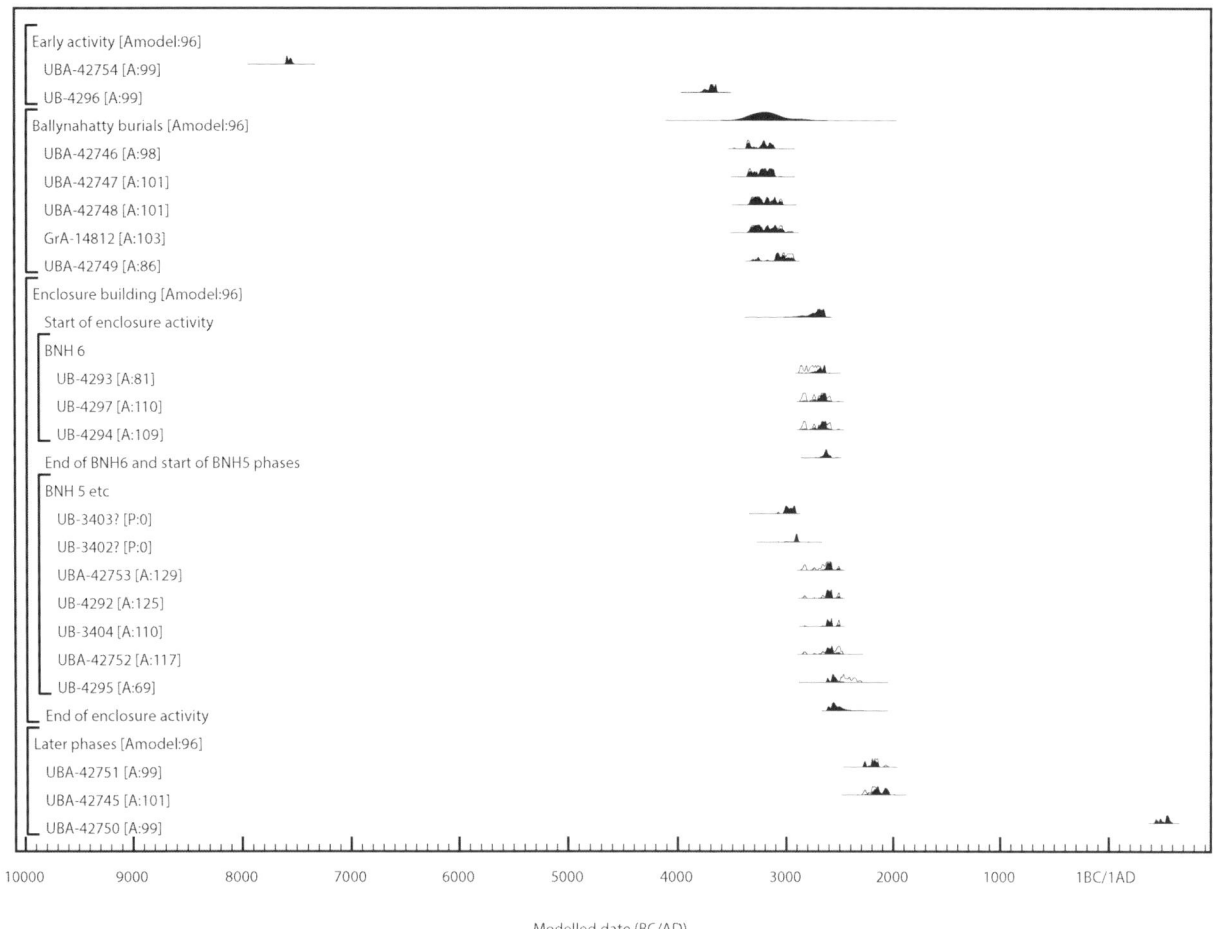

Figure 10.1 Radiocarbon dates from the Ballynahatty complex calibrated and modelled using kernel density estimates and sequence and phase constraints in OxCal (Bronk Ramsey 2009; 2017). Filled areas represent the modelled posterior probability of the dates. Note that UB-3403 and UB-3402 have been tagged as outliers for the BNH5 phase and the software has estimated they have effectively zero probability as belonging in that phase.

time should be used as evidence for a separate phase of activity or whether they can contribute to the knowledge of the timber circle complex, partly because taphonomic processes have likely heavily filtered what survives and partly because the archaeological traces of funerary activity at the site are fragmentary and difficult to interpret. It is also appropriate to consider also the dating of the nearby 1855 tomb and whether there is a degree of contemporaneity between it and the mortuary activity at the timber circle complex. Seven radiocarbon determinations (including repeat measurement of one sample, which can be combined for statistical analysis) obtained by Schulting *et al.* (2012) are available from the 1855 tomb, which have been recalibrated for use with the chronological model (Table 10.2). Kernel density estimation has been used in the model in Figure 10.1 to determine the overall distribution of these funerary events in time and to illustrate how they pre-date the main events of the timber circle complex, with activity peaking between approximately 3300 and 2900 cal BC.

Turning to the chronology of the timber circle complex itself, the statistical spread of dates can be used to estimate if their distribution in time is consistent with the priority of BNH6 in the sequence of events at the timber circle complex. It should be noted that the Bayesian model cannot prove that this sequence is correct, only that it may be wrong in the sense that the model is not plausible in the light of the evidence. The analysis presented here is indeed consistent with the notion that BNH6 was constructed before BNH5 but should not be read as definitive independent proof that this is so. Table 10.3 summarises the dating of this sequence, as calculated in *OxCal*. Here, the dates from BNH6 are considered as being originally deposited uniformly, in a distinct phase pre-dating those from BHN5. The dates from BNH5 are similarly treated, although dates UB-3402 and UB-3403, on samples of bulk charcoal, were tagged as outliers. The OxCal algorithms have estimated that there is effectively zero probability that these samples belong in the sequence of events that constrain the other radiocarbon dates (Fig. 10.1). Instead, they must reflect either masses of old wood enveloped within the bulked charcoal samples dated here, significant residual material from the mortuary phase of the site's use, or both.

Table 10.1 Radiocarbon dates from the Ballynahatty excavations.

Lab. identifier	Determination BP	Method	Calibrated date (95.4%) cal BC	Phase modelled date (95.4%) cal BC	Material	Fill	Cut	Interpretation
UBA-42754	8537±42	AMS	7600–7520	7600–7520	Charred twig	509	443	Residual Mesolithic activity in BNH6 (inner ring)
UB-4296	4915±37	LSC	3775–3635	3775–3635	Bulk charcoal	927	926	Residual Early Neolithic activity at centre of BNH6
UBA-42746	4547±33	AMS	3370–3100	3370–3100	Burnt poss. human bone	635	678	Residual burial activity in BNH6 (inner ring)
UBA-42747	4513±28	AMS	3360–3100	3355–3100	Burnt human bone	1016	1017	Human cremation deposit
UBA-42748	4467±29	AMS	3340–3025	3335–3030	Burnt human bone	943	–	Human cremation deposit
GrA-14812	4460±40	AMS	3350–2935	3340–3025	Burnt human bone	–	591	Human cremation (Coarse Ware)
UBA-42749	4400±31	AMS	3310–2910	3320–2915	Burnt human bone	588	–	Human cremation deposit
UB-4293	4152±24	LSC	2875–2630	2860–2585	Bulk charcoal	373	417	BNH6 activity
UB-4297	4114±26	LSC	2865–2575	2840–2585	Bulk charcoal	644	948	BNH6 activity (Four-Poster)
UB-4294	4106±27	LSC	2860–2570	2840–2580	Bulk charcoal	634	635	BNH6 activity
UBA-42753	4083±30	AMS	2855–2495	2650–2495	Charred hazel	828	783	BNH5 Entrance Chamber activity
UB-4292	4070±26	LSC	2845–2490	2635–2495	Bulk charcoal	62	31	BNH5 eastern setting activity
UB-3403	4355±26	LSC	3075–2905	Outlier	Bulk charcoal	–	43	BNH5 activity (old or residual material)
UB-3402	4293±30	LSC	3010–2875	Outlier	Bulk charcoal	–	64	BNH5 activity (old or residual material)
UB-3404	4062±21	LSC	2835–2490	2630–2500	Bulk charcoal	–	22	BNH5 activity
UBA-42752	4051±38	AMS	2845–2465	2635–2495	Charred hazel	827	783	BNH5 Entrance Chamber activity
UB-4295	3950±46	LSC	2575–2295	2630–2460	Bulk charcoal	819	799	BNH5 activity
UBA-42751	3762±26	AMS	2285–2045	2290–2130	Charred hazelnut	635	634	Bronze Age activity in BNH6 Inner Ring
UBA-42745	3751±39	AMS	2290–2035	2210–2030	Burnt human bone	1544	–	Bronze Age burial activity in south Annexe
UBA-42750	1544±28	AMS	cal AD 430–595	AD 430–595	Charred grain	1453	1452	LIA/early medieval activity in southern Burning Pit

Measurement instruments are Accelerator Mass Spectrometry (AMS) or Liquid Scintillation Counting (LSC). Dates are calibrated here using IntCal20 (Reimer *et al.* 2020) and phase modelled using OxCal 4.4 (Bronk Ramsey 2009; 2017).

Table 10.2 Dates from the Ballynahatty '1855' tomb.

Lab. identifier	Determination BP	Calibrated date (95.4%) cal BC	KDE modelled date (95.4%) cal BC	Material
UB-7194	4587±34	3510–3105	3495–3100	Human tooth (man. LM2)
UB-7521	4584±37	3505–3100	3495–3100	Human tooth (mand. RM1)
UB-7248	4507±36	3360–3090	3355–3095	Cremated human bone
UB-7059	4465±38	3345–3020	3340–3025	Human tooth (max. M1, 'Ballynahatty 2')
UB-7247a	4452±33	3335–2930	3335–3020	Cremated human bone (repeated measurement)
UB-7247b	4440±33			

All dates are from Schulting *et al.* (2012) but recalibrated using IntCal20 (Reimer *et al.* 2020) and modelled using kernel density estimation (KDE) in OxCal 4.3 (Bronk Ramsey 2017) with the dates on human remains from the area of the timber enclosures. Not included in this list is an unexpectedly late result from the mandible of the same individual as UB-7521 which Schulting *et al.* (2012, 12–13) argue is probably the result of contamination with preservative (UB-6723, 4165±36 BP, 2885–2630 cal BC). The presence of a calibration plateau in the IntCal20 curve hinders more precise dating of archaeological materials from these centuries.

Table 10.3 Summary of events present in the Bayesian model of the site's chronology shown in Fig. 10.1.

Event	Modelled estimate (68.2%) cal BC	Modelled estimate (95.4%) cal BC
Start date of BNH6	2760–2630	2940–2600
Start date of BNH5	2650–2600	2700–2510
End date for enclosure activity	2610–2490	2620–2420
Duration of BNH6	0–120 years	0–310 years
Duration of BNH5	0–120 years	0–240 years

Highest posterior probability density intervals calculated using OxCal (Bronk Ramsey 2009).

Based on this Bayesian model, the main phase of use of the timber circle complex likely spans approximately 2700–2500 cal BC. Considering the outer margins of the highest posterior probability density intervals, activity at BNH6 could have begun as early as 2940 cal BC, and activity at BNH5 could have lasted until 2420 cal BC. The Bayesian model can also estimate the duration of activity associated with the dates from BNH6 and BNH5 respectively, which in both cases perhaps spans less than 120 years (Table 10.3).

10.5 Radiocarbon dates from comparable Irish sites

Developer-funded fieldwork has contributed enormously to our understanding of the prehistoric past in Ireland. This is especially true for the years that have elapsed since the Ballynahatty fieldwork was done; there are now at least 20 (Carlin 2016, 123) excavated timber circles and a growing awareness that they are part of a gamut of Late Neolithic timber structures. Drawing on the radiocarbon information compiled by McLaughlin (2020), there are at least 37 radiocarbon dates associated with these building traditions in Ireland (Table 10.4). Through KDE modelling the density of these dates in time, it is possible to study whether the Ballynahatty enclosures fit into a tradition of building with timber. Research on these sites has not been exhaustive, and there are likely to be other excavated examples, although it is not expected that the overall chronological patterns in the KDE analysis will significantly change.

10.6 Discussion

10.6.1 Early activity

The Mesolithic-period charred twig from BNH6 dates to an early peak in Mesolithic activity in Ireland around 7600 cal BC, identified by Chapple *et al.* (2022). It can be seen as a signal of the early occupation of this landscape, but little else can be said as no associated Mesolithic artefacts were recovered from the excavation. However, an Early Mesolithic axe and blade were found just 700 m away during field walking along the R. Lagan at Edenderry (see Section 2.3.4).

Early Neolithic activity at the site occurred around 3775–3635 cal BC in the vicinity of what would become BNH6. This activity was detected at Ballynahatty through other lines of evidence. Plunkett *et al.* (2008 and Chapter 3) found clear palynological evidence for activity in the immediate vicinity of the site in the sediment core extracted from the nearby Ballynahatty Bog. This included an expansion in Gramineae pollen and that of *Plantago lanceolate* (often seen as signals of disturbance), the appearance of Cerealia-type pollen, and mineral content in the profile indicating increased soil disturbance. Early Neolithic activity is also attested archaeologically at the site by the presence of Carinated bowls. This is a time of considerable activity through the landscape in both settlement and funerary contexts (McLaughlin *et al.* 2016), and as such, it is not surprising that a signal of this wider pattern is present at the site.

Table 10.4 Radiocarbon dates from Late Neolithic sites in Ireland with timber structures and postholes.

Lab. identifier	BP determination	Date cal BC (95.4%)	Site	Interpretation	Material	Ref.
SUERC-21320	3995±30	2570–2465	Annaghilla, Tyrone	House	Alder charcoal	1
SUERC-21642	4055±40	2845–2470	Annaghilla, Tyrone	House	Hazel charcoal	1
SUERC-21331	4060±30	2840–2475	Annaghilla, Tyrone	House	Oak charcoal	1
SUERC-21655	4080±40	2860–2475	Annaghilla, Tyrone	House	Oak charcoal	1
SUERC-20760	4080±30	2850–2495	Armalughey, Tyrone	Structure	Oak charcoal	1
SUERC-20654	4100±35	2865–2500	Armalughey, Tyrone	Structure	Oak charcoal	1
SUERC-20790	4020±30	2620–2470	Armalughey, Tyrone	Timber circle	Alder charcoal	1
SUERC-20784	4030±30	2620–2470	Armalughey, Tyrone	Timber circle	Oak charcoal	1
SUERC-20779	4040±30	2660–2470	Armalughey, Tyrone	Timber circle	Hazelnut shell	1
SUERC-20774	4045±30	2830–2470	Armalughey, Tyrone	Timber circle	Hazelnut shell	1
SUERC-20786	4060±30	2840–2475	Armalughey, Tyrone	Timber circle	Willow charcoal	1
SUERC-20777	4070±30	2845–2490	Armalughey, Tyrone	Timber circle	Willow charcoal	1
SUERC-20794	4080±30	2850–2495	Armalughey, Tyrone	Timber circle	Oak charcoal	1
SUERC-20795	4110±30	2865–2575	Armalughey, Tyrone	Timber circle	Alder charcoal	1
SUERC-20785	4110±30	2865–2575	Armalughey, Tyrone	Timber circle	Oak charcoal	1
SUERC-20778	4135±30	2870–2585	Armalughey, Tyrone	Timber circle	Oak charcoal	1
SUERC-20770	4135±30	2870–2585	Armalughey, Tyrone	Timber circle	Oak charcoal	1
SUERC-20789	4155±30	2875–2630	Armalughey, Tyrone	Timber circle	Oak charcoal	1
SUERC-31603	4145±30	2875–2625	Ask, Wexford	Associated with structure	Hazelnut shell	2
UB-13157	4060±25	2835–2475	Ballynacarriga 3, Cork	Structure	Oak charcoal	3
UBA-14224	4078±24	2845–2495	Glassdrummond 7, Down	House	Hazel charcoal	4
UBA-14226	4116±25	2865–2575	Glassdrummond 7, Down	House	Hazel charcoal	4
UBA-14227	4162±29	2875–2630	Glassdrummond 7, Down	House	Hazel charcoal	4
UB-11128	4008±25	2570–2470	Grange 4, Meath	Posthole	Hazelnut shell	5
UB-12902	4096±23	2850–2505	Kilmainham 1C, Meath	Structure	Pomoideae charcoal	5
UB-12081	4138±27	2870–2585	Kilmainham 3, Meath	Structure	Hazelnut Shell	5
UB-12092	4261±27	2910–2785	Kilmainham 3, Meath	Structure	Hazelnut Shel	5
SUERC-31935	4005±30	2575–2470	Lagavooren 7, Louth	Timber circle	Burnt bone (pig)	6
SUERC-31931	4050±30	2835–2475	Lagavooren 7, Louth	Timber circle	Burnt bone (unidentified)	6
SUERC-31930	4205±30	2895–2675	Lagavooren 7, Louth	Timber circle	Burnt bone (unidentified)	6
SUERC-16970	4395±35	3260–2915	Parknahown 5, Laois	Posthole	Oak charcoal	7
UBA-15437	3989±27	2570–2465	Paulstown, Kilkenny	Timber circle	Ash charcoal	8
UBA-8747	4070±31	2850–2475	Prumplestown Lower 2, Kildare	Timber circle	Oak charcoal	9
Beta-246958	4020±40	2830–2460	Raynestown 1, Meath	Posthole	Oak charcoal	10
Beta-247032	4180±40	2885–2630	Skreen 2, Meath	Structure	Oak charcoal	11
GrN-24734	3990±45	2620–2350	Whitewell, Westmeath	Timber circle	Uncertain	12
GrN-25726	4040±40	2840–2465	Whitewell, Westmeath	Timber circle	Uncertain	12

Calibrated using IntCal20 (Reimer *et al.* 2020) and rowcal (McLaughlin 2019). Refs: 1. Dunlop and Barkley 2016; 2. Kelly *et al.* 2012; 3. Lehane *et al.* 2011; 4. Dunlop 2015; 5. IAC pers. comm.; 6. IAC n.d.; 7. NRA n.d.; 8. Coughlan and Brick 2009; 9. Bolger *et al.* 2015; 10. Elder 2009; 11. O'Sullivan *et al.* 2013; 12. Grogan 2007.

10.6.2 Late Neolithic Ballynahatty

The phase of mortuary activity at Ballynahatty spanning 3300–3000 cal BC is earlier and distinctive from the activity associated with the timber complex. This reflects a wider reduction in the archaeological visibility of funerary monuments at this time. Drawing on the analysis of dates from megalithic tombs by McLaughlin (2020), we can plot the density of dates from Ballynahatty compared to the overall signal of activity at megalithic sites elsewhere in Ireland (Fig. 10.2). In this model, the intensive use of 'developed' passage tombs like Knowth and Newgrange causes a peak around 3000 cal BC that quickly fades. At this time, activity in other Irish timber circles and buildings begins to increase in intensity. As Figure 10.2 illustrates, the KDE model for the dates at Ballynahatty matches the tempo of this wider dynamic.

Focusing on the Bayesian model of the BNH6–BNH5 sequence, we can begin to infer change at a generational level, despite the uncertainties that remain. Figure 10.3 plots the posterior probability distribution of the events modelled in Figure 10.1 in detail and, for the purposes of illustration, suggests that the activity attested by the radiocarbon dates may have spanned only 3–4 generations of people. The end of the timber circle complex before 2500 cal BC is quite well resolved by this model. It also matches the dynamic seen at other sites.

10.6.2.1 The Giant's Ring henge

Given the well-resolved sequence of events from the Ballynahatty timber circle, it is natural to ask where the

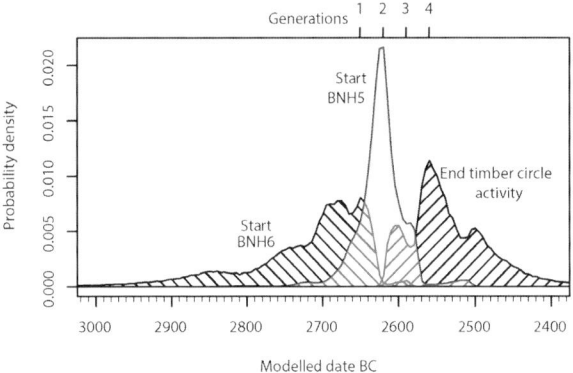

Figure 10.3 Posterior probability density of the date of the phases of activity at BNH6 and BNH5. The generations shown in the upper x-axis are speculative (placed at 30-year intervals) but serve to illustrate the relatively short chronology of the site.

adjacent 'Giant's Ring' henge fits into the sequence. This feature has not been directly dated, but a comparison with other similar dated features elsewhere could help. Unfortunately, in Ireland, solid chronological data from henge enclosures are extremely scarce. A radiocarbon date from a henge ditch at Tonafortes, Co. Sligo (Danaher 2007) places the enclosure in the Early Bronze Age (Beta-199778, 3840±50 BP, 2460–2145 cal BC), but this seems very late compared to their Late Neolithic chronology in Britain. Other Irish dates associated with Late Neolithic enclosures are a curving ditch around a Drumlin at Carnbane 9, Co. Armagh (UBA-14837, 4209±29 BP, 2895–2675 cal BC; Dunlop 2015), and a pit at likely Middle Neolithic enclosure at Kilskeagh 3, Co. Galway (UBA-17312, 4018±29 BP, 2615–2470 cal BC; O'Neill 2013). None of these provides a good proxy for the Giant's Ring, however. A 26 m diameter henge excavated at Carrowreagh, Co. Down, some 12 km northwest of Ballynahatty, remains undated at the time of writing (Historic Environment Record of Northern Ireland pers. comm. 2022). Turning to Britain, the closest dated parallel to the Ballynahatty complex is the Mount Pleasant complex in Dorset. Here, through careful Bayesian analysis of some 59 radiocarbon dates, Greaney et al. (2020) have determined that the henge at the complex was built in 2610–2495 cal BC (95.4%), perhaps 2580–2530 cal BC (68.2%). If Ballynahatty conforms to this pattern, the Giant's Ring may closely post-date the timber circle complex or at least the BNH6 enclosure.

10.6.3 Later activity

Two subsequent phases of dated human activity have resulted in later archaeological features becoming superimposed on the timber circle complex. The first of these occurred in the Early Bronze Age, with two statistically indistinguishable dates around 2285–2035 cal BC from the centre of BNH6 and the southern Annexe. The BNH6 date is difficult to interpret, but the sample of burnt human bone from the southern Annexe clearly indicates that the

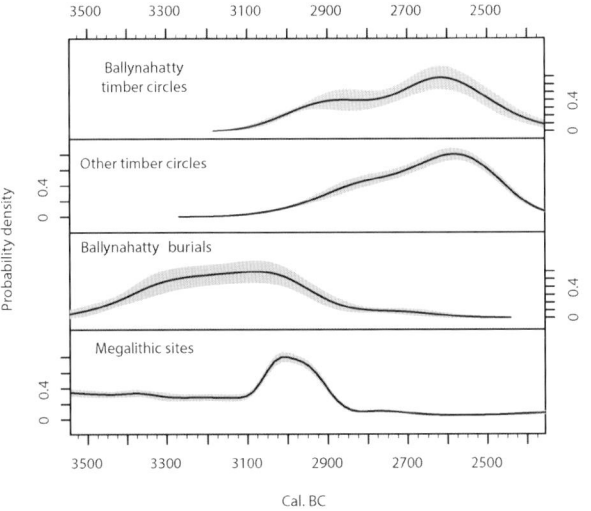

Figure 10.2 Kernel density models (McLaughlin 2019) of activity levels at Ballynahatty, compared to activity level estimates derived from ensembles of radiocarbon data from similar archaeological contexts elsewhere in Ireland. The KDE model of timber circles was calculated using Table 10.3; the model of activity at megalithic sites was calculated as described by McLaughlin (2020). Note the similar dynamic shared by Ballynahatty and the other sites and how the decline in mortuary activity at Ballynahatty occurred at the same time as activity at megalithic sites dwindled.

site was, by this stage, once again the focus of funerary activities. More widely, this was a time of Bronze Age expansion across the British and Irish isles, and in Ireland, there was an identifiable peak in funerary activity around 2000 cal BC (McLaughlin 2020).

The charred grain from the 'southern burning pit' dates to the 5th or 6th centuries AD, the very early medieval period. Density models of similar archaeological finds of cereals from elsewhere in Ireland reveal this to be a time of rapid agricultural expansion matching the proliferation of various kinds of settlements (Hannah and McLaughlin 2019). Other, as yet unexcavated, features at Ballynahatty, including a possible ringfort, likely date to this time.

10.7 Conclusions

Whilst the main phase of activity at the complex is in the 3rd millennium cal BC, the 18 radiocarbon measurements from the site span some 6000 years. The Early Mesolithic, Early Neolithic, Early Bronze Age and early medieval activity detected in the radiocarbon samples from the site all date to identified peaks in human activity in the Irish landscape. Finding these 'early' cultural phases at Ballynahatty illustrates how the archaeological excavation of any site also serves as a sampling exercise capable of detecting signals of large-scale patterns.

For the site itself, what emerges from the analysis of the radiocarbon dates is one of a place of mortuary activity becoming a monumentally enclosed space. This matches the sequence of events at the adjacent Giant's Ring. Bayesian tools help resolve generational change, as there is a distinct possibility that the BNH5 enclosure was built by the children of those who erected BNH6 and that the complex was, in turn, abandoned during the time of their children or grandchildren.

This opens fresh perspectives on how the site was experienced. According to the rule-of-thumb that timber posts will erode at a rate of 3–4 mm each year (Wainwright 1989, 155), posts at the timber circle complex would have lasted about 100–150 years, this fits well with the Bayesian chronology presented here. The more massive 1m posts at the Outer Façade may have stood for 300 years (also cf. Greaney *et al*. 2020). If these survived, the likely burning event at the end of the timber circle could have been extant features in the landscape, along with the Giant's Ring and passage tomb, when the site was once again selected for funerary purposes by a new wave of people. These Bronze Age farmers are now understood to have little or no genetic ancestry with the Neolithic residents of Ballynahatty, but their presence speaks to the enduring prominence of the monumental features of the Neolithic landscape that also includes the nearby passage tomb and Giant's Ring enclosure.

Part 3

What does it all mean?

11

Interpreting the excavation results in the wider context of prehistoric Ballynahatty

11.1 Introduction

This chapter attempts to interpret the excavation results in terms of an overall chronological narrative of activities in prehistoric Ballynahatty, discussing the development of the ceremonial complex and its relationship to other monuments in the wider landscape, particularly the Giant's Ring henge and the passage tomb inside it. The main focus will be the Late Neolithic complex of timber structures, but the first part of the chapter (Sections 11.2–11.5) presents an overview of the evidence for activities spanning the Early Neolithic to the early medieval period.

11.2 Early and Early to Middle Neolithic activity

An early date from the Ballynahatty excavations comes from a radiocarbon charcoal sample taken from the central structure of BNH6, of *3785–3639 cal BC* (Table 10.1; UB-4296). This does not date the structure itself, which is several hundred years later. Instead, it is probably a legacy of an earlier occupation during the Early Neolithic, represented by just a handful of artefacts, all from uninformative contexts. These artefacts consist of four sherds of Early Neolithic Carinated Bowl pottery (discussed in Chapter 6); two leaf-shaped arrowheads, plus a fragment of another, unfinished, and a possible roughout for this kind of arrowhead (Chapter 7); and a fragment of imported Arran pitchstone, probably from a microblade (Chapter 8). There may be other elements of the lithic assemblage that are also of Early Neolithic date: the fragments of porcellanite axeheads and the other stone axeheads, for example, fall into this category. Three of the Carinated Bowl sherds are from the topsoil, while a fourth came from the plough-disturbed uppermost level of the fill of posthole OR 1 in the BNH5 Outer Enclosure (C64) and its presence there will be as a residual object. The leaf-shaped arrowheads and the complete and fragmentary axeheads are mostly from the topsoil, while the pitchstone fragment was found in the upper fill of one of the West–North–West postholes; if its identification as a fragment of a microblade is correct, it must have been residual in its findspot.

These artefacts attest to some kind of activity in the area during the first half of the 4th millennium BC, and it may well be that some of the stone artefacts that have been discovered in the area as stray finds or as a result of systematic fieldwalking (as discussed in Chapter 2) are also of Early Neolithic date. Without associated structural evidence, it is impossible to say whether an Early Neolithic settlement existed in the area, but this is quite possible. It remains to be seen whether the putative cursus monument (BNH16), identified from aerial photographs and mentioned in Chapter 2, is of Early Neolithic date, like its counterparts in Scotland (Thomas 2007a); only excavation could determine that.

One structure that is highly likely to be of Early Neolithic date is the simple passage tomb inside the Giant's Ring henge (BNH2). As discussed in Chapter 1, this survives as five orthostats enclosing a chamber 1.5 m across (Fig. 11.1 (1); Section 5.2). One massive capstone is in place, and another has slipped and now blocks the remains of a short entry passage defined by one surviving orthostat (Plate 85). Harris (1744) reported the existence of a stone kerb (Section 1.2.1), though later commentators have disputed this, and it is unclear whether the monument ever had a mound; there is no surviving evidence that would help in determining this. In 1917 Lawlor excavated beneath the chamber and found burnt bone in an area of modern disturbance (Section 1.3.1). Though incomplete and much abused, this site demonstrates the main attributes of a simple passage tomb: a chamber defined by a series of orthostats and roofed by overlapping capstones, often corbelled. There is a distinct entrance which may be developed into a passage and there may have been a kerb of stones surrounding this. While this monument is undated, it has been argued (Sheridan 1986) that simple passage chambers and closed megalithic chambers in general in Ireland are likely to date to around the beginning of the 4th millennium BC. Support for the view that these are Early Neolithic monuments is offered by the *termini ante quos* radiocarbon dates in the third quarter of the 4th millennium cal BC obtained by dating fragments of bone and antler pins from two such monuments in the Carrowmore cemetery, Co. Sligo (Bergh and Hensey 2013) and by the radiocarbon dating programme of human and animal remains from several larger and more elaborate passage tombs in the Carrowkeel cemetery, Co. Sligo, which has indicated an earliest phase of use in that cemetery (attested in the deposition of burnt red deer antler) of *3641–3382 cal BC* to *3366–3165 cal BC* (Kador *et al.* 2018, 236).

A phase of activity whose chronological position is best described as 'Early to Middle Neolithic' is represented by the 14 'Globular Bowl' sherds (which, as explained in Chapter 6, are likely to represent the kind of modified Carinated Bowl pottery in use between *c.*3600 BC and *c.*3300 BC) and by the flint javelin head. Once again, there may be other lithic material that belongs to this phase of activity. The 'Globular Bowl' sherds were found in the primary and secondary fills of the postholes in the Entrance Chamber and in the fill of a charcoal-rich pit (C16 in Area 45 at the rear of BNH6). Those from the Entrance Chamber appear to belong to a single vessel. It is highly likely that the construction of the Entrance Chamber disturbed the remains of centuries-old activity, and this will account for the apparent association between some 'Globular Bowl' sherds and Grooved Ware in the postholes.

11.2.1 Middle Neolithic activity

This is almost exclusively represented by evidence for funerary practices, mostly associated with the use of heavily decorated Coarse Ware, and consists of the following (including the monument discovered in 1855):

1. The 'subterranean chamber' or compartmented cist-like structure, found in 1855 in the field immediately to the northwest of the Giant's Ring, containing cremated and unburnt bone and Coarse Ware pots (MacAdam and Getty 1855; Fig. 11.1 (2); Section 1.2.2), together with another similar structure found in a neighbouring field earlier in the 19th century but not investigated;
2. The underground chamber, C591, located near the eastern end of the ridge running south of the entrance to the Annexe (Section 5.2; Figs 5.4, 11.1 (3));
3. Four deposits of cremated human bone, in a line and spaced 1 m apart, just 7 m away from C591 (Section 5.2). The best-preserved of these is 'cist' C1014 (Fig. 11.1 (4)), which consisted of a pit with two flat stones on its base, a transverse stone on the northern side, with an upright Coarse Ware bowl containing cremated remains resting on the flat stones. Small split stones provided an edging of symbolic orthostats. A cluster of small stones suggested that the position of the deposit had been marked by a miniature cairn. The other deposits of cremated remains – C1012, C1016 and C1018 – had been badly plough-damaged, but a cluster of small stones survived at C1016, and it is possible that all these deposits had originally been marked by miniature cairns;
4. Two deposits of cremated human bone and (in one case) associated lithics (C588 and C943: Section 5.3), possibly associated with a pit containing just stones (C998), and discovered during the excavation of the Late Neolithic timber enclosure BNH6;
5. Cremated human bone found in C635, the fill of C678 – residual material found in the primary fill of a posthole in the Inner Ring of timber enclosure BNH6;
6. Finds of Coarse Ware pottery and associated stone artefacts in shallow scoops and features (including C931) containing dark soil and split stones in Areas 45 and 46 to the immediate west of BNH6 and close to the edge of the excavated area; there may have been further features in the unexcavated area beyond. These features are heavily truncated, and it is unclear whether they ever contained any human remains;
7. Other finds of Coarse Ware, some in the topsoil and ploughsoil overlying Late Neolithic postholes in Areas 33, 37 and 44–46, and the rest from C876 and C777 in the northern Annexe and from Area 42;
8. Finds of flint hollow scrapers, several found in the vicinity of the much later, early medieval 'Burning Pit' C1453 – and note that some of the petit tranchet derivative arrowheads may be of Middle Neolithic date.

Several radiocarbon dates have been obtained for the Middle Neolithic phase of activity at Ballynahatty, all of them definitely or probably from human remains (Section 10.4, especially Tables 10.1, 10.2 and see Schulting *et al.* 2012). Seven – including one duplicate date and one that was rejected on the grounds of probable contamination by consolidant – were obtained from cremated and unburnt human bone from the 1855 subterranean chamber (*ibid.*), while five more come from calcined bone found during the excavations: in chamber C591; in deposits C1016, C588 and C943; and in C635, the primary fill of one of the Inner Ring postholes of the BNH6 enclosure. Rowan McLaughlin's modelling of all the dates (except the 'reject' date from the 1855 structure) concluded that the peak period of this burial activity ranged between 3300 cal BC and 2900 cal BC (Section 10.4; note that the presence of a plateau in the IntCal20 calibration curve prevents a more precise dating estimate). It may be that there are further examples of residual Middle Neolithic cremated bone among the undated fragments found in the timber complex but only further dating can establish whether this is the case.

With regard to the funerary structures, the predominant mortuary practice is cremation, although in the 1855 compartmented chamber, as explained in Chapters 1 and 5, unburnt human bones were added in a second phase of activity. MacAdam and Getty's account (1855) of this structure describes the 'subterranean' chamber as being radially segmented into six segments around its circumference, stone-lined, and having a corbelled roof and an entrance facing east (Section 1.2.2; Figs 1.4 and 11.1 (1)). The primary activity involved the deposition (probably sequential) of the remains of at least nine cremated individuals, possibly many more, with some of the bones contained within three or four Coarse Ware bowls (see Section 1.2.2.1, for a discussion of the pottery.) One calcined sheep bone was also present among the human remains (Section 9.2.1). The secondary activity involved the deposition of unburnt human and animal remains. Access to the monument, therefore, had to be achieved on several occasions. This is not an uncommon practice and may represent an attempt to promote the legitimacy of occupation by associating the burials of one social group with the ancestral remains of another (Thomas 1991, 40) – although in this case, there may not have been a long interval between the deposition of the cremated and the unburnt bones, according to the radiocarbon dates obtained by Schulting *et al*. (2012 and see Chapter 10, Table 10.2). The chamber was covered by small stones, as though to form a marker cairn, but which allowed access by preventing the entrance from collapsing. MacAdam and Getty (1855) reported that another compartmented cist had been found in a neighbouring field earlier in the century. (Note that MacAdam and Getty also report that 'several' cists were found on Bodel's land in 1855, with a stone slab at the base and another as a lid, being 'shorter than a man' and containing 'urns'; whether these are comparable with 'cist' C1014, or are of Early Bronze Age date, it is impossible to tell.)

As for the funerary features found during the 1990–2000 excavations, while these do not match the 1855 structure in their design, nevertheless they are connected to it by the presence of cremated human bone; by the similarity in radiocarbon dates (as discussed in Chapter 10); and, in some cases, by the presence of Coarse Ware pottery. All the excavated features are situated along the top of the ridge closest to, and in full sight of, the passage tomb (BNH2) inside the Giant's Ring some 250 m to the south. It is tempting to suggest that they, and the 1855 subterranean chamber, may have been inspired in some way by the Early Neolithic passage tomb, which appears to have acted as a nodal point for the subsequent evolution of the ceremonial landscape of Ballynahatty. (See Fig. 11.1 for a comparison of the ground plans of these structures in comparison to that of the Early Neolithic passage tomb.) The type of pottery associated with the 1855 structure and with several of those excavated during 1990–2000 seasons is the same as the pottery found in passage tombs elsewhere in Ireland (and traditionally referred to as 'Carrowkeel Ware').

The variously-shaped Middle Neolithic funerary features at Ballynahatty join others in Ireland that are associated, directly or indirectly, with Coarse Ware or with other artefacts of types found in passage tombs (Brindley and Lanting 1990, 6) but which cannot be described as 'passage tombs' (*i.e*. as megalithic, above-ground monuments with a chamber, passage and, usually, a round cairn). These non-megalithic graves tend not to attract much attention in discussions of the passage tomb tradition (*e.g*. Herity 1974), and yet they are an important element in the narrative of Middle Neolithic funerary practices in Ireland. Moreover, an unexpected additional connection to the passage tomb tradition has recently come to light as a result of the DNA study of

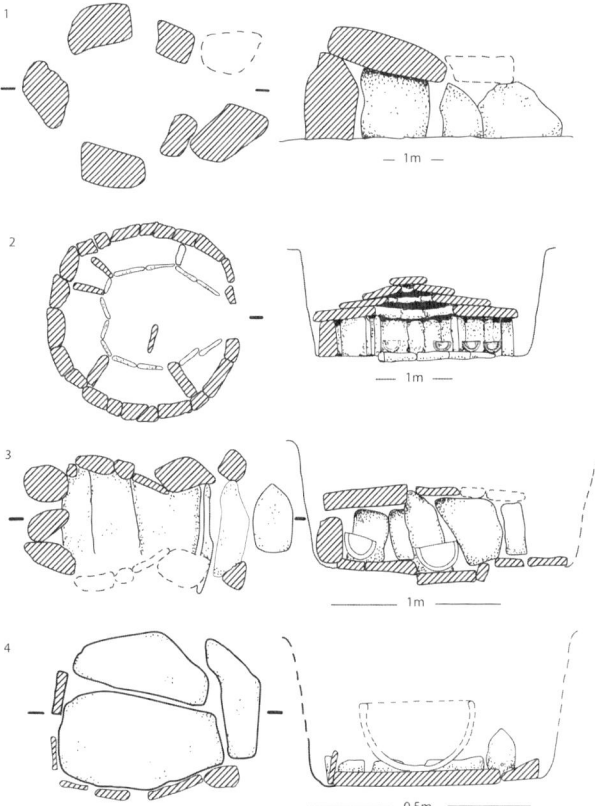

Figure 11.1 Comparative plans and sections through passage tomb BNH2 (1), and related subsurface chambers: 1855 (2), C591 (3) and C1014 (4). (1) is after Collins 1957; (2) section is reconstructed from a description in text and plan by MacAdam (MacAdam and Getty 1855). Scale variable to emphasise the striking resemblance in plan of (3) and (4) to BNH2 despite their difference in size.

Irish Neolithic human remains by Cassidy *et al.* (2020). This has shown that the individuals buried at a long (6 × 1 m) communal stone burial 'cist' surrounded by slabs decorated in a style reminiscent of (but not matching) passage tomb art at Millin Bay, Co. Down (Collins and Waterman 1955; Schulting *et al.* 2012), are genetically linked to much earlier individuals buried in passage tombs at Carrowmore, Carrowkeel and to an individual buried at Newgrange – and form part of a group that is genetically discrete from other Neolithic individuals in Ireland. The term 'monument with passage tomb affinities' has been proposed to describe the 'cist' at Millin Bay and the 1855 'subterranean chamber' at Ballynahatty (Schulting *et al.* 2012), and perhaps 'Middle Neolithic funerary practice with passage tomb affinities' is an appropriate way to group the disparate set of non-megalithic funerary sites in Ireland. A brief review of this evidence is offered here.

At Millin Bay, in addition to the long 'cist' containing disarticulated unburnt human remains and cremated remains, several smaller stone cists were found containing cremated human remains. This invites comparison with the small 'cists' at Ballynahatty. A sherd from a Coarse Ware bowl was found on the ground surface under the mound that covered the Millin Bay cists and a broken flint petit tranchet derivative flint arrowhead was also found. The radiocarbon dates obtained by Schulting *et al.* (2012) from unburnt and cremated bone from Millin Bay suggested a date range for depositions there between *c.* 3380 and 3050 cal BC (*ibid.*, 37) – a range that is affected by the same issues with the radiocarbon calibration curve as that pertaining to the Ballynahatty dates. Also in Co. Down, an unusual monument at Ballynoe that included a boulder-built chamber measuring 1.5 × 0.6 m, set at the east end of an oval mound partly surrounded by a closely-set circle of stones, was found to contain cremated human remains and a sherd of Coarse Ware (Groenman-van Waateringe and Butler 1976).

Elsewhere, at the Mound of the Hostages, Tara, Co. Meath, three small cists containing cremated remains and (between them), Coarse Ware bowls, beads, pendants and pins were found built against the outside of the slabs of a passage tomb (O'Sullivan 2005, 68–79). Further deposits of cremated human remains were found outside the passage tomb, including a series of 17 around the perimeter of the cairn, some in unlined pits, others in pits with a stone setting or protective slabs (*ibid.*, 63). Modelled radiocarbon dates suggest that the cists were erected *3285–3075 cal BC (95% probability)*, probably the 32nd century cal BC, and the perimeter deposits were probably laid down around 3100 cal BC (Bayliss and O'Sullivan 2013, 42–47, fig. 12) – once again, a comparable date range to that for the Middle Neolithic funerary activity at Ballynahatty.

Simple deposits of Middle Neolithic cremated remains, with or without pottery, have been found at Townleyhall A, Co. Louth (Liversage 1960), and inside the henge at Monknewtown, Co. Meath (Sweetman 1976).

At Townleyhall, it appears that the site may have started as a settlement, with a lightly-built house surrounded by a bank and ditch; after burning down the structure, two deposits of cremated remains – one in a small pit and covered by a stone, the other deposited as a spread and associated with Coarse Ware – were deposited and the whole was covered over by a mound. At Monknewtown, a Coarse Ware pot containing cremated human remains was found 'placed in a small shallow depression in the gravel surface close to the tail of the [henge] bank', just inside the bank (Sweetman 1971, 28–29 and pl. iii). Sadly, it is impossible to say whether it was deposited before or after the henge was constructed. This is all the more regrettable since the henge offers a parallel to the Giant's Ring, whose dating is as uncertain as that of Monknewtown. The question of the date of the Giant's Ring will be discussed below.

11.3 Late Neolithic activity: the timber enclosures BNH5 and BNH6

The complex of Late Neolithic structures will be described and discussed in detail below; suffice it to summarise the key aspects of these structures here (Fig. 11.2). The structures consist of an oval enclosure (BNH5) defined by a double ring of posts and with an entrance at the southeast, within which is a 'square-in-circle' structure (BNH6) consisting of a Four-Poster arrangement of large posts, surrounded by a double ring of posts with an entrance to the southeast. In its centre, there is a squarish arrangement of close-set posts that has been interpreted as having supported a platform, perhaps for the laying out of the dead for excarnation. A small trapezoidal (wedge-shaped) setting of posts within the circuit of the BNH5 enclosure has been called the Eastern Setting (ES), and there are indications that another similar structure had faced this so that the two effectively flanked the approach to the enclosure. Opposite the entrance to the BNH5 enclosure is a rectangular structure described as an Entrance Chamber, with four large posts in its interior and entrances at its west and east; the east entrance is flanked by a façade. Additional features in this area are parallel rows of posts running in various directions – east to west (north and south) and northwest to southeast which together define an area labelled the 'Annexe'. The excavated features are only a small part of what must have been a more extensive complex.

There is stratigraphic evidence to suggest that not all the elements were constructed at the same time; the three construction phases are tabulated (see Table 11.2) and described below. It appears that BNH6 and the Eastern Settings were built first (with the Four-Poster in the square-in-circle BNH6 structure being erected before its encircling rings). The oval enclosure BNH5 was then added. These episodes are referred to below as Phase 1. Phase 2 saw the realignment of the entrance of BNH5

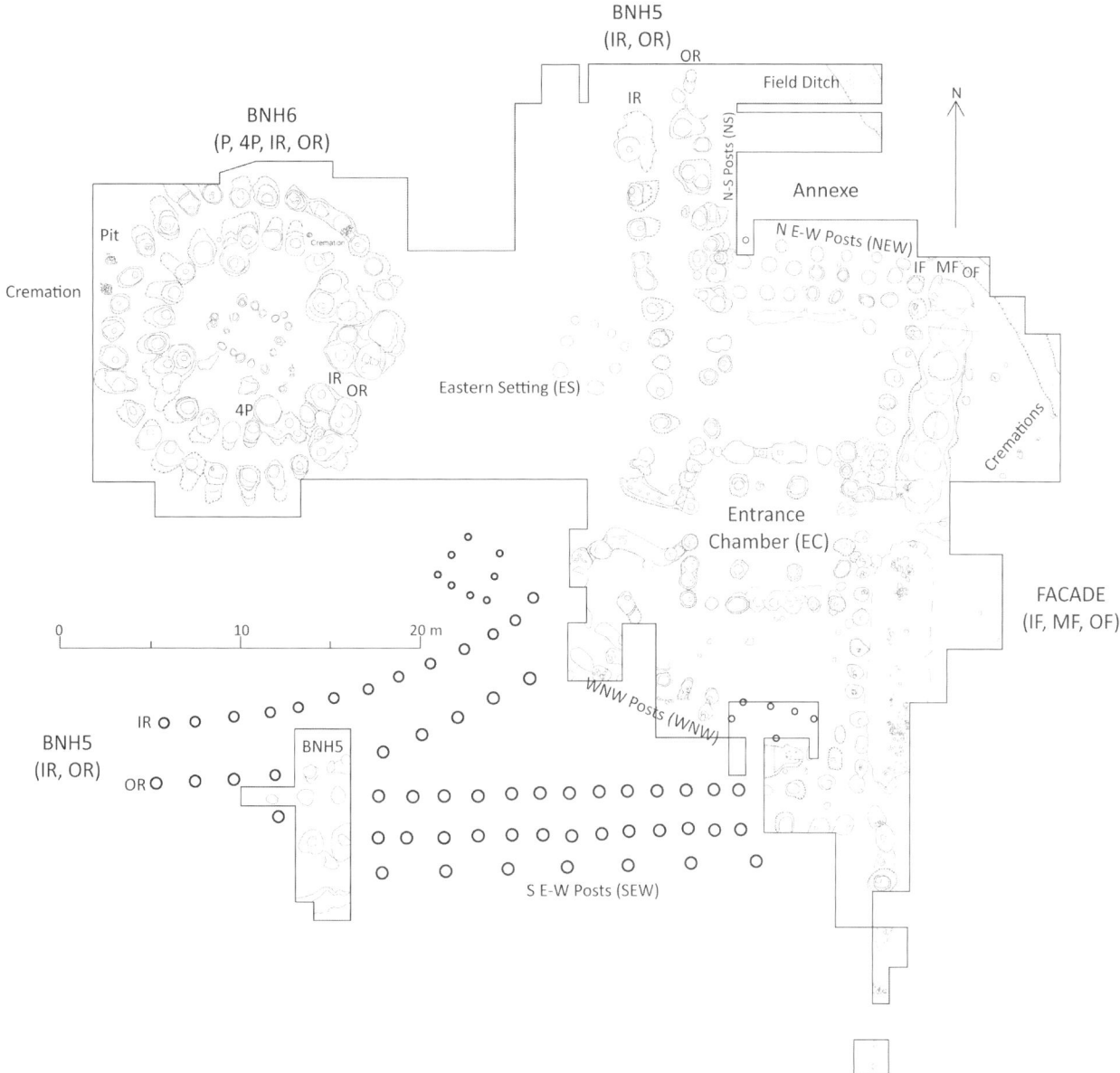

Figure 11.2 Site plan of excavation showing main feature groups. P: Platform; 4P: Four-Poster; BNH6 IR: Inner Ring; BNH6 OR: Outer Ring; BNH5 IR: Inner Ring; BNH5 OR: Outer Ring; NS: North–South Posts; NEW: Northern East–West Posts; WNW: West–North–West Posts; SEW: Southern East–West Posts; IF: Inner Façade; MF: Middle Façade; OF: Outer Façade.

to the east and the addition of a facade and part of the Annexe, while Phase 3 saw the extension of the Annexe, the erection of the Entrance Chamber, the erection of some stand-alone posts and the enhancement of the facade by the addition of a line of more imposing posts. A fourth phase is defined by the deliberate burning down of all the timber structures.

The timeline for these activities, as defined by the modelling of the suite of radiocarbon dates (presented and discussed in Chapter 10), is as follows: BNH6, the 'square-in-circle' structure, has a 95.4% probability of having been erected within the date range *2940–2600 cal BC (2760–2630 cal BC at 68.2%)*. The oval enclosure BNH5 has a 95.4% probability of having been erected within the date range *2700–2510 cal BC (2650–2600 cal BC at 68.2%)*. The end date for activities at BNH5 is estimated to be *2620–2420 cal BC at 95.4% probability, 2610–2490 cal BC at 68.2%*. Taking a minimalist view of the duration of activities, it is possible that neither BNH6 nor BNH5 was used for longer than 120 years and that the main phase of activity in the complex fell within the bracket 2700–2500 cal BC.

As discussed below and in Chapter 13, it is possible that the function of this theatrical landscape of imposing timber structures was to act as the locus for ceremonies connected with the dead – thereby continuing, albeit in a much more monumental fashion, the use of the area for dealing with the dead as seen during the Middle Neolithic.

If the interpretation of the central square setting of postholes in BNH6 is correct then, during Phase 3, the dead would have been brought up to the site, passing through the monumental façade and the Entrance Chamber, then through the entrance to the BNH5 enclosure, then through the double ring of BNH6, to be placed on an exposure platform at the heart of BNH6 for excarnation. What then happened to the defleshed remains – whether they were scattered or cremated – is not known. The scale of the construction work suggests the involvement of many people and it is likely that seasonal ceremonies to commemorate and honour the dead occurred.

The question of whether the Giant's Ring – another structure whose creation would have included a considerable input of communal effort – was already standing when the timber complex was erected is addressed below. (Whether the standing stone at Ballynahatty, or any of the sub-surface features identified through the various kinds of survey undertaken, could have been of Late Neolithic date is another question, but one that cannot be addressed without excavation.)

11.4 Early Bronze Age activity

There are indications, from artefactual finds, radiocarbon dates and palaeo-environmental evidence, that people were active and farming in the area during the Early Bronze Age and, once again, it may well be that their activity was funerary in nature. The artefactual material from the excavations comprises sherds of a pot which, as discussed in Chapter 6, was initially identified by the late Professor Derek Simpson as a Beaker, but which is more likely to be a Vase Food Vessel of ApSimon's 'Irish-Scottish' type (Fig. 6.8); and three sherds that are most likely to belong to a Collared Urn (Fig. 6.9). In addition, the suite of Ballynahatty radiocarbon dates includes one, from a sample of 'charred hazelnut' within the inner ring of BNH6 (C635), of *2290–2130 cal BC* (3762±26 BP, UBA-42751), and another, from calcined human bone found in the southern Annexe (C1544), of *2210–2030 cal BC* (3751±39 BP, UBA-42745).

The Collared Urn sherds came from the ploughsoil overlying the square-in-circle monument BNH6 – the same general area as the dated hazel. The date obtained from the hazel is, however, arguably too early for this material to have been associated with the Collared Urn, for which a currency in Ireland of 1850/1830–1700 cal BC has been proposed by Anna Brindley (2007, 328). There are several finds of Collared Urns in the area around Belfast (*ibid.*, fig. 34), so the presence of one at Ballynahatty is consistent with a broader pattern. As for the probable date of the Vase Food Vessel (whose sherds were found 'broadly associated' with posthole BNH5 OR13, C1711 – the point where the WNW posts join the BNH5 enclosure): according to Brindley's typo-chronology of Irish Early Bronze Age pottery, Vases of this general type (her 'Stage 3' Vases) should date to *c.* 1830–*c.* 1740 cal BC (*ibid.*, 328). Both Collared Urns and Vases are normally found associated with human remains, so it is quite possible that this had been the case with these two pots as well.

Elsewhere in the Ballynahatty area, there is plentiful evidence of funerary activity, and at least some of the many cists, 'urns', and 'vast quantities' of human remains that were reportedly found during the 19th century (reviewed in Section 1.2.3) are highly likely to be of Early Bronze Age date. The aerial imagery has shown a number of ring ditches associated with graves across the landscape, including a cluster of four (BNH9–12, Fig. 2.6). These probably date to the Bronze Age, although not necessarily to the Early Bronze Age (Fig. 4.1, Table 4.1), The aerial photographs also show many more marks produced by human interference in the soil, but these are largely uninterpretable and undatable.

11.5 Early medieval activity

This is represented by the remains of the suspected ringfort (BNH 3), by the Burning Pit C1452 (Section 5.7), and possibly also by the tiny beads, suspected to be of glass (Chapter 8), that were found in Late Neolithic contexts (the postholes in the northwest part of BNH6) and which may have arrived there as a result of bioturbation. Cremated human bone, along with burnt animal bone and two small beads (one of glass, one possibly of faience), was found in the 'Southern Burning Pit', suggesting that it was connected to some kind of funerary activity – possibly even *in situ* cremation – and a sample of charred grain from the pit produced a radiocarbon date of *cal AD 430–595* (1544±28 BP, UBA-42750).

The early medieval period was a time of agricultural expansion and a number of ringforts were constructed in the Lagan Valley. There are at least three in a 1.5 km radius of the BNH3 ringfort (see Plate 2, Figs 1.4, 2.2), which sits at the western end of the ridge. Burnt grain found in the ditch of BNH3 suggests that this was a farming settlement. The presence of human remains in the Burning Pit extends the period of mortuary use of the Ballynahatty landscape to the 5th/6th centuries AD.

11.6 Further consideration of the Late Neolithic timber monument complex

The rest of this chapter focuses on the construction, use, meaning and significance of the Late Neolithic timber monument complex.

11.6.1 *The scale of construction*

In its final form, the timber monument complex at Ballynahatty contained an estimated 471 posts. Species identification from the charcoal shows that oak predominated, especially in the inner enclosure (BNH6) and the Annexe, although hazel was present in the Platform postholes and

Table 11.1 Species identifications of charcoal.

Context	Feature group	Identification
3	BNH5 I	Alder (*Alnus*) × 3
48	BNH5 I	Birch (*Betula*) × 2
24	BNH5 O	Birch (*Betula*) ×1
57	BNH6 I	Oak (*Quercus*) × 6
126	BNH6 O	Oak (*Quercus*) × 6
79	BNH6/Four-Poster	Oak (*Quercus* × 1
X1	Platform	Hazel (*Corylus avellana*) × 3
4	Eastern Setting	Birch (*Betula*) × 3
27	North–South Posts	Birch (*Betula*) × 2
134	Northern Annexe	Oak (*Quercus*) × 6

Note: charcoal was recovered from all features but not all of it was identifiable. Dominance of any one species in a particular feature group cannot be assumed and may represent infill rather than the posts themselves. Identification by David Brown, Palaeoecology, Queen's University Belfast.

birch in the Eastern Setting and N–S postholes. The Outer Enclosure (BNH5) contained birch and alder (Table 11.1).

Assessing the scale of this enterprise is fraught with difficulties. The variables, such as estimating the density of prehistoric woodland and the catchment area for the timber supplies, will only ever be an informed guess. However, with this caveat, it is worth attempting to give an impression of what was achieved. The depth of the postholes and the size of the ramps indicate tall and substantial posts. Their above-ground height will always be problematic but it can be reasonably estimated as three times the depth of the posthole. Post lengths vary from 3.00 m to 9.00 m, and diameters from 0.15 m to 1.00 m+. This would have entailed felling and trimming over 470 trees using stone-bladed axes and transporting an estimated 550 tonnes of timber to the site. The largest – the posts of the Outer Facade – could individually have weighed over 5 tonnes.

Stands of such trees with long straight poles would have been readily available for felling in the Lagan Valley and would have been ideal candidates for the many postholes of BNH5 and 6 (Plate 86). Although some would have been available locally, logs may still have had to be brought some distance and may even have been floated down the Lagan. All this would have had a considerable impact on the local environment. Moving the timber would have created its own problem but draught animals were available. An investigation of nearly 4000 cattle bones from the Middle Neolithic site of Kilshane, Co. Dublin, has shown that a small proportion were kept beyond the normal age for slaughter to between 40 months and 20 years. Some of the larger males were possibly oxen and showed pathological changes consistent with their use for traction. Such animals would have been essential to move heavy logs out of the forest and bring on site (Pigière and Smyth 2023). Evidence of cattle was found in both Coarse Ware and Grooved Ware contexts at Ballynahatty.

11.6.2 Working to plan: Phasing the build

Four phases in the sequence are discernible (excluding a possible preliminary, single ring of posts at BNH6) with an overall trend towards aggrandisement, control of movement within and outside the complex and, presumably, the elaboration of rituals (Table 11.2). Although the construction of 1a and 1b may have been separated in time, they are part of a unified concept and so are bracketed together. It is possible, with reasonable certainty, to establish a sequence of construction and, ultimately, decommissioning at Ballynahatty. More difficult is determining its period of use, when additions were made and if any of the feature groups in the sequence were deconstructed before its final immolation.

The earliest focus of activity in the complex seems to have been at the BNH6 enclosure. BNH6 is a double concentric circle of posts, 16 m in overall diameter, set in 2 m deep postholes with radiating ramps. The Outer Ring has been flattened on either side of the entrance to create an impressive façade. The paucity of linking stratigraphy has already been discussed, although it is possible to identify sequential steps during the construction. McLaughlin's Bayesian analysis of the radiocarbon dates from Ballynahatty (Table 10.3) confirms the priority of BNH6 in the sequence and, at highest probability (95.4%), activity could have commenced at 2940 cal BC and finished with the decommissioning of BNH5/6 by 2420 cal BC. However, given the longevity of earthfast timber, the most likely scenario is that the site was used for less than 120 years within the range of 2700–2500 cal BC. The evidence from the postholes tends to confirm this, with most of the posts having been completely removed, some with traces of the lower post-butt in place, and some having completely rotted *in situ*. Larger posts would have lasted longer than smaller ones, as would those erected in the latest phase. Comparison of the later NEW Annexe

Table 11.2 Construction phases represented at Ballynahatty.

Phase	Activity
Phase 1a	BNH6, Eastern Settings & entry segment
Phase 1b	BNH5 enclosing BNH6 & with the SE entrance based on entry segment
Phase 2	Annexe defined by Inner & Middle posts (NEW, WNW & Façade) with entrance realigned to the east
Phase 3	Extension of Annexe to the south (SEW) & outer stand-alone posts, including outer line of Grand Façade & insertion of Entrance Chamber
Phase 4	'Cremation' of the complex, deconstruction, and memorialising

postholes with those of the original BNH6 entrance reinforces this point. It is not possible to ascribe specific dates to these phases but, in Figure 10.3, McLaughlin speculates that if BNH6 was started by the first generation of builders, BNH5 could have been started by the second, and the fourth generation decommissioned the complex. Its foundation and history would have been well within the collective memory of the fourth generation.

11.6.3 Phase 1a BNH6: the planning and sequence of construction

It has been proposed that there was an initial single timber circle based on the surviving six postholes between the Inner and Outer Ring of BNH6 on either side of the entrance (see Section 5.4.1), possibly similar to that found at Knowth (Eogan and Roche 1997, 101–123). If so, this was then developed into the complex double-row building BNH6, described here as Phase 1a.

11.6.3.1 Planning the circle

The timber circle was not any random arrangement of posts in circles but conceptually sophisticated architecture, elegant in its simplicity and executed with remarkable precision to a standardised unit of measurement. There was a clear intent to control movement and carefully arrange viewpoints for dramatic visual impact. Given the size of the postholes and the posts they supported, the precision and order with which this was achieved is remarkable. The plan of BNH6, and there must originally have been one, was based on two conjoined circles centred on points A and B (Fig. 11.3) of a central axis C–A–B. The main feature groups are arranged on a series of regularly spaced notional concentric circles around A (A1–7). The four corners of the central group of posts are laid on A1, and the Four-Poster structure on A3. The Inner Ring is on A4, the northern deposit of cremated remains on A5 and the Outer Ring on A6. The distance between the circles approximates to the distance between the posts along the circumference. Point C lies on circle A7. This was a small pit filled with split stones. From here, arcs could be drawn (C1) to set out the flattened façade of the eight outer posts of BNH6. Any of the central posts on one side could be matched by the arc intersecting the same circle on the other side, although these could also be placed by the projection of the diagonals through the central square.

A second circle (B3) of the same diameter as A7 lies on the C–A axis at centre B. Four posts are set out on either side of B3, positioned by another arc projected from C (C2), which passes through B and intersects with the outer circle at points D. Lines from these two points (D1 and D2) passing through B to the other side of the circle create two segmented structures with posts laid out along two further, similarly spaced, circles, B1 and B2 – the Eastern Settings. Extending the C–A–B axis to the east, an open-ended segment is created on the circles B1–B3,

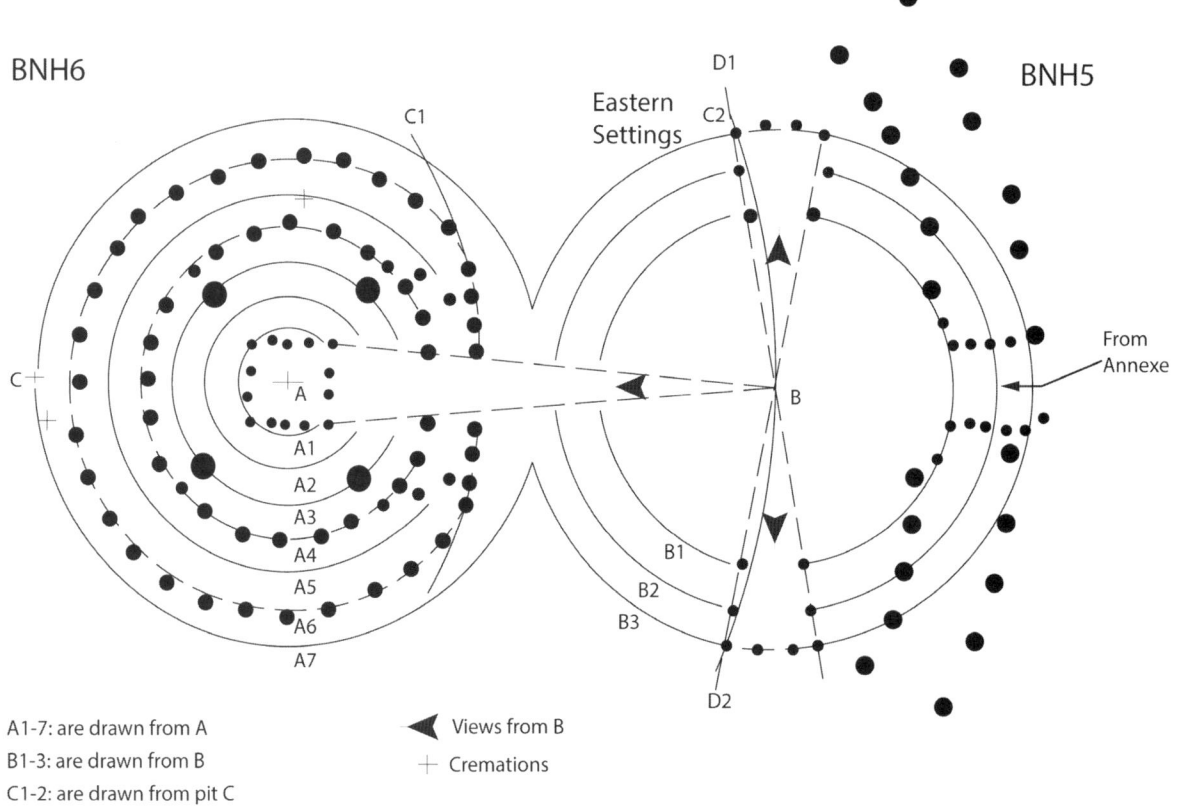

Figure 11.3 Construction plan of BNH6 and 5.

which marks the direction of approach along the axial line to the inner enclosure BNH6 and becomes the entry point through the outer enclosure, BNH5. This establishes the most important position in the whole complex – point B. Because of their segmental shape, the Eastern Settings have a narrow entrance and a wider rear wall. Visibility of the interior is, therefore, restricted, and a full view of both is only available from point B. Looking forward into BNH6, B is the first point at which the entirety of the central platform can be seen through the entrance facade. Working to a plan, all this could have been set out with a measuring stick, two stakes and a length of rope. Because the arc C2 could not have been made with the posts of BNH6 in place, the Eastern Settings must have been part of the original concept and construction.

Immediately within BNH6 were four taller posts – the Four-Poster structure – arranged in a square. It is difficult to see what structural benefit these would have given other than as markers for the position of a significant activity which could be located, but not seen, from outside the enclosure. Understanding this activity, which was the focus of the entire structure, depends on the interpretation of the 14 central postholes.

This structure had a distinct orientation, being formed by two opposing sides of four postholes (EP5–8, EP1 and 12–14) facing the entrance and the two other sides of five postholes (EP1–5 and EP8–12). There is no clear entrance, and if there was a narrow central door on the side facing the entrance to BNH6 (between EP13 and 14), as could be expected in a plan which displays a strong central axis, the rear wall might be expected to have the five-post organisation of the other sides. It could be argued that the central posts (EP3 and 10) of the 'five hole' sides supported a ridge beam with rafters running down to a wall plate carried on each of the 'four hole' sides. However, there is no variation in the depth of postholes, and a deeper hole would be expected to stabilise the taller ridge posts necessary to carry the weight of a roof. Rather than a hut, it is much more likely that the regular height of these posts supported a platform (Fig. 11.4).

11.6.3.2 Constructing the circle

Most of the postholes were dug discretely into the subsoil, but there is considerable overlapping of the cuts at the Four-Poster postholes and the entrance (see Fig 5.19). Figure 11.5 suggests the order in which the postholes in this section of BNH6 were dug, with the Inner Ring of holes after 4P2. It is probable that the holes of the Inner Ring were not dug consecutively but alternately, which would have allowed more room to dig the hole, raise each post, and backfill.

A similar sequence is found at the other postholes of the Four-Poster structure. In the north, the ramp of 4P3 is cut by IR17, which also clips its primary fill (Fig. 11.6). The relationship with IR16 is more ambiguous because of difficulties in identifying respective fills. In the east, the primary fill of 4P4 is cut by IP21 and IP22. Finally,

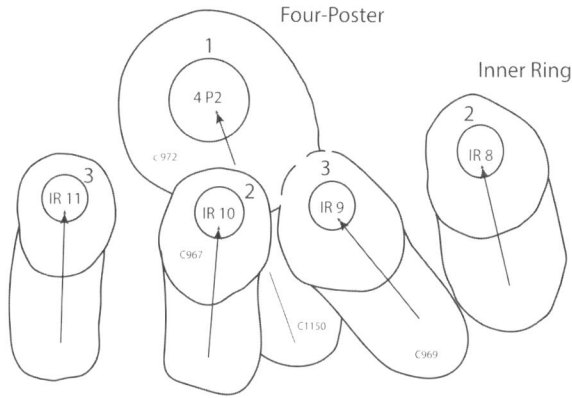

Figure 11.5 Suggested construction sequence (1,2,2,3,3) of 4P2 and IR8–11. Arrows show the direction in which the posts were erected.

Figure 11.4 Reconstruction of the central post setting as a platform. Posts EP5-8 on the northwest side, and EP1 and 12–14 on the southeast side carry the four supporting beams for the platform itself.

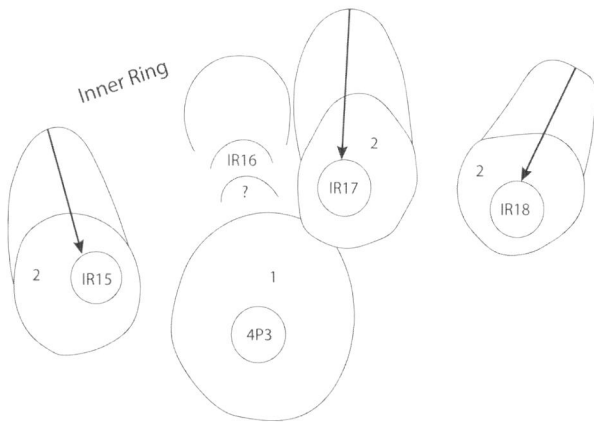

Figure 11.6 Possible construction order of postholes BNH6 4P3 and the Inner Ring postholes IR15–18.

in the south, IR4 and IR5 both cut into the ramp of 4P1, and IR3 cuts both IR4 and the ramp (Fig. 5.19).

The Four-Poster structure was, therefore, erected before the Inner Ring of BNH6. The ramps for these four large posts, like those of the BNH6 rings, radiate outwards, and the ramp cuts of the Four-Poster posts were cut through when the Inner Ring posts were erected. After their removal, the position of the posts was marked by boulders and cobbles. Initially, the quantity of stone in the upper levels of this secondary fill must have intentionally formed a surface cairn, but this progressively collapsed into the hole as the contents gradually settled or were subsequently removed by later cultivation. Burnt pig bone was recovered from the secondary fill of C453.

The orientation of the ramps on the Inner Ring shows that construction would not have been possible if the Outer Ring of posts was already in place. At the southern side of the entrance (Fig. 11.7), IR1 was cut by OR1; therefore, the Inner Ring preceded the Outer Ring. In the Inner Ring, IR1 was cut by IR2, and a slot suggests planking between them. IR4 is cut by IR3, which is probably later than IR2.

The Outer Ring also shows how the digging of the postholes and erection of posts was achieved alternately, with OR1 and OR3 completed and stable before OR2 and OR4 were dug out. Evidence of slots at the surface of the posthole again suggests a planked infill. Figure 11.7 also shows the relationship with the intermediate posts (A), which are earlier than the Inner and Outer Rings and, as already noted, may belong to an earlier phase of building. Comparison to the recently discovered Four-Poster Enclosure at Newgrange suggests that they may be an integral part of the building (as discussed below, Section 11.8.1).

Of the Four-Poster postholes, 4P2 was cut by IR9 and IR10 (Fig. 11.5), and 4P3 was cut by IR17 and probably IR16 (Fig. 11.6), so they preceded the Inner Ring. At 4P2, it was possible to see a sequence in the construction of the Inner Ring, with alternate postholes probably being dug to facilitate access. This can also be seen where IR3 is dug after IR2 and IR4 and in the Outer Ring, where OR2 and OR4 are dug after OR1 and OR3. The sequence of construction at BNH6 was, therefore:

1. A possible earlier single (inner) ring based on the diagonal intermediate posts;
2. The platform;
3. The Four-Poster structure;
4. The Inner Ring;
5. The Outer Ring and façade

With the exception of 1, this was probably a single, continuous building operation.

11.6.3.3 BNH6: to roof or not to roof?

Whether or not timber circles were roofed is an issue that has received considerable attention and has arguably not yet been resolved (Musson 1971; Gibson 1992; 2005; Bourke 1997). The presence or absence of a roof, however, would seem a fundamental factor in the function of the sites and would certainly be linked to the nature of the ritual performed within; it would seem primary to the experience of the monument and perhaps would even preclude them from functioning in the same way. The problem arises because little evidence to support or deny the presence of a roof is ever found. The reconstruction of remains often comes down to speculation and personal preference (Gibson 2005, 140). The problems are well illustrated by the excavation of the timber circles at Knowth and Ballynahatty. There are similarities between their excavated remains which are likely to be of a similar date. The BNH6 circle is larger than the Knowth structure, which has a maximum diameter of just over 9 m, and Ballynahatty is arguably more elaborate. At Knowth, the presence of a roof is favoured, based on the posthole arrangement (Bourke 1997, 293; Fig. 11.8).

Alternatively, BNH6 argues the case for an unroofed structure (Hartwell 1998, 40; 2002, 528; Fig. 11.9). No evidence for roofing was recovered during the excavation, but this does not conclusively prove that it had never existed (Hartwell 1998, 40; 2002, 528). However, the structure does not lend itself to being roofed. If it were, each rafter would need to be at least 10 m long to span the 8 m distance from the Outer Ring to the centre and give a sufficient pitch to the roof for water runoff. The combined weight of rafters and thatch would be unsustainable without support and would have required a central post and/or four or more posts in a midway position. At Ballynahatty, the central square setting of 14 posts would be unnecessary as a support for rafters, would have impeded any other central structure, and removed four square metres of space from use. The positioning of the Four-Poster posts so close to the Inner Circle of posts excludes them as a supportive element and would actually impede a roofed structure.

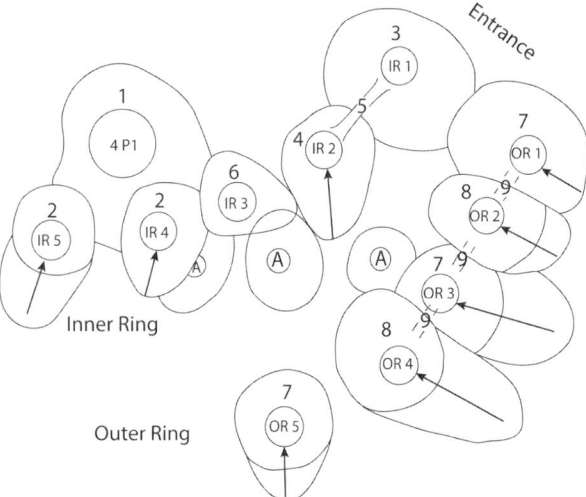

Figure 11.7 Postholes of BNH6 south entrance showing a possible construction order. In practice, all the Inner Ring holes would have been dug before the outer ones. The three intermediate postholes (A) are earlier.

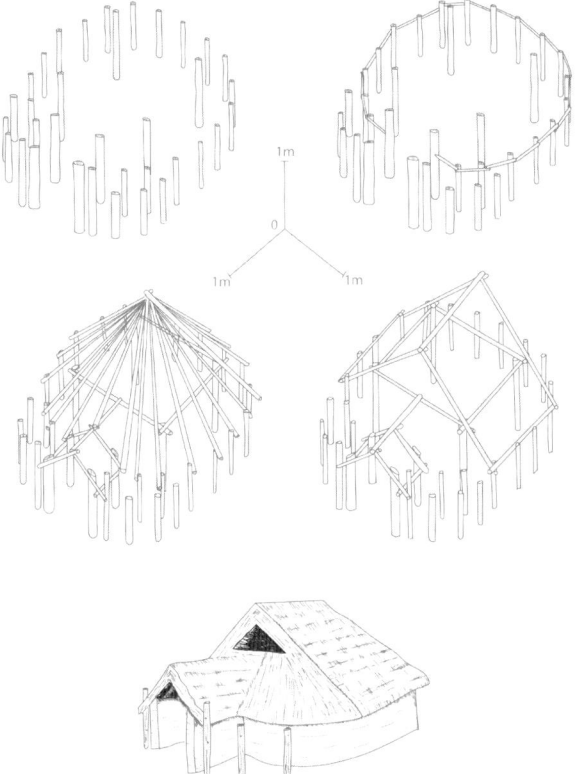

Figure 11.8 Reconstructions of Knowth timber circle. The fourth example is favoured by Bourke and is also shown in the artist's impression (after Bourke 1997, 287–288, fig. 1-2).

If the interpretation of a central platform is accepted, this also rules out a roofed structure on the basis of function.

The preference for unroofed structures appears to have been prevalent at other excavated timber circle sites. At the early excavation at Woodhenge, Wiltshire, the excavator proposed that the circle was not roofed and drew parallels with Stonehenge (Gibson 2005, 129–130). Cunnington (1929) also felt, despite inconclusive evidence either way, that the remains excavated at the Sanctuary, Wiltshire, were unroofed, although it has been argued elsewhere that the structure may have been roofed (Pollard 1992, 216). The excavator of Arminghall, Norfolk, also favoured an unroofed interpretation (Clark 1936, 13). Musson, having studied the evidence for the two circles at Durrington Walls, Wiltshire and the remains from the Sanctuary and Woodhenge, found it difficult to see a convincing developmental sequence (Musson 1971, 377) and concluded that the excavated remains probably represented freestanding posts, perhaps with lintels (Mercer 1982, 157; Richards and Thomas 1984, 197; Fig. 11.10). At Sarn-y-bryn-caled, Powys (Gibson 1994) and Balfarg, Fife (Mercer 1982, 153), unroofed interpretations have also been favoured. The span of a roof at Sarn-y-bryn-caled is likely to have been too great for the size of the upright posts (Gibson 1994, 211). The reconstruction of that timber circle demonstrated, however, the need for some horizontal element to add visual order (*ibid.*; fig. 3.46).

11.6.4 Phase 1b BNH5: enclosing the enclosure

As already discussed, the entrance segment is demonstrably formed on the C–A–B axis at the eastern side of the second structural circle (B) on which the Eastern Setting sits. This formal, rigid layout implies that any other approach is blocked. The BNH5 enclosure strains to include the northern Eastern Setting but swings round to incorporate seamlessly the entrance segment. BNH5 has therefore been built to surround and encapsulate the existing BNH6 and is later in the building sequence but probably part of the same phase. It functions to restrict access and to focus the formal approach on the entrance segment. Morphologically, BNH6, the Eastern Settings, the entrance segment and the BNH5 enclosure can be linked as a single construction event.

Structurally and functionally, the perimeter posts of both enclosures are identical. The outer enclosure, BNH5, defines a space nearly 100 m long by 70 m wide, defined by *c.* 470 posts arranged in pairs. There is a second possible break in the wall to the north, shown by a slight irregularity. The small kettle lake, Ballynahatty Bog, is downslope from here. The south side of the enclosure sits on the southern edge of the ridge, deliberately placed to emphasise its height and presence from below. The structuring of access, the careful choreographing of movement, and the restriction of sightlines, which are so clear in the design of the building, only make sense if the enclosure walls provided a solid barrier as opposed to being perforated with gaps and views between each free-standing post.

The remains of charcoal-filled horizontal slots and a charred wood plank were recovered between the two entrance posts of the Inner Ring and the four entrance posts of the Outer Ring of BNH6 on either side of the entrance. This suggests that there may have been planked walling between posts giving a clearer, more structured focus to the entrance area. It is also possible that lintels capped the posts of the inner and outer circles, giving emphasis and definition to the circular shape, which may otherwise have appeared ragged. Without lintels or infill, the structure would have appeared simply as a mass of posts, as demonstrated by the reconstruction of the Sarn-y-bryn-caled timber circle (Fig 11.10). However, aerial structures, such as horizontal bracing between the posts or lintels, would leave no trace (Hartwell 2002, 528; Gibson 2005, 138).

The carefully structured entrances, spaces and sight lines only make sense if the enclosures presented a solid wall. Analysis of the entrances at Ballynahatty shows that, in each case, the inner and outer rings functioned as a single unit and not as a replacement or enlargement. A simple wattle fence could suffice for infill but would not necessitate such substantial supporting posts or match their durability. Planking or split posts are possibilities, as planks had been used on the façade, but neither of these would require a double fence. Planking could have

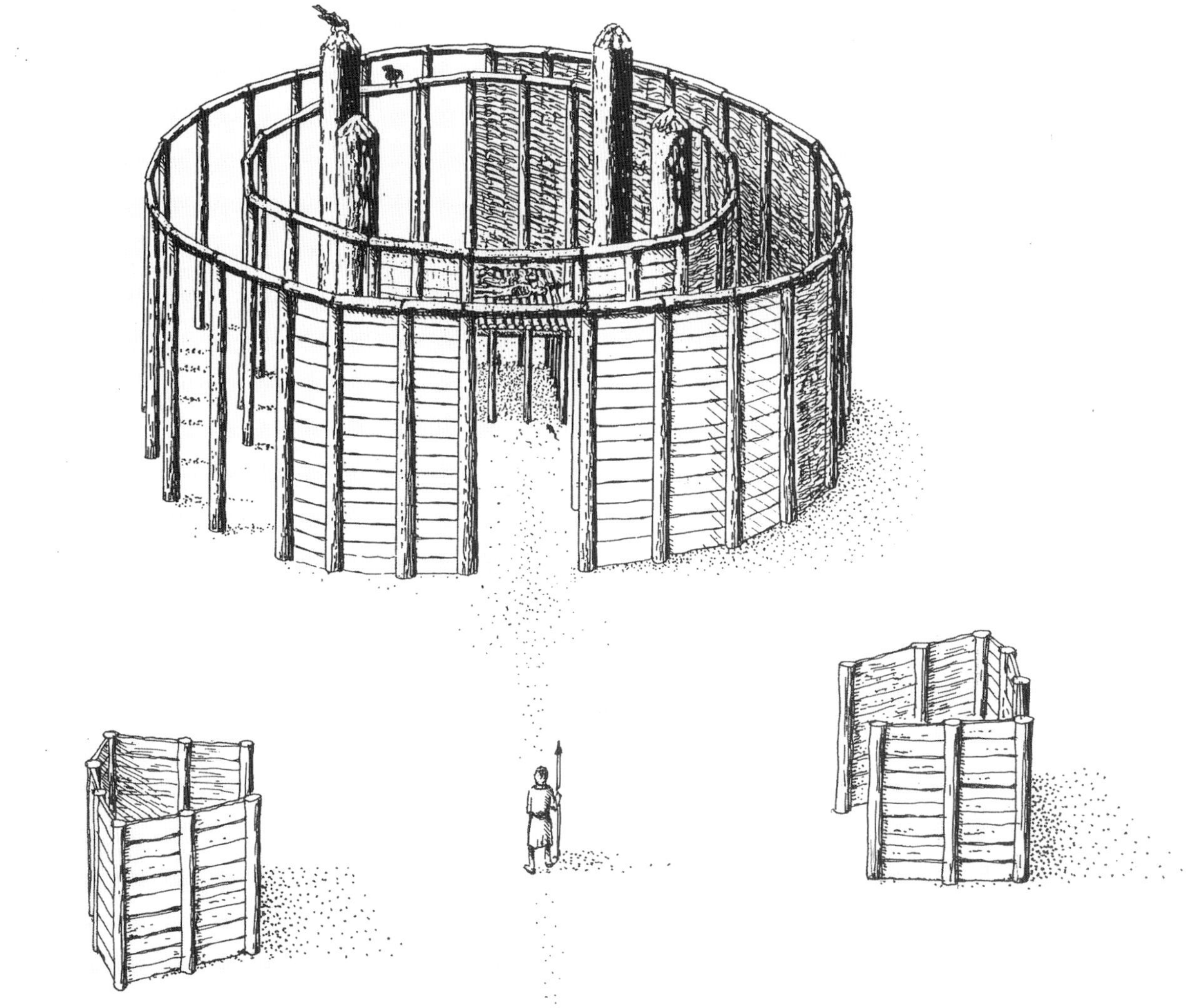

Figure 11.9 An early reconstruction of BNH6 showing the effect of infilling with wattlework panels and planking (from Hartwell 1998, 41, fig. 3.5).

Figure 11.10 Reconstruction of Sarn-y-bryn-caled (after Gibson 1998a, 109, figs 94–95).

been used to retain an earth wall, but the amount of soil required would necessitate a ditch or at least a continuous shallow quarry 'scoop'. There is no evidence for this or, indeed, a bank. What would satisfy the purpose of exclusion and longevity and require a double row is if the posts functioned as retainers for a wood infill. The most available source and the most elegant solution for this 'sandwich' would have been the branches and trimmings of the trees themselves. Analysis of a felled mature oak from the Blenheim Palace estate, Oxfordshire, showed that the branchwood from beyond the straight timber trunk represented 58% of the volume of the tree (Sylva.org). The pairing of posts and repetition every 1.4 m (up to 2 m in BNH5) would provide ample wood for infill. In such a way, an impenetrable barrier could readily have been erected (Fig. 11.11). The heavy weight of the branches would require strong and secure earth-fast posts to prevent splaying, especially if the intended height was 6 m.

Evidence for possible infilling between posts has been found at other timber circle sites across Ireland and Britain. Dunlop and Barclay (2016, 40 and 42) propose that the unique four-post pits that form the imposing outer ring (Circle A) at Armalughey, Co Tyrone, were designed to hold 'horizontal timbers or planking, creating a solid wall

Figure 11.11 Timber wall: an impenetrable barrier created by infilling branches and trimmings between upright posts.

around the structure'. Gibson (1994) has suggested that Sarn-y-bryn-caled may have had wattle and daub panels. Carbonised planking from both Machrie Moor I, Arran and North Mains, Perth and Kinross may indicate that screens or barriers were erected between the posts (Gibson 2005, 114). At Bleasdale in Lancashire, the presence of birch poles laid parallel to each other in a narrow trench may represent the remains of a fence or screen which was supported by the larger upright posts (*ibid.*, 153), and it has been suggested that reeds could have been used as screening material at Woodhenge (*ibid.*, 112). As with the evidence for roofing, the indicators for infilling are not conclusive.

11.6.5 Phases 2 and 3, Annexe and entrance: impress and control

Phase 2 saw the construction of the Annexe, which is really an elaborate entrance structure defined by the Northern East–West posts (excluding NEW1–6), West–North–West, Inner and Middle Facade feature groups. The paired postholes were similar in form to those of the inner and outer enclosure walls but were not as deep (at least 1.40 m) and at closer intervals (*c.* 1.40 m). The distance between the paired posts was more variable (1.00–2.00 m), but, importantly, the NEW and WNW posts articulated with the existing Outer Ring posts of BNH5. This was a functional join confining entry into the Annexe to the main entrance through the Inner and Middle Facade. The entry was lined between the Inner and Middle Façade posts by at least three vertical posts in a slot. On the clearer north side, this is shown by EC34–35b (Figs 5.38 and 5.42), although the eastern end of the slot had been cut by the secondary fill of the Outer Facade. This produced a wide entrance nearly 4.00 m across into an unrestricted Annexe interior. The paired posts would have had a fill similar to that in the main enclosures BNH5 and 6, producing a physical and visual barrier to the interior, although at 4.00 m, not as high. The precision with which this was accomplished was of a more variable standard than BNH6 and suggests that this was neither in the original plan nor executed by the same builders. This is clearly a transitionary precinct for activities/rituals separate from the outside world and preparatory for entry into the main enclosure. This is where most of the pottery was found and evidence of scattered hearths.

The most distinctive feature of this phase is the change in the orientation of the complex. The Phase 1 entrance through BNH5 was from a southeasterly direction, and although this is retained, the Annexe allows a new outer entrance facing east along the line of the ridge. The main approach towards BNH5 changes from non-specific to strongly directional in Phase 2 with movement up the ridge from the east. This change is reflected in the shape of the Annexe with the northern limit, the NEW posts, orientated east–west, whereas at the southern limit, the WNW posts are orientated towards the southeast.

Phase 3 saw the development of the Annexe to produce a grandiose architectural statement designed further to control movement through this space. A straight wall of paired posts, similar in construction to the BNH5 and BNH6 enclosures, was constructed at the south end of the Annexe running east–west to join the southern tip of the Façade to the BNH5 wall (the SEW posts). This structure does not fully join the enclosure wall at the west end and was probably cosmetic, increasing the apparent length of the BNH5 enclosure when viewed from the henge and by people moving through the lower ground between BNH5 and the embankment. The final embellishment was a line of stand-alone posts, arranged in four groups of seven, immediately outside the Annexe walls. These are the NEW1–6 posts (with an additional post seen in the excavation baulk), the outer SEW posts seen in the aerial photographs and the seven posts on each side of the entrance forming the Outer Façade. The north and south lines were of similar height to the walls they fronted, whereas the Outer Façade posts were the largest in the whole complex. These posts measured about 1.00 m in diameter, and the depths of the postholes indicated a height of as much as 6.00 m. The importance of this grand façade cannot be overstated; neither can the difficulties of acquiring, erecting and removing such large, 5–6 tonne logs. The break of slope at the eastern end of the ridge occurs a few metres from the Main Façade, and when viewed from below, the façade would have risen majestically and impressively on the skyline. The façade, when viewed from the outside, is slightly concave. Because of the care with which the site was laid out, this is unlikely to have been accidental, and the aim may have been to provide more focus on the central entrance and passage. The alignment of the façade was important because it is orientated on the passage tomb, showing that, even at this late stage in the development of the complex, the founding tomb was a controlling presence (see Plate 36). The north end of the façade is downslope, and the tomb cannot be seen from this position, but it comes into view once the entrance is reached on the brow of the ridge. It was probably at this time, if not in the previous phase, that an extension was created at the southern end of the Middle Façade designed to draw a sight line from the façade down to the passage tomb to emphasise this continuing link. Conversely, from the passage tomb, the actual Annexe entrance could not be seen, although a procession toiling up the slope would be visible, only to disappear suddenly through the façade into the Annexe.

At the bottom of the slope is a depression, which occasionally holds water, although it tends to be ploughed over today. It remains to be investigated but could be a stage in the ceremonial progression through the area. The final ascent was then enhanced by the towering glory of the Outer Façade and the climactic entry into the complex.

Considerable energy was expended on the entrance façade. There is no exact parallel for the excavated Annexe area at Ballynahatty, but similar features have been uncovered elsewhere. Both Armalughey, Co. Tyrone, and Glenshane, Co. Derry, have entrance enclosures. The entrance façade at Armalughey is of particular note as it creates a distinct 'funnel-shaped' entrance into the timber circles (Dunlop and Barkley 2016, 43). As at Ballynahatty, the postholes of this entrance façade were massive, measuring between 1.00 m and 1.50 m in diameter, and they were sunk up to 0.80 m in the ground (*ibid.*). Similarly, dramatic entrances have been noted at a number of British sites. At Avebury in Wiltshire, for example, the large entrance stones mark the henge entrance (Plate 87). It also appears that access to timber circles was often not direct, and entranceways and access routes were elaborate (Thomas 1999, 56; Gibson 2005, 107–112).

A façade was uncovered in association with Durrington Walls's southern circle (Wainwright and Longworth 1971, 27; Fig. 11.12), and at this site, the northern circle was approached by a timber avenue, and again through a façade (*ibid.*, 42; Fig. 11.13). At the component timber circle site, excavated at Lugg, Co. Dublin, a timber avenue approached the circle (Kilbride-Jones 1951, 322). An annexe-type structure was also found associated with structure 2 within palisade 2 at West Kennet, Wiltshire (Whittle 1997, 77; fig. 3.43). This may have had a similar

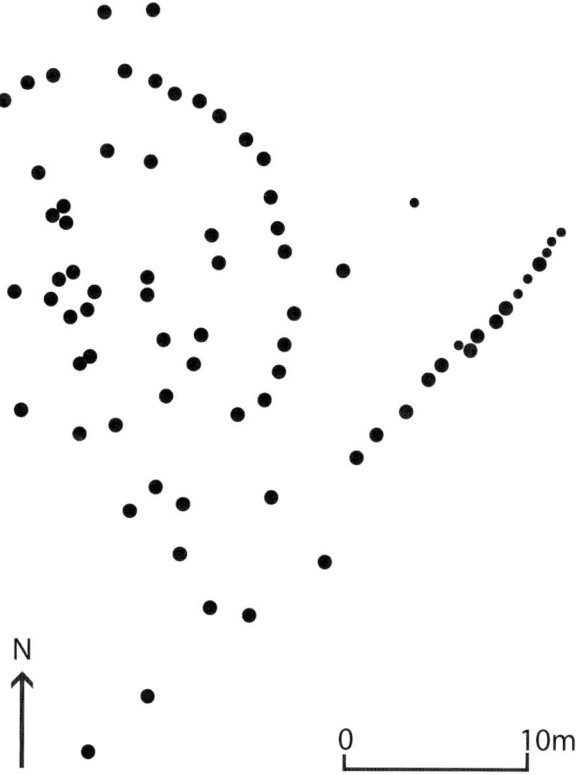

Figure 11.12 Durrington Walls, Southern Circle, Phase 1 (after Wainwright and Longworth 1971, fig. 11).

function to the annexe at Ballynahatty, perhaps to screen the interior from view or to act as a transitional area between the interior and exterior.

The entrances through the façade and BNH5 and the BNH6 entrance are not on the same alignment, resulting in the interior not being visible on the initial entry to the Annexe area. Indirect entrances have also been noted at Sarn-y-bryn-caled (Gibson 1994; 2005, 108), at Mount Pleasant, Dorset, the Sanctuary (Pollard 1992, 224) and North Mains (Barclay 1983, 182). Evidently, the façade screened from view the areas beyond which were not meant to be universally viewed.

The features immediately beyond the entrance in the Annexe area represent more than one phase of construction. It is clear from the remains that the Annexe had an important function, perhaps as a place where ceremonies could be prepared or performed, prior to moving through to the interior of BNH5, and so was a zone of transition, vital to choreograph movement through the structure (Hartwell 2002, 530). BNH6 would still not have been visible until passing through the BNH5 entrance. The architecture at BNH5/6 was carefully laid out to ensure that people moved through the structure in the prescribed fashion and to ensure a clear division between those who were allowed into the Annexe and those who were not. There was even a division between those who were allowed to pass beyond the Annexe into the main BNH5 enclosure and those who were completely excluded.

Although the full extent of the timber complex had now been reached, there were further developments within the Annexe. A 9.00 m square setting of posts, the Entrance Chamber, was created (Phase 3). A modification of this size was an important consideration because three alterations had to be made to accommodate it. Two of these have already been discussed – the reorientation of the BNH5 entrance with the Entrance Chamber exit and the removal of a BNH5 Outer Ring post to enable the insertion of the northwest corner post (EC22) of the Entrance Chamber. On the east side, there is also a distortion of the square caused by the existing line of the Inner Façade such that the south side of the Entrance Chamber is slightly shorter. The entrance to the Chamber is the same width as the exit, leaving a narrow gap of c. 1.40 m, possibly closed by a door. Together with the narrower gap in the Outer Façade, this created a small vestibule c. 3.40 m square. There was a trace of a small posthole alongside EC36 which could have matched EC4 and EC5, and together this may be evidence of a door to close the entrance. The closing of the gaps between the Entrance Chamber and the Vestibule would seem to suggest that the Entrance Chamber was also a closed unit. If so, a different method of enclosing the space would have been necessary in the absence of paired posts. Planking would be appropriate here, although there is no other evidence for it. Gaps would have been necessary to allow movement into the Annexe spaces to the north and south. The central space is dominated by four taller posts. These four central postholes were deeper and the posts taller than the Entrance Chamber itself and would have marked a point of significance. They form a rectangle allowing passage between and perhaps a pause to allow rituals to take place in the Annexe (see Fig. 5.38).

An important aspect of this realignment was the effect on viewpoint. Standing in the Entrance Chamber passage, the sightline through the exit bypassed the Eastern Setting and the entirety of BNH6 so that only the internal expanse of the Outer Enclosure could be seen. The final section of the journey to the Inner Enclosure was only revealed when the entrance to BNH5 was reached.

11.6.6 Phase 4: destruction and removal or conversion and preservation?

Much of the excavated remains relate specifically to the destruction of the site. The abundance of charcoal associated with many of the features shows that the primary ceremonial function ended with widespread burning. One hundred and thirty-nine charcoal samples were taken for the purposes of identification and possible dating. Large quantities of oak were present, but also birch and alder. In addition, 282 bulk soil samples (1976.00 kg) from BNH6 were wet sieved in the laboratory – although weighing nearly 2 tonnes, this was a tiny proportion of the total secondary deposit – and 10.60 kg of charcoal from it was extracted and dried. This sandy subsoil, 0.54%

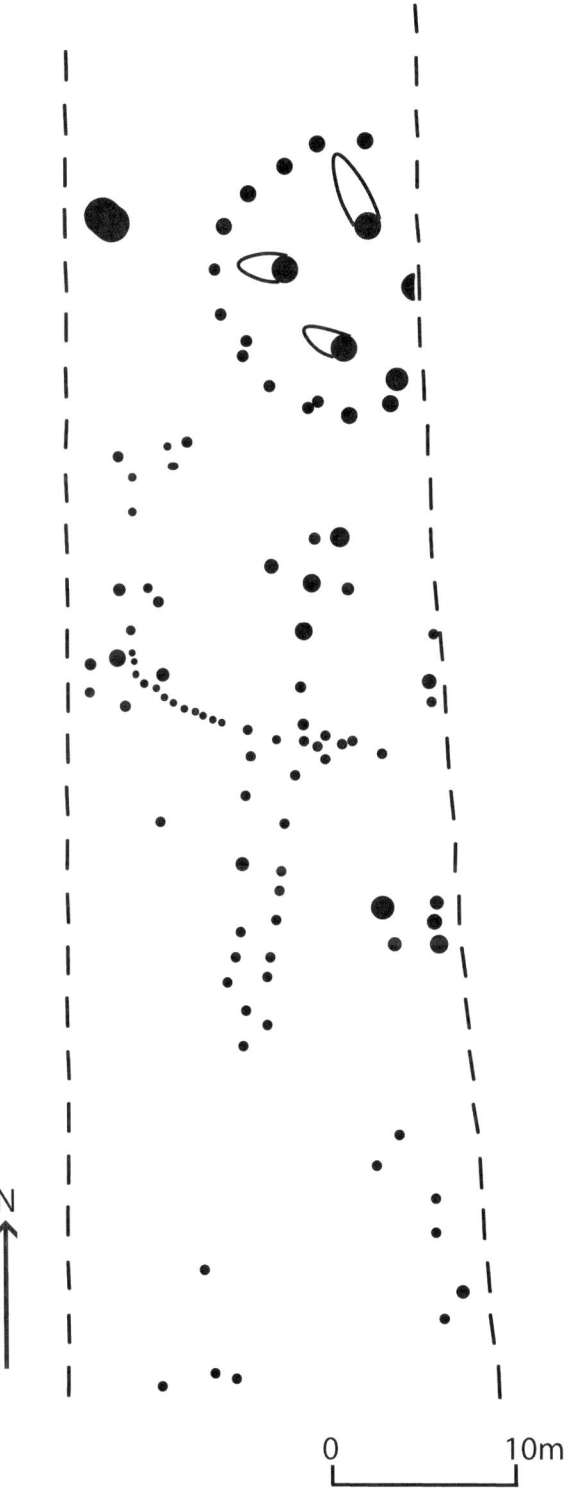

Figure 11.13 Durrington Walls, Northern Circle (after Wainwright and Longworth 1971, fig. 3).

Thomas (1999, 60) has proposed that the architecture apparent at these sites is an indication that the ordering and division of space had become more elaborate than the simple inside–outside division evident in earlier periods or, indeed, the first phase of BNH5/6.

by weight, equates to approximately 3.5% by volume (Daniel Taylor, pers. comm.) or nearly a quarter of a cubic metre from the soil samples alone. This was not evenly distributed: greater amounts occurred in the Four-Poster postholes (0.93% by weight), but this was concentrated in the secondary fill of posts 4P1, 2 and 3, with most in 4P2 and negligible amounts in 4P4. The primary fill of all the postholes contained a little charcoal, and this may simply relate to the initial clearance of the site or possibly the charring of the post to prevent rotting. The few examples of postpipes, where posts had rotted *in situ*, had concentrations of charcoal flecks or a band at the top, often in the shape of a small funnel where the sediments had settled and been drawn down, but with relatively little further down the pipe.

Like the layout and the rituals that the monument must have enabled, the end appears to have been very much prescribed – but was this an act of transformation? The closeness of the postholes rules out selective burning of individual or groups of posts, so the conflagration which engulfed the enclosures must have ended in the total incineration of the upstanding structure. Because a substantial amount of charcoal is found in all feature groups (with the exception of the central platform), it is likely that all these groups were still upstanding at the end.

The question remains as to whether this was accidental or deliberate burning. If deliberate, was this iconoclastic or an act of ceremonial sympathetic transformation? Burning of the structure would have stopped at ground level due to a lack of oxygen to feed the flames and would have left a charred stump. Had the stumps been allowed to stay and rot, this might have indicated that the burning had been accidental; but here we have ample evidence of the complete removal of posts. This includes not only the larger and, presumably, most significant posts, such as those of the Four-Poster structure in BNH6, the Entrance Chamber and the Annexe façade but also smaller posts of the Annexe. The size of the Four-Poster posts demanded a more invasive approach to their removal, and a funnel-shaped pit was dug into the primary fill around each post to loosen it before it was pulled out. Ash, charcoal, burnt soil and the primary fill were then redeposited in the hole but a substantial component was usually cobbles and larger stones and these could often be found deep in the remains of the postpipe. Indeed, there were so many stones that they could originally have formed marker cairns. In the postholes of the north Annexe (NEW), there were no surviving extraction funnels but the secondary fill took the form of a multi-layered sandwich of charcoal-rich soil and stones. The inescapable conclusion is that the intention was to remove totally as many posts as possible. Only the poor state of survival of the post butt would have hindered this. However, attempts at retrieval would cause considerable disruption and mixing of the posthole contents, with some posts being removed cleanly, some requiring a funnel-shaped pit to loosen the post, while others would have been only partially removed or had completely disintegrated in the ground. Even a post removed cleanly would have caused damage to the top of the postpipe as it eventually became unstable and toppled. More mixing and disruption would have been caused as soil, stones and ash were scraped and shovelled back into the holes and trampled underfoot. All the time, a blizzard of ash whipped up by the wind would have spread charcoal over the surrounding area. This would explain the problems encountered during the excavation of defining edges in the upper layers and distinguishing between primary and secondary fills and decayed posts and why some postholes were so disturbed that they could only be recorded as profiles.

There does remain a practical problem of how to extract a post from the ground with sufficient dexterity to allow the pipe to be preserved so that it could be refilled with stone. As explained previously, the erection of a timber post was facilitated by a ramp, a tipping point, gravity, rope and muscle. A burnt post was only a quarter of its original weight (the section left in the ground), but removal was now against gravity, and there was the added problem of friction – the 'grip' of the primary fill. The obvious suggestion was to lever the post vertically out of the ground, but could this be achieved in practice? The only way to demonstrate this as a possibility was to test it experimentally, and this is described in some detail as it also has a bearing on the effort taken to construct the enclosures.

Initially, it was decided to dig a test hole, with the only determinant being the depth of 2 m. The presumption here is that the depth of the hole is related to the height of the post it is meant to support – three times the depth. The reason for the length and width of the hole was an unknown but access and the size of the post must have been major constraints. The digging implements were a simple, unbranched, single-tined 2-year-old red deer antler and a bucket.

Red deer had been introduced into Ireland in the earlier Neolithic and, although their bones are not common on archaeological sites of the period, they were presumably widespread by the end of the Neolithic. At Newgrange, red deer remains represented 3% of the total bone assemblage by 2500 BC. Deer will have been valued not only for the meat and skins they provided but also for their antlers, which were used as tools and for tool making. The male stag, with an average lifespan of 16 years, will produce useful antlers from its second year until it dies. Older stags produce antlers with more tines. Antlers are initially cartilaginous but undergo a process of ossification before rutting and are shed annually in May and June. Through its lifetime, a stag can produce as many as 30 hard but highly durable antlers, which could be used or modified as picks. No antler was recovered from the excavation but bone generally does not survive at Ballynahatty unless burnt. The use of antler is implied by its utility.

The sandier levels of posthole BNH6 OR8 produced evidence of striations caused by a pointed digging tool, presumably an antler pick.

Safety considerations determined the use of an aluminium ladder to access the developing hole. A running log was kept of progress and the task of digging the hole was given to Billy Dunlop, a long-standing and very energetic volunteer, then in his mid-70s. The utility of the antler pick quickly became apparent when the hole hit the ubiquitous, indurate, stony layer in the upper subsoil. Using the basal burr as a hammerhead it was possible to smash the surface progressively and then scrape away the loosened material into the bucket using the tine. This was far more effective than using a modern metal pick which tended to stick in the surface.

It soon became obvious that the length of the hole is directly related to access and represents the distance between the top of the head and the seat of the person digging the hole. Similarly, the width is determined by the width of the digger plus elbow room. Even so, digging a hole must have been an uncomfortable experience. The total time taken for one person to dig this hole was 21 hours. No attempt was made to dig a ramp but this was estimated to take a further 5 hours.

Next, a discarded 5 m length of telegraph pole was inserted, and the hole gradually backfilled. The displaced subsoil fill was then periodically rammed down to ensure a tight hold on the post (Plate 88). To try and remove the post without digging it out, two vertical timber 'levers' were placed on either side of the post and all three very loosely tied together by a loop of rope at the bottom. A fulcrum was placed on either side in the form of two large stones and the levers were brought down in opposite directions to rest on them (Plate 89). This had the effect of twisting the rope and provided a vice-like grip on the post. By continuing to push down on the far end of the levers, the near end was raised above each fulcrum and the post moved upwards by 0.10 m. Simply leaning against the post provided enough friction against the fill to stop it from sliding back down the hole. The levers were then raised, which loosened the rope, and slid down to ground level, and the process repeated. The post progressively came out of the ground, although it caused a lot of disturbance at the top of the hole as it toppled over (Plate 90). However, a 2 m ranging rod was put in the postpipe to show that it was still open and then filled with alternate layers of stones and soil (Plates 91 and 92).

As a proof of concept, the experiment showed that it could have been possible, using the simplest of materials, to lift a 5 × 0.30 m post out of the ground and replace it with stone and a secondary fill. Against this, most of the actual timber posts were larger, more irregular and naturally tapered outwards at the base. It may have been necessary to dig around parts of the post and to rock it back and forth to ease the passage resulting in general disruption. It would have been too difficult to lever out a complete, 0.5 × 8 m, 1–2 metric tonne (Huber's formula – a measure of log volume used in forestry) post vertically against gravity, whereas the stump, at a quarter of its weight, remains a possibility. If the post butt had partially rotted this would have facilitated removal and may provide circumstantial evidence for the longevity of the timber enclosures.

Although wood itself is time-limited and will ultimately rot in the ground, it could have survived beyond the lifetime of its creators. The use of mature timber imbues the structure with instant age. The selection of massive timbers for the Façade also gives legitimacy by utilising components which were already ancient. Whereas stone is lifeless and therefore of no apparent age, trees were living and growing with roots in the land.

Ballynahatty is by no means the only timber circle to have experienced this type of 'ritual deconstruction'. Many of the Irish timber circles appear to have been subject to some form of ritual deconstruction, where the posts were either burnt or deliberately removed before being backfilled. To date, Ballynahatty is the only Irish timber circle where there is evidence of posts having been burnt *in situ*, although evidence of burning has been recorded at Dorchester on Thames Site 3, Oxfordshire (Gibson 2005, 158). However, many of the sites display evidence of the post having been extracted. At Armalughey, Co. Tyrone, Balgatheran (Buildings 1 and 3), Co. Louth, Bettystown, Knowth, Co. Meath, Lowpark, Co. Mayo, Paulstown, Co. Kilkenny. Prumplestown Co. Carlow and Scart, Co. Westmeath (Eogan and Roche 1997; Eogan 2000; Monteith 2008; Elliott 2009; Ó Drisceoil 2009; Bolger *et al.* 2015; Dunlop and Barkley 2016; Gillespie n.d.) there is evidence of posts having been deliberately extracted and the voids backfilled with artefact-rich material. The extraction of posts has been noted at some British sites, including Dorchester 3, Oxfordshire (Gibson 2005, 124) and the Southern Circle at Durrington Walls (Thomas 2007b).

11.6.7 What do the artefacts tell us?

Distinguishing between accidental incorporation and intentional deposition of artefacts is problematic. Ritual deposition would be more convincing if there was an obvious pattern or uniformity of pottery or lithic density. It could equally be argued that activity areas shown by the inevitable casual discard of broken pots and lithics had been dug through by the post builders resulting in the accidental incorporation of this cultural material. The original activity may have been ritual and, therefore, the artefacts were ritually derived, but this residual material does not constitute 'ritual deposition'. This can be demonstrated with respect to burnt bone. Apart from individual deposits of cremated remains, small quantities of burnt bone appeared in nearly 150 samples from scattered contexts across the site, strongly suggesting that burnt bone was present over the entire surface through which the postholes were dug. Similarly, the presence of an isolated artefact in a posthole can only be considered a 'token' deposit if there was a clear positioning showing an intention to incorporate it. Unless there was obvious patterning, such as repeat deposition, this should be considered accidental incorporation.

At least 31 Grooved Ware vessels are represented at Ballynahatty, with the majority being of Brindley's Knowth 1 style. Four sherds of Knowth 3 style Grooved Ware, regarded as the earliest of the Irish Grooved Ware styles, were found in the BNH6 area but are not necessarily related to it. These may provide the link in the use of the area between the Middle Neolithic funerary activities and the later Grooved Ware/timber enclosures. Grooved Ware is not found in the primary posthole deposits of BNH6 (Phase 1a), which suggests that, for the builders, this was a new location. Knowth 1 style pottery makes a very limited appearance in the construction of the larger BNH5 enclosure (Phase 1b). Most of the Grooved Ware from primary contexts comes from the Entrance Chamber and the Annexe. Knowth 1 and 2 style Grooved Ware, including sherds of a single Knowth 2 pot, were distributed amongst four postholes in the southwest corner of the Entrance Chamber beside the exit. Whereas a plausible case could be argued that these postholes were dug through the residual remains of a broken pot, the presence of a sherd from the same pot being found in a SEW posthole over 12 m away strongly suggests this is deliberate deposition. This also ties the construction of the Entrance Chamber to the Southern East–West posts in Phase 3.

Nearly 20% of the Grooved Ware came from the northern Annexe and the primary deposits of the adjacent Entrance Chamber postholes. In this case, it is likely that the postholes had been dug through the residual remains of at least seven Knowth 1-style Grooved Ware vessels in the Phase 2 Annexe. This was not simply a passageway but an activity area with traces of hearths. Some sherds had carbonised deposits on the inside but not on the breaks and, together with a general lack of abrasion, indicates that these Grooved Ware cooking pot sherds were not lying around on a floor where they would be subject to wear but had fairly soon become incorporated into protective deposits after breakage. A stony layer, interpreted as a floor, covered some of these pots and may relate to the Phase 3 elaboration of the Annexe with the insertion of the Entrance Chamber. Very little pottery comes from Phase 4, the secondary destruction deposits. Seven sherds of Knowth 1 and 2 style were found in the entrance postholes and in two of the Four-Poster postholes of BNH6.

Continued general occupation of the area after BNH5/6 was destroyed is shown by five sherds of a single Food Vessel found close to the southern side of the BNH5 entrance but not stratigraphically related to it. Three sherds of a Collared Urn were also found in topsoil clearance, adding to the ample evidence of an Early Bronze Age presence, as discussed above (Section 1.4).

There was a large lithic assemblage at Ballynahatty (5036 pieces) that included flakes and blades, but 25% is angular shatter, the product of knapping, and 20% of that is burnt. The commonest artefact type is small scrapers. These vary in shape and are mostly round, but 17 are hollow scrapers, probably used as spokeshaves; these are most likely to relate to the Middle Neolithic activity and will have been residual in the timber monument complex. Some tools were expedient, having multiple functions such as cutting and piercing. Knives and projectiles are rare but, where present, petit tranchet and petit tranchet derivative arrowheads dominate. Ten per cent of the artefacts are modified core tools which suggest that these were being made for a specific purpose, possibly as wedges. Edge damage shows that these tools were being used on a variety of materials, including bone. Overall, this represents an unusually large collection for a mortuary site, with the modified tool assemblage covering cutting, scraping, piercing and splitting functions singly or in combination.

Understanding how these tools were used at Ballynahatty is problematic. Carbonisation on the pottery and hearths in the Annexe and the presence of calcined cattle, pig and sheep/goat bones indicate the preparation and consumption of food. Where identifiable, all but one of the cattle and pigs had achieved full size, and the meat cuts were a mix of qualities suggesting that the animals were slaughtered nearby and consumed on-site. A substantial part of the lithic assemblage was the result of fabrication, so knapping of flint was another activity, and much of this was burnt, possibly in the final burning of the monument. This would seem to point to ceremonial feasting during which 'raw products' in the form of animals and flint were brought to the site, the tools were made, and the animal was butchered, cooked and consumed as food. Another possibility is that tools were created to be used in the excarnation process. Fragments of human bone were found in the primary fill of a BNH6 Inner Ring posthole which presumably pre-dated the site but some were also present in postholes in the north and east sides of the Entrance Chamber and in a Southern East–West posthole. This was calcined bone (unburnt bone does not survive), so a cremation process was also in operation here. The possibility remains, however, that this could be residual bone from the Middle Neolithic funerary activities.

Variability in the lithic distribution has produced a more distinct pattern than the pottery. A third of the finds are from the topsoil, spread across the site but concentrated over BNH6, the northern Annexe and BNH5. Most tools are found in primary deposits, especially in Phases 2 and 3 of the Entrance Chamber and north Annexe. Burnt lithics, especially debitage, are found in Phase 4 secondary deposits, mostly in the BNH6 entrance, mostly in the Inner Ring, and 4P1. These tend to be zones of transition.

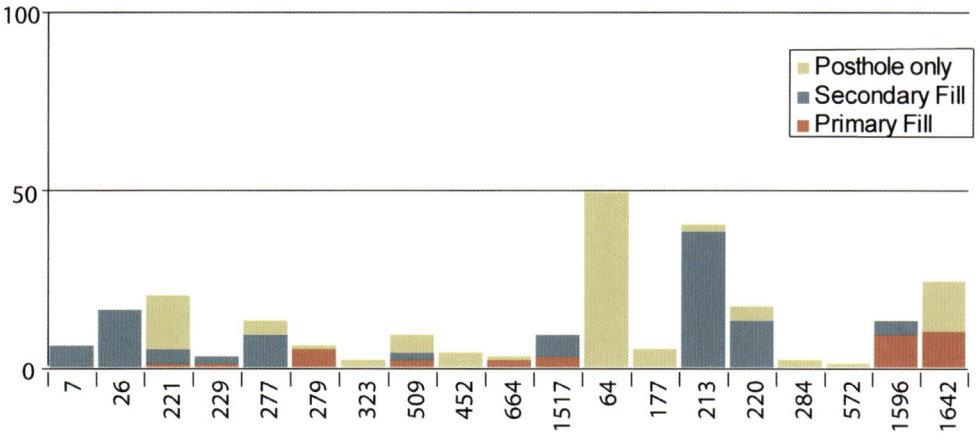
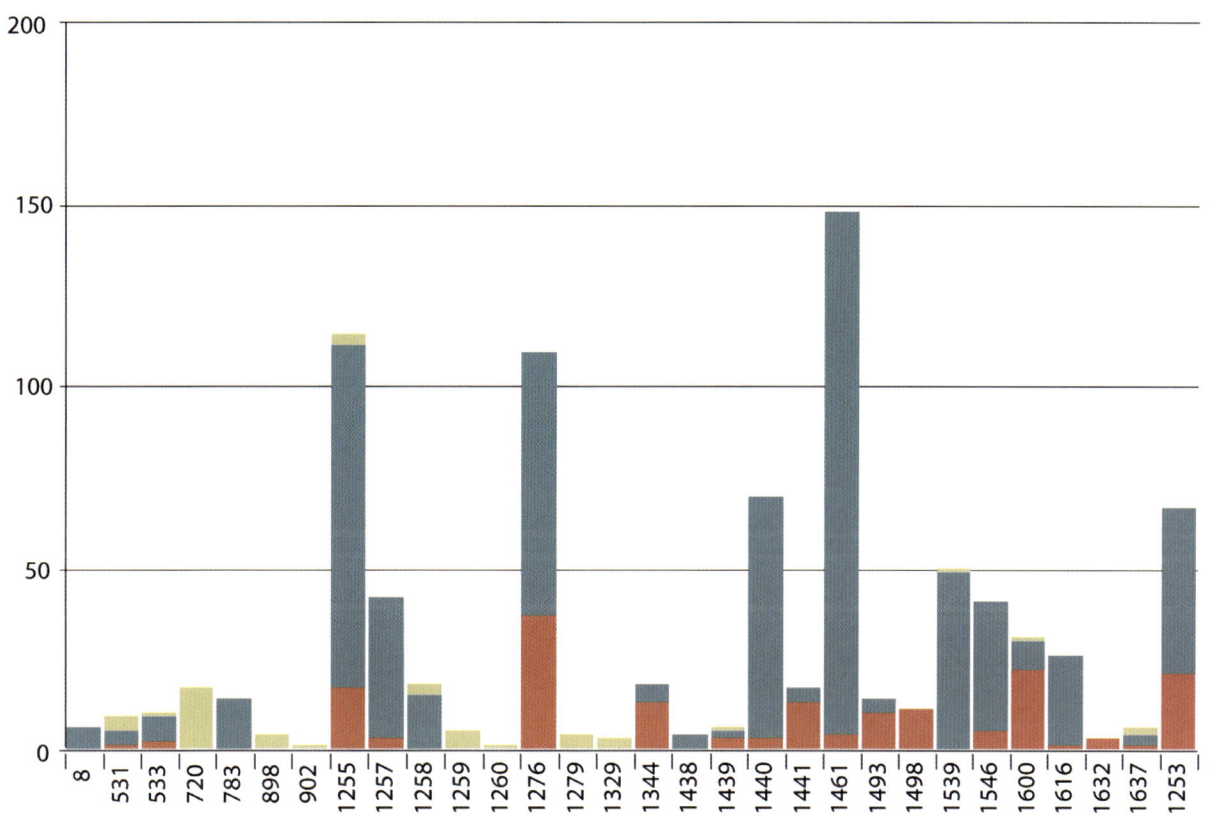
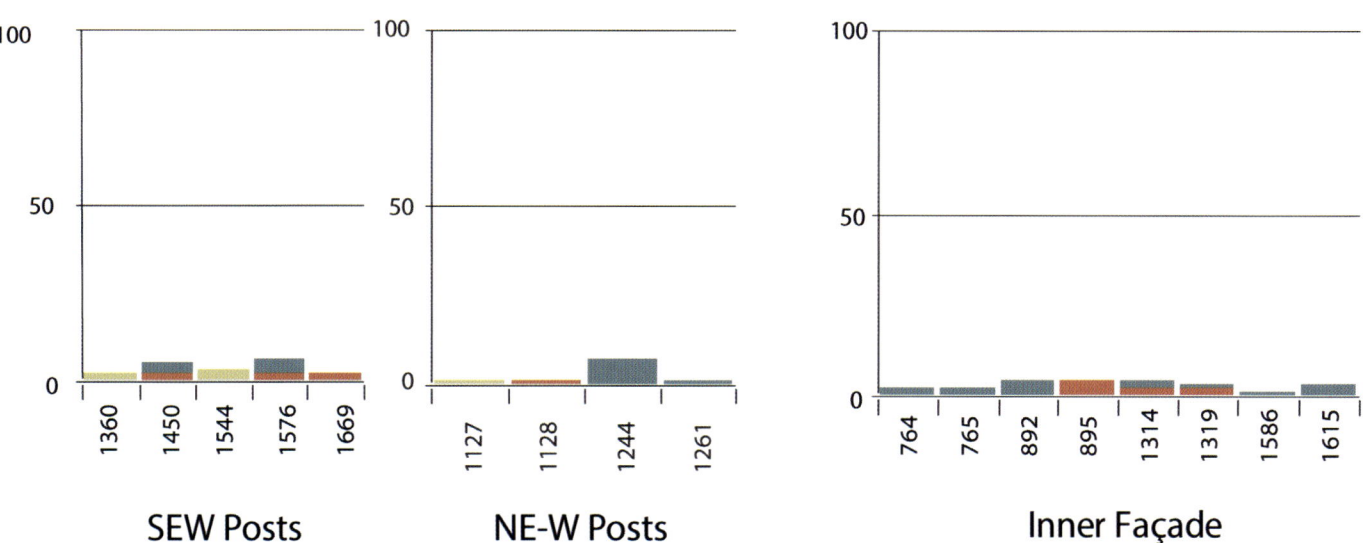

Plate 74 Distribution of lithics in the primary and secondary fills of postholes in each feature group. X-axis shows the posthole number and the Y-axis shows the number of lithic artefacts recovered.

Plate 75 Burnt flakes recovered from Ballynahatty.

Plate 76 Two intact stone axeheads recovered during excavations: a) 6111; b) 6164.

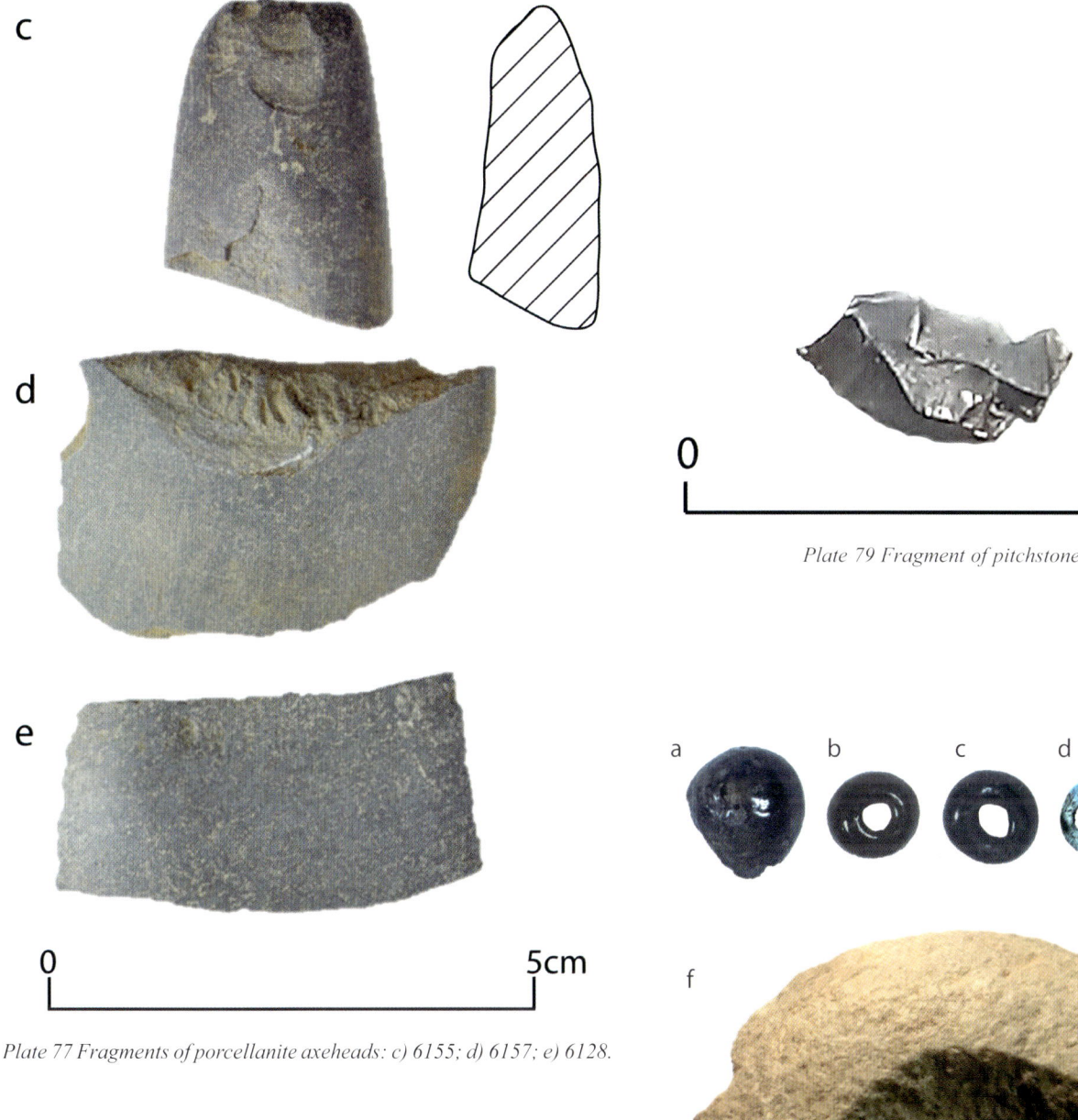

Plate 77 Fragments of porcellanite axeheads: c) 6155; d) 6157; e) 6128.

Plate 79 Fragment of pitchstone artefact.

Plate 80 Beads a–c, e: glass; d: 'faience'; f: stone.

Plate 78 a) flint 'ball' (6117); b) stone 'ball' (6109).

Plate 81 Possible bead of lignite.

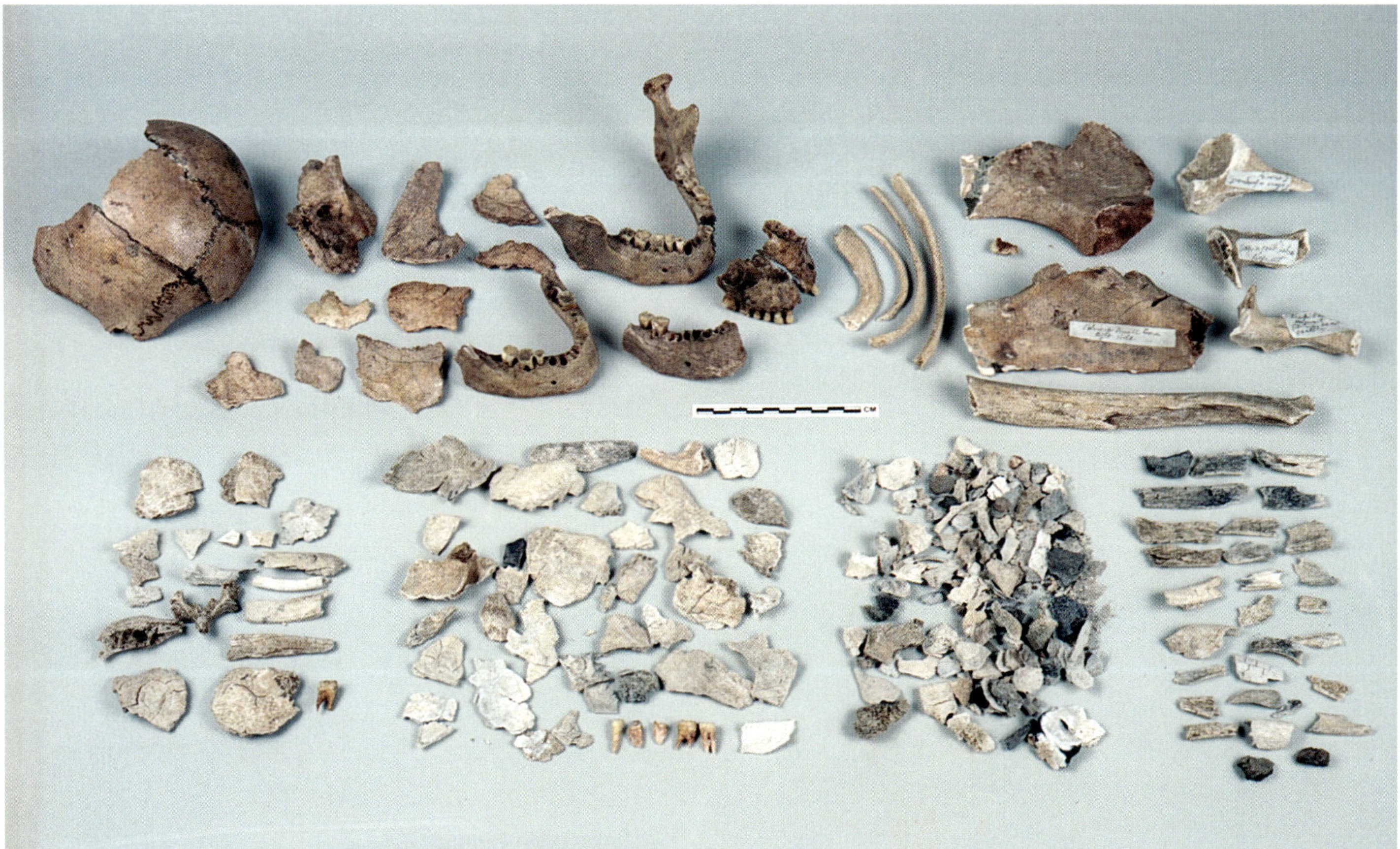

Plate 82 Contents of the 1855 tomb in the collection of the Ulster Museum, with the exception of the two skulls recorded in MacAdam and Getty's report. The bone fragments were a later insertion, placed on the cremated bone and separated from it by a layer of sand.

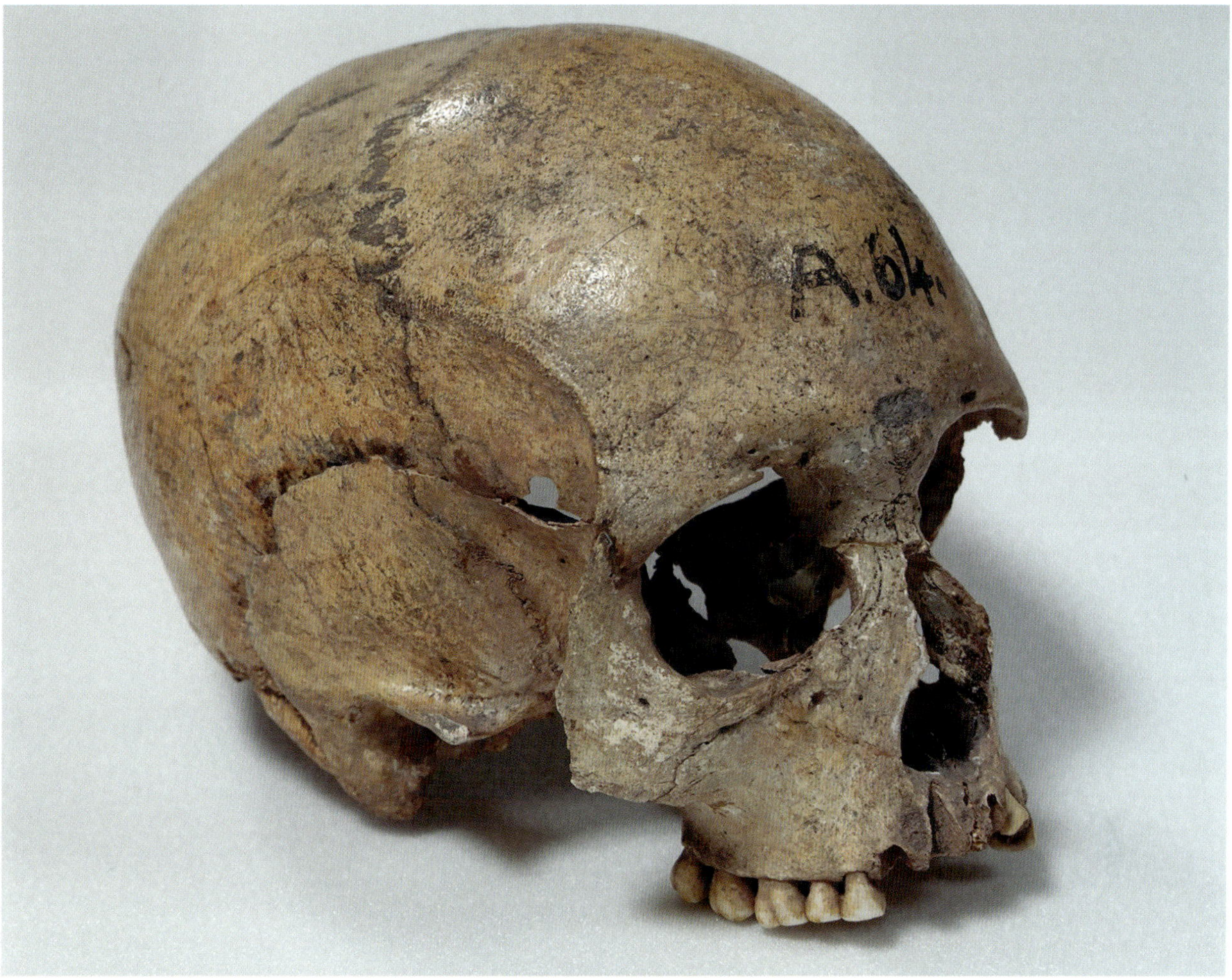

Plate 83 Skull A64, one of the two published skulls (both female aged 20–25 years) published in 1855. Popularly known as 'Ballynahatty Woman', she was the subject of further research undertaken in the early 21st century (see Fig. 13.12).

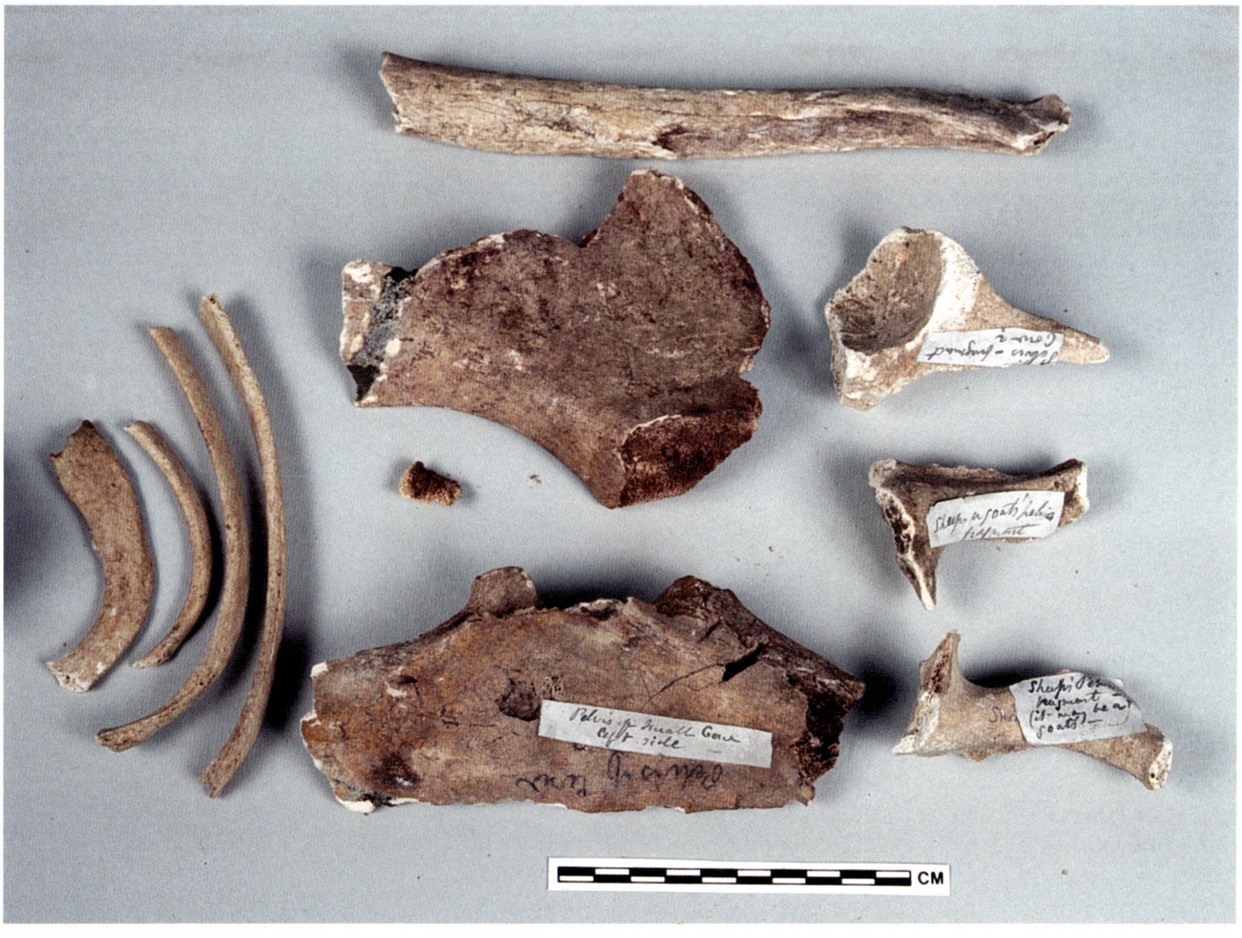

Plate 84 Animal bones from the 1855 tomb.

Plate 85 The passage tomb (BNH2) with blocked passage to the left, looking east to the entrance of the henge (BNH1) (photo: Barrie Hartwell).

Plate 86 A modern example of two oaks growing in a densely wooded environment at The Argory in Co Armagh with diameters at shoulder height of 60 cm and 50 cm and straight poles with trunks of over 13 m persisting high into the crown (photo: Barrie Hartwell).

Plate 87 The entrance to Avebury (photo: Barrie Hartwell).

Plate 88 Ramming the primary fill around the post (photo: Barrie Hartwell).

Plate 89 Levering out the post (photo: Barrie Hartwell).

Plate 90 The butt of the post topples as it is removed, causing considerable disruption to the surface (photo: Barrie Hartwell).

Plate 91 The hollow post mould shown by the ranging rod (photo: Barrie Hartwell).

Plate 92 Post extraction experiment: dropping stones into the void to create the secondary fill (photo: Barrie Hartwell).

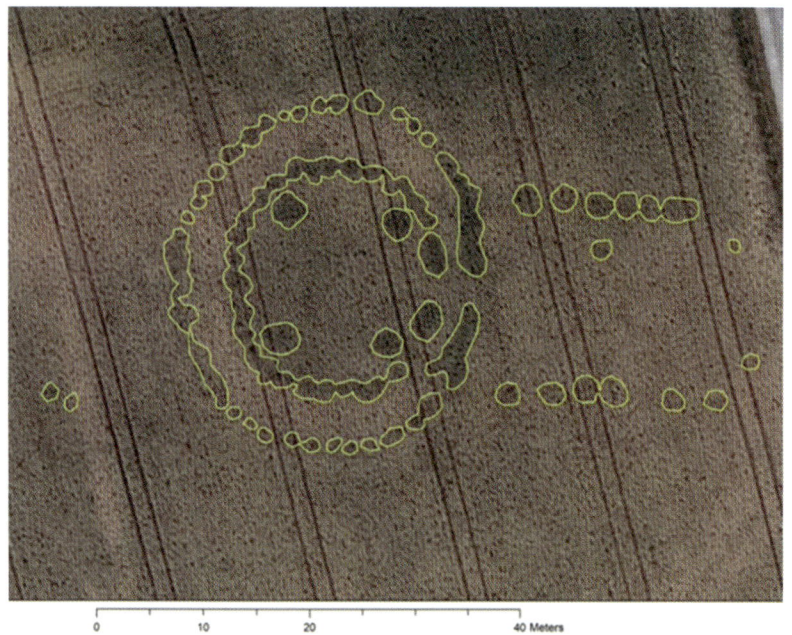

Plate 93 Newgrange 'Four Poster Enclosure' (Condit and Keegan 2018, fig. 23). Image © Department of Housing, Local Government and Heritage; base image © Bluesky International Ltd.

Plate 94 Newgrange 'Four Poster Enclosure' and encircling bank (Condit and Keegan 2018, fig. 27). Image © Department of Housing, Local Government and Heritage; base image © Bluesky International Ltd.

Plate 95 Giant's Ring: the northeast entrance and straight east section which runs along a steep slope (photo: Barrie Hartwell).

Plate 96 Newgrange Henge A (Condit and Keegan 2018, fig. 27). Image © Department of Housing, Local Government and Heritage; base image © Bluesky International Ltd.

Plate 97 3D approximation of the timber circle in Ballynahatty (image: Robert Barratt).

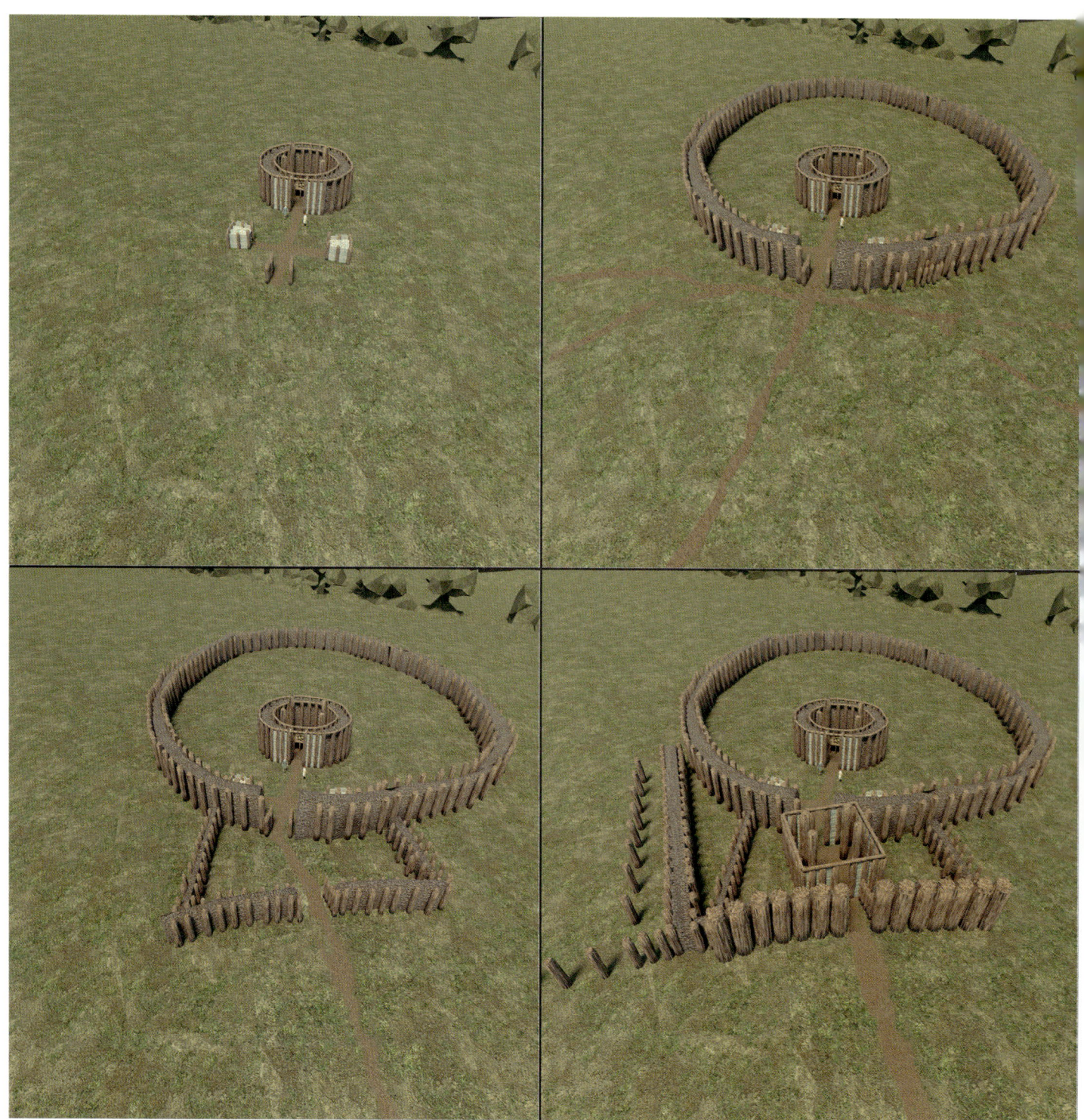

Plate 98 Four different versions of the model, showing choronology of the site: top left: phase 1a; top right: phase 1b; bottom left: phase 2; bottom right: phase 3 (image: Robert Barratt).

Plate 99 Racks of skulls in the Eastern Settings – the ancestors (image: Robert Barratt).

Plate 100 Platform structure portrayed as an excarnation platform with hypothetical elements used to convey the experience of visiting the site in use (image: Robert Barratt).

Plate 101 Early development of the model showing testing of different types of fills (image: Robert Barratt).

Plate 102 The timber 'Temple' from the south. The SEW line of posts accentuate the length of the structure, visually unifying BNH5 and the Annexe (image: Robert Barratt).

Plate 103 Approach to the Grand Façade from bottom of the ridge. The short southerly extension to the Façade (left) provides a visual balance to the northern mass (right) of the larger enclosure and places the Façade and entrance in central position (image: Robert Barratt).

Plate 104 The Entrance Chamber, looking into the interior of BNH5. BNH6 cannot be seen (image: Robert Barratt).

Plate 105 The tall timbers of the Entrance Chamber draws focus on the sky (image: Robert Barratt).

Plate 106 Inside the Annexe (image: Robert Barratt).

Plate 107 *View into the Temple (BNH6) (image: Robert Barratt).*

Plate 108 Decaying cadavers on the central Platform open to the skies (image: Robert Barratt).

Looking in greater detail at the distribution of lithics in the postholes of the two greatest concentrations – BNH6 and the Entrance Chamber – interesting patterns emerge (see Fig. 7.6). There is a clear emphasis on deposition at the front of BNH6. Simply dividing the enclosure in half, front and back, shows a 71% to 29% distribution of lithics. If the site is divided on either side of the axial line running through the entrance, the distribution is 55% on the left (southwest) to 45% on the right (northeast). A division into southern and northern halves gives a more revealing distribution of 60% to 40%. The distribution of lithics in postholes within BNH6 may also be taken as a proxy for the relative importance of the posts they contain. The Inner Ring holds 62% of the lithics against 11% in the Outer Ring. The Four-Poster postholes hold 26% and the Central Platform less than 1%. During extraction of the posts, the upper layers of the Four-Posters and the adjacent three postholes of the Inner Ring would have had a very messy interface, so if each is taken as a single composite unit, the Four-Posters would hold 58% of the lithics. At the other end of the distribution, three areas stand out for their low lithic counts. The Central Platform, because of its relatively small size, would not have lasted for the lifetime of BNH5/6 and was probably not present at its end. The Outer Ring, despite having six more postholes, has only 11% of the lithics, and ten of these postholes contain no lithics at all. Finally, 4P3, the most northerly, has less than 4% of the lithics. A similar but stronger pattern is evident in the Entrance Chamber, with 80% of the lithics in the southern half and an emphasis towards the southwest. The east–west division (56% west) is not as pronounced as the front/rear division of BNH6 because this is an open-ended routeway to the interior and not the destination. There is a greater emphasis on the Four-Posters (60%), especially to the south.

In both areas, there is a strong emphasis on the southern side, presumably demonstrating the continuing influence of the henge and the passage tomb. The Four-Posters continue their role as markers of significant spaces, but the entrance and Inner Ring of BNH6, which face into and define the interior, mark this as a space of core importance – an inner sanctum.

Apart from the flaked lithics, there was also a small coarse stone component. Considering the amount of woodworking which had taken place, there were remarkably few axeheads found (Section 8.3.1). As noted at the beginning of the chapter, it is unclear whether these axeheads (and fragments thereof) date to the Late Neolithic phase of activities. One shows no obvious signs of use. Another, heavy and blunt, is certainly utilitarian. In addition, there were fragments of three polished porcellanite axeheads and a porcellanite flake, probably from a further axehead. A flint 'ball' (from BNH5 OR6) and a stone 'ball' (from BNH6 IR1) are of uncertain significance. Less fine than the smooth stone balls found in Middle Neolithic passage tombs, a possible use as hammerstones should be borne in mind. The most intriguing artefact is a very large oblate perforated object of lignite, which may well be a large bead (Section 8.3.5), found in the entrance to the Entrance Chamber. The presence of this unusual piece in such a prominent situation indicates deliberate deposition.

11.7 Relationship between the timber monument complex and the Giant's Ring

The question remains as to the chronological relationship between the timber monument complex and the Giant's Ring. This question is crucial to our understanding of the sequence and tempo of monumental activities at Ballynahatty, not least because it is clear that the construction of the henge – a huge, flattened-circular embanked enclosure around 180 m in diameter, possibly originally with two entrances, at the northeast and southwest – will have involved a massive communal effort, just as the construction of the timber monument complex will have done.

The excavations by A.E.P. ('Pat') Collins (1957) established that the monument consisted of a number of structural elements, including (from the outside of the monument inwards) an outer marker bank, 1.50 m high; the bank itself, 4.00–5.00 m high; a stone revetment; a berm; and a ditch, 15.00 m wide by 1.00–1.50 m deep, which was the source of the bank material. To reconstruct the original shape of the bank, three separate slopes need to be considered: the present interior and exterior slopes and the projected slope of the internal stone façade recorded in the lower bank section. Natural collapse and settlement would most likely produce a steeper exterior slope angle, and this was confirmed with reference to the recent excavation spoil heap, which existed for several years and came to a similar natural angle of rest at 35°. However, the interior slope is only 20°.

Producing a slope angle from two irregular stones is subjective but building a stone façade on the inside wall of the bank could have increased the angle from the present 20° to about 70°. The high stone content gives inherent stability, but it is possible that within a few years of its initial construction, slippages would have reduced the interior closer to 35°, the external bank angle. Figure 11.14 shows how the bank may have originally appeared. This would have been even more impressive with the open 1.00–2.00 m deep quarry trench in front of it. 'Replacing' the soil from the trench to the bank gives it a broader, flatter top. With the quarry trench re-established, the interior focus drops from 180.00 m to a more restricted 150.00 m in diameter and further increases the visual separation between the central area and the bank.

How should we view this vast enclosed area? Was the point of the banks to exclude views of the exterior and create an area set apart, as suggested by earlier commentators of the Ring? In this model, the bank contains the participants, protecting them and the rituals they perform from the view of the greater population. The inner façade

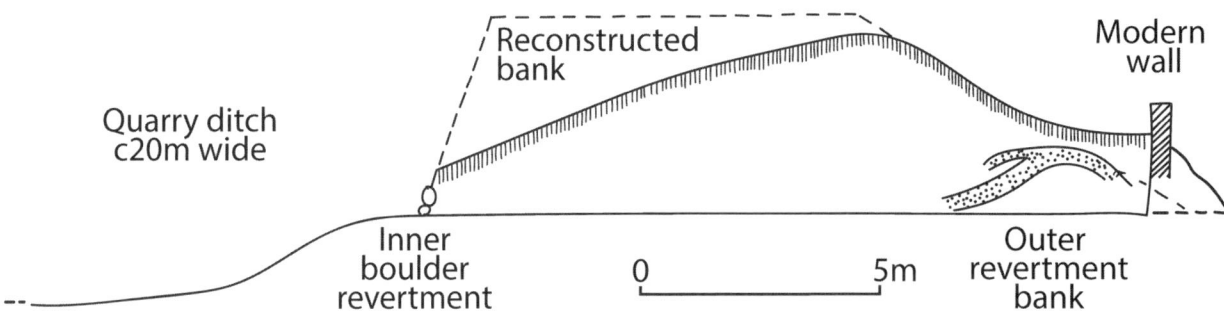

Figure 11.14 Conjectural reconstruction of the Giant's Ring bank (Barrie Hartwell).

of the bank would have been an immensely impressive 4.00–5.00 m tall stone wall, functioning as an inward looking barricade strictly defining the interior space – not as a barrier against the outside world. The bank would not have created a serious impediment to a human and was unnecessarily high to simply constrain larger animals, such as cattle, but perhaps the point was to isolate the ceremonies and rituals within and to contain the results. The interior thus becomes a dangerous place.

In an alternative model, the interior is still a vast stage on which these rituals are conducted but it is also a public spectacle, acted out between 'actors' and 'audience'. The size of the monument – and the fact that many people will necessarily have been involved in its construction – suggests that it was designed with large numbers of participants in mind. If it was to hold a viewing public, allowing a generous square metre per person, 650 people could stand in a single, continuous line on the bank top. However, the top has been made deliberately broad and could accommodate at least six rows, so a maximum number could have been as much as 3900. Such large numbers are not unreasonable considering the size of the henge.

Whatever the actual function of the Ring, it was a staggeringly impressive piece of construction and remains the most impressive surviving henge in Ireland. Quantifying its construction is fraught with difficulty, and the variables seem endless, but it is worth emphasising how big a task this was. We may assume a basic unit of a 0.20 m cube (about the amount which can be easily carried as a basket of soil or a lump of stone *i.e.* 1/125 of a cubic metre). Each basket would need to be moved from the quarry ditch to bank position, a distance of 5.00–30.00 m, and requiring a climb of between 2.00–5.00 m. We can assume an average distance of 17.50 m and height of 3.50 m required for each basket moved. Experimentation shows that this return journey, with a load, could be done in 3 minutes, provided there was no delay. Generating the required amount of soil/stone with an antler pick from the ditch would take about 5 minutes. A 1.00 m wide section through the reconstructed bank equates to 50 m³. Therefore, it would take one person digging, and another moving the load, 6250 journeys to construct a 1.00 m

section of the bank or 4 million journeys for the complete 640.00 m circumference. With an average 5 minute turnaround per journey, this would take 333,333 hours. If we then factor in an average of 8 hours of daylight, this would have spread the time taken to 41,666 days for our hardworking pair (83,332 person days). At that rate, it would take 100 people 833 days to complete the job. Most likely, the work would be undertaken between summer harvest and spring, from November through to March, over a period of 5–10 years.

As for the question of 'When was the Giant's Ring built?', this is difficult to answer, and all that can be stated with confidence is that it was constructed long after the passage tomb which it encircles – possibly as long as 1000–1500 years. There is no dating evidence from the monument itself and, as pointed out by Rowan McLaughlin in Chapter 10, the dating evidence from other Irish henges and similar-looking monuments does not offer any convincing pointers. As noted above (Section 11.3), the evidence from Monknewtown henge, Co. Meath (Sweetman 1971) is frustratingly inconclusive: while it is generally assumed that the evidence for Beaker-associated activities inside the enclosed area post-dates its construction, the evidence relating to a Coarse Ware ('Carrowkeel') bowl containing cremated human remains is ambiguous. That pot was found in a shallow scoop within the monument, close to the 'tail' of the bank. Stylistically, it is most likely that the pot dates to between 3300 and 2900 cal BC, by analogy with dated examples of this kind of pottery elsewhere, but there is no direct stratigraphic link between the pot's findspot and the bank. Even if the bank had been created by scraping earth from the henge's interior, this does not prove that the pot post-dates the henge, as it could have been buried at depth.

In 2004, on the basis of comparative dating of henges in Britain, Sheridan proposed that the Giant's Ring is most likely to have been constructed between 3000 and 2600 cal BC (Sheridan 2004, 31). Since then more dating evidence in Britain has been obtained and, in 2020 Susan Greaney and colleagues summarised the current state of knowledge regarding large henge monuments in southern England (Greaney *et al.* 2020).

It appears that the earliest securely dated henge monument in Britain and Ireland is the circular, single-entrance example at Stones of Stenness, Orkney (Ritchie 1976), constructed during the 30th century cal BC (Bayliss *et al.* 2017). This is considerably smaller than the Giant's Ring, however, with an estimated bank diameter of *c.* 65 m. The larger, *c.* 90 m wide circular monument consisting of a bank with external segmented ditch at Stonehenge – not a henge according to a strict definition and having more in common with Neolithic 'enclosed cremation cemeteries' – has been dated to *2995–2900 cal BC* (Parker Pearson *et al.* 2020, 536).

A number of henges seem to have been constructed around the middle of the 3rd millennium BC and these include the very large 'mega-henges' found in southern England, including Mount Pleasant, Dorset (Greaney *et al.* 2020). This irregular oval monument, 370 × 320 m across, with several entrances in its bank and external ditch, appears to have been constructed during the *26th century cal BC* (*ibid.*, 213–8), and the henge bank at Durrington Walls, Wiltshire, seems to have been constructed early in the *25th century cal BC* (Parker Pearson *et al.* 2013, 171), while that at Avebury has a provisional construction date of *2580–2470 cal BC (95% probability*, cited in Greaney *et al.* 2020, 224). Smaller henges were also being constructed around this time, with a circular, single-entrance example being built at Forteviot, Perth and Kinross (*c.* 50–60 m in diameter, *25th–23rd century cal BC*: Brophy and Noble 2020, 147). Beaker pottery was already in use in Britain during the 25th century and indeed was present in primary contexts at the Forteviot henge, so technically the examples constructed after *c.* 2500 cal BC can be regarded as belonging to the Chalcolithic period. Other henges – particularly the slightly oval, two-entrance henges – were built during the Early Bronze Age, with North Mains, Perth and Kinross (Barclay 1983), having a *terminus post quem* of *2140–1960 cal BC* (Sheridan 2003; *cf.* Bradley 2011 on other Early Bronze Age henges in Scotland).

This, therefore, presents a wide comparative chronological framework – Late Neolithic, Chalcolithic and Early Bronze Age – into which to slot the Giant's Ring. Ideally, of course, the dating of Irish henges should be undertaken by dating Irish examples independently of their British comparanda but in the absence of such a framework, it could be suggested that the most likely date for the Giant's Ring may be the centuries around 2500 cal BC – that is, around the time when the very large henges of southern England were being constructed. The Giant's Ring is considerably smaller than the Wessex mega-henges, but it is nevertheless nearly three times as big as the early 3rd-millennium example at Stones of Stenness. If its builders had seen the mega-henges of southern England (and, as argued in Sheridan 2004, this is a possibility), then it is easy to imagine how they might have sought to emulate them by constructing a huge and impressive monument at Ballynahatty.

If that hypothesis is correct, it raises the possibility that the timber monument complex at Ballynahatty was deliberately destroyed (around 2500 cal BC) in order to be replaced by a more enduring monument – and one whose location drew on the sacredness of the ancient monument, the Early Neolithic passage tomb, which it enclosed. Only further excavation at the Giant's Ring, and some good luck, could show whether that is the case or not.

Regardless of order, there is little doubt that there is a direct relationship between the timber monument complex and the Giant's Ring, either as contemporaneous sites or as a permanent replacement of timber for earth and stone. The area between them is a clear space in a 'busy' landscape, and the SEW posts of the final phase of the Annexe directly overlook the northern embankment of the henge. A line of posts run around the west side of the henge but stops at the ridge, thereby defining the western edge of the complex. To the south and east, it is limited by the edge of the plateau. The ridge marks the northern side, and BNH5/6 sits on top of it, overlooking the whole complex. The southerly extension of the façade creates the visual link back to the passage tomb. Therefore, they should be regarded as a single entity.

However, based on the physical development of the BNH5/6 enclosures, it is possible to outline a plausible developmental relationship. Robert Barratt (Chapter 12) shows that, in its earliest phase, BNH6 conforms to no readily identifiable orientation, other than a common southeasterly direction, and bears no obvious relationship to the henge or passage tomb. The easterly reorientation of the later Annexe and its great façade shows that passage up the ridge from the east became the prescribed grand ceremonial approach. This begins at the lower end of the ridge and, therefore, on the very edge of the eastern side of the plateau. The northeast entrance to the henge is placed in the most unlikely position considering general access because it is also jammed against the eastern edge of the plateau. This entrance must be for ceremonial purposes only because access to the bank can be from anywhere around the perimeter. From the ceremonial exit of the henge to the ceremonial start of the climb to what might be referred to as the timber 'Temple' is a walk of 100 m along the edge of the plateau. Because of the fixed locations of the ridge and the passage tomb, the entrances to the henge and the 'Temple' could not be put any closer together. The probability, therefore, is that the henge was not in existence during the initial phases of BNH5/6 but that its construction was tied to the final reorientation of the Annexe. These two sites are inextricably linked and should not be viewed as two separate sites but as elements of one great, integrated building project. A more tentative link is offered by Collins's observation, during his 1954 excavation, that the surface of the internal henge quarry ditch was covered with charcoal flecks. During and after the conflagration that ended the Temple, ash and charcoal would have blown around the Ballynahatty plateau for days until settled by rain (Collins 1957, 44). The models of inclusion (henge) and exclusion (BNH5/6) suggest functional interdependence and duality rather than replacement – the living below and the ancestors above, the theatre and the mortuary temple, public ceremony and intimate ritual, the present and eternity, the earth and the sky.

The construction of the Giant's Ring henge will also have post-dated the Middle Neolithic funerary activity at Ballynahatty, as well as referencing the Early Neolithic passage tomb in its design. Elsewhere, especially regarding the Boyne Valley, it has been argued that the end of the practice of constructing passage tombs around 3000 cal BC, and the start of different monument-building traditions, marked a dramatic cultural change or even an influx of people from elsewhere (Eogan 1991; Stout 1991; Eogan and Roche 1997). Stout has argued that this change from passage tomb construction to the construction of earthen embanked enclosures is comparable with the changing trends in Britain dated to around 2500 cal BC and proposes (for the Boyne region) that although the mechanism for this change is difficult to determine, it may be the result of an influx of newcomers (Stout 1991, 253, 255). Eogan and Roche similarly propose that the end of the passage tomb tradition signifies the beginning of a new complex which 'was not something that resulted from indigenous development' (1997, 221). Others, principally Neil Carlin (2017) and Alison Sheridan (2004; 2014) interpret the evidence differently with Sheridan, for example, citing a complex interplay of influences between the Boyne Valley and Orkney, and between Ireland and southern England, as major factors shaping developments in that part of Ireland over the first half of the 3rd millennium.

At Ballynahatty, the construction of the timber monument complex was indeed a radical departure from previous practices: the elaborate BNH6 square-in-circle structure, with excarnation at its core, seems to burst fully formed onto this landscape and the scale of community involvement, including the development of the BNH5 enclosure and the Giant's Ring, shows a degree of organisation and a spiritual momentum at odds with that reflected in the Middle Neolithic funerary practices that preceded their construction. However, it is not necessary to posit the arrival of new people to account for this; internal social dynamics, perhaps featuring the emergence of an elite capable of persuading large numbers of people to work together to create the impressive timber monument complex, could account for the observed changes (*cf.* Cooney and Grogan 1994, 90). The people who organised this major building project (and that of the Giant's Ring) appropriated the by-then ancient passage tomb in their ideology, siting the timber monuments on the ridge overlooking the passage tomb, erecting a line of posts leading south from the Middle Façade towards it, and then constructing the Giant's Ring around it.

11.8 Relationship with other Irish timber circles, and with 'square-in-circle' structures in Britain

Along with Knowth, Ballynahatty was one of the first timber circles to be discovered in Ireland. While timber circles have been well documented in Britain since the early 20th century, when Woodhenge was excavated between 1926 and 1928 by the Cunningtons (1929), it was not until the 1990s that Knowth (Eogan and Roche 1997) and Ballynahatty (Hartwell 1998) were uncovered by research-led excavations. Since these two initial sites, practically all of the known Irish timber circles have been discovered through development-led archaeology. These excavations have greatly increased the number and range of timber circles across Ireland, and continued research is uncovering more significant sites, such as the 'Four-Post enclosure' cropmark, discovered at Brú na Bóinne, Co. Meath, during a recent aerial survey (Condit and Keegan 2018)

To date, there are just over 20 potential timber circle sites within Ireland (Fig. 11.15). There is a high concentration of these sites in the east of the country, with an especially high concentration around the Boyne Valley in Co. Meath. However, excavations have revealed a much wider distribution, with examples also known from Countys Armagh, Carlow, Cork, Derry, Down, Kilkenny, Louth, Mayo, Westmeath and Wexford. Generally, the circles appear singly, although they are often set within a larger ceremonial landscape, as is the case at Ballynahatty. There are, however, a number of clustered sites. The clustered circles tend to have smaller, less elaborate structures and, in some cases, these have been interpreted as domestic structures (as discussed below). All the Irish circles fall into Darvill's 'Small Circle Tradition', which encompasses circles of <30 m (Darvill in press, 80). The Irish examples range from 17.50 m at Armalughey, Co. Tyrone, to just over 4.00 m for Building 3, Balgatheran, Co. Louth, which means that Ballynahatty is one of the largest Irish examples (Fig. 11.16). The variation in size may signify scales of purpose, with the smaller ones designed to be used by individual families or small groups, while the larger, more elaborate structures are likely to have served as foci for dispersed communities, drawing people together for important events.

There is a great deal of conformity between Irish timber circles, with a range of design elements that were variously adopted and adapted to meet the needs of the builders/users. Typically, the Irish timber circles consist of a sub-circular ring of posts surrounding a central rectangular four-post structure – a 'square-in-circle' arrangement. Many of them also include other elements, such as enhanced entranceways or 'porticos' and entrance screens/façades, as observed at Ballynahatty. In a few instances, the circles are fronted by enclosures, as is the case at Armalughey, Ballynahatty and Glenshane, but this does not appear to be a common feature. While all the recently excavated sites have contributed substantially

11. Interpreting the excavation results in the wider context of prehistoric Ballynahatty

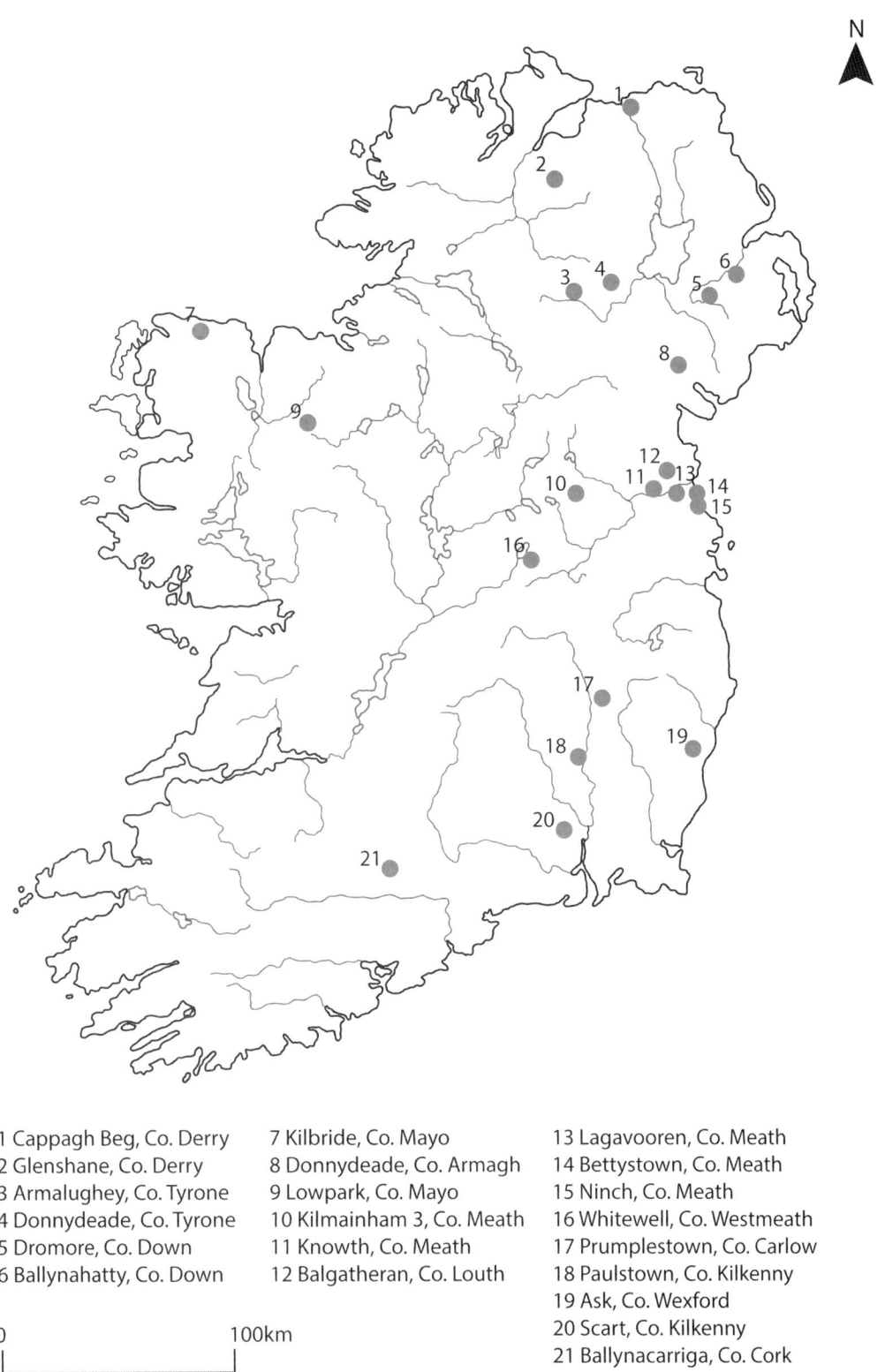

1 Cappagh Beg, Co. Derry
2 Glenshane, Co. Derry
3 Armalughey, Co. Tyrone
4 Donnydeade, Co. Tyrone
5 Dromore, Co. Down
6 Ballynahatty, Co. Down
7 Kilbride, Co. Mayo
8 Donnydeade, Co. Armagh
9 Lowpark, Co. Mayo
10 Kilmainham 3, Co. Meath
11 Knowth, Co. Meath
12 Balgatheran, Co. Louth
13 Lagavooren, Co. Meath
14 Bettystown, Co. Meath
15 Ninch, Co. Meath
16 Whitewell, Co. Westmeath
17 Prumplestown, Co. Carlow
18 Paulstown, Co. Kilkenny
19 Ask, Co. Wexford
20 Scart, Co. Kilkenny
21 Ballynacarriga, Co. Cork

0 ⊢——————⊣ 100km

Figure 11.15 Definite and possible Irish timber circle sites in Ireland.

to our understanding of the timber circle phenomenon in Ireland, they have also raised a lot of questions regarding how and why they were erected. It is obvious that some of the Irish examples involved a great deal of complexity and effort to construct. As we have already discussed, there is increasing evidence from the Irish examples that great care was also taken in their deconstruction, with evidence of posts being deliberately removed or, in some instances,

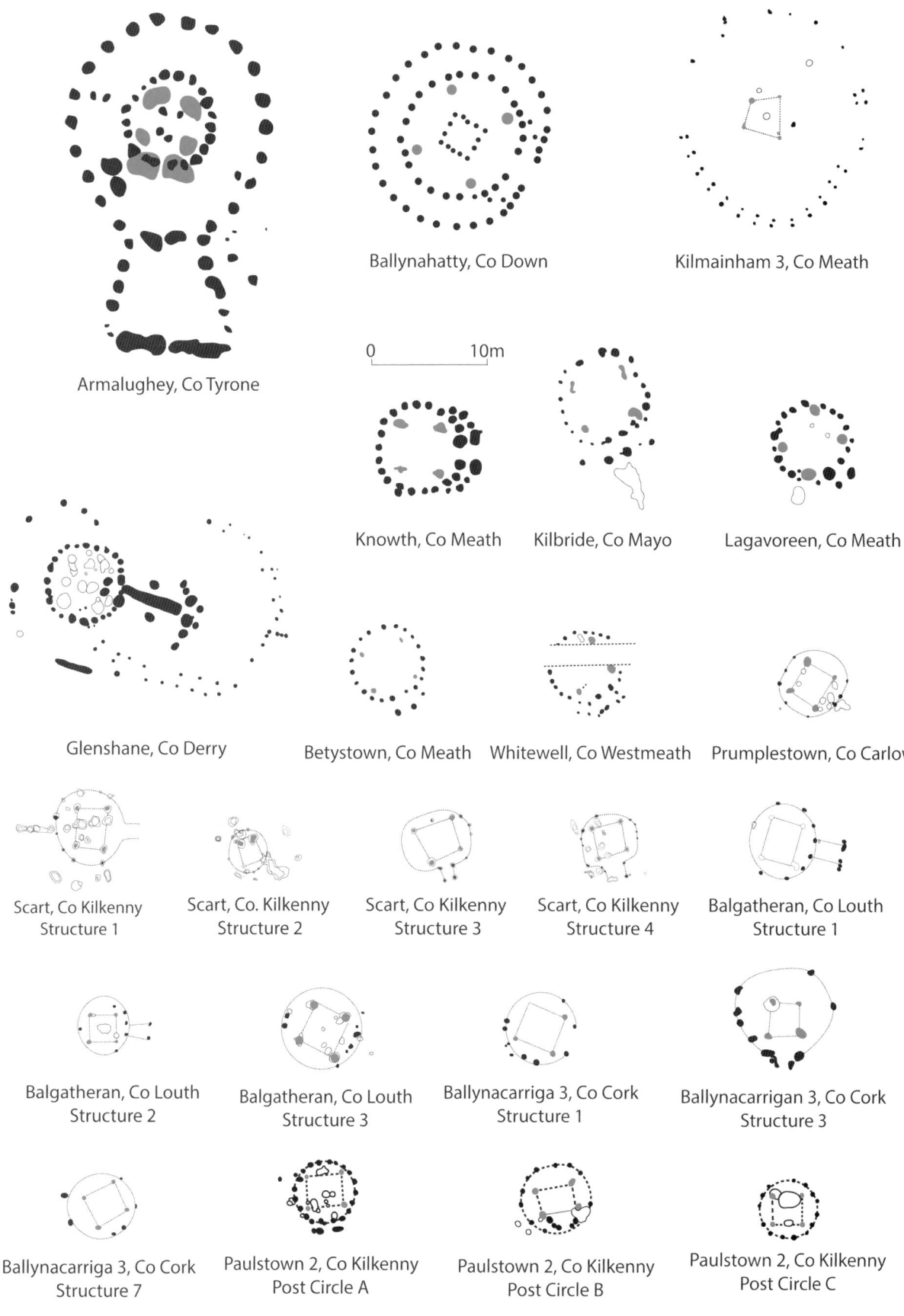

Figure 11.16 Plans of possible Neolithic Irish timber circles.

burnt before the postpits were backfilled with material, including deliberately deposited artefacts. It appears that these timber circles had defined life cycles, part of which called for the sites to be ritually dismantled.

While it is difficult to imagine that some of the larger structures, such as Ballynahatty and Armalughey, were anything other than ceremonial, it is often much harder to determine the nature of some of the smaller Irish sites. One of the main reasons for this is that timber circles and Late Neolithic houses are archaeologically very similar. Examples of Late Neolithic houses are relatively rare in Ireland but the known examples are similar in scale, form and construction to the smaller timber circles (Smyth 2011, 22). Material recovered from timber circle sites tends to be domestic in appearance, with lithics, Grooved Ware pottery, charcoal, burnt bone and cereals frequently recovered from these sites (Carlin 2016, 200). All this serves to blur the lines between our understanding of the domestic and the ritual at the end of the Neolithic and makes it difficult to be definitive. Examples of this can be found at Ballynacarrigan, Co. Cork (Lehane *et al.* 2011; 2019), Balgatheran, Co. Louth (Ó Drisceoil 2009) and Scart, Co. Westmeath (Monteith 2008; Laidlaw 2017), where the excavators have discussed the issues surrounding the classification of these sites. In summing up his discussion, Monteith (2008, 64) conceded that the structures at Scart may 'actually lie somewhere between these two interpretations, as the line between domestic and ritual may not be as distinct as often believed'. It is entirely possible that concepts of domestic and ritual were much more fluid, with some structures having served multiple functions, or their functions may have changed and transformed over the course of their lives (Thomas 2010).

It may also be that timber circles represent a deliberate monumentalisation of the domestic realm, with the structures intended to replicate contemporary houses (Bradley 2003, 13, Thomas 2010, 12; Carlin and Cooney 2017, Laidlaw 2017). Indeed, in discussing the 'square-in-circle' structure at Durrington Walls, Julian Thomas has argued for its status as a possible cult house (Thomas 2007b). As already noted, Irish timber circles are often found encircling a central rectangular structure defined by four posts. This 'square-in-circle' format is observed across Ireland and Britain during the Late Neolithic, at sites that have been ascribed both domestic *e.g.* Slieve Breagh, Co. Meath (de Paor and Ó h-Eochaidhe 1956), Greenbogs, Aberdeenshire (Noble *et al.* 2012), Wyke Down, Dorset (Green 2000) and ritual functions *e.g.* North and South Circles, Durrington Walls, Wiltshire (Darvill 2006), Machrie Moor, Arran (Haggarty 1991). An up-to-date distribution map of the 'square-in-circle' sites of Ireland and Britain has recently been published by Greaney *et al.* (2020, fig. 17).

Carlin and Cooney (2017, 46) suggest that timber circle builders sought to create and emphasise group identity within the local community by incorporating aspects of the domestic and everyday into the ritual realm; hence the similar forms and materials used. One of the main differences in form that has been used to distinguish domestic sites from timber circles is the lack of a central internal hearth (Smyth 2013, 320; Laidlaw 2017, 46). However, Smyth (2014, 94) has suggested that the central four-post structures, found within many timber circles, may be symbolic of the four sides or corners of the central hearths seen in Late Neolithic houses.

This ritualisation of the everyday may also explain the distinctly domestic nature of the artefacts found in association with timber circles, with quotidian artefacts being imbued with ritual significance through the practice of deliberate deposition. Instances of such are a common feature at Irish timber circles and are particularly associated with the deliberate destruction of the monument. As already discussed (see Section 6.3.4.3), Armalughey, Ballynahatty, Bettystown, Balgatheran, Lagavooren, Lowpark and Scart all display evidence of Grooved Ware deposition associated with the later stages or end of the site. This trend is replicated at post-circle A, Paulstown, Co. Kilkenny, although there, the pottery belongs to the Beaker tradition (Carlin 2011). This is also observed with lithic material at many sites, including Ballynahatty (see Section 7.3.3). It would also appear that certain postholes, especially the central Four-Poster setting, the entrance structure posts and the rear post, directly opposite the entrance, received a disproportionately large quantity of material (Eogan and Roche 1997; Eogan 2000; Cotter 2006; Phelan 2007; Stafford 2012).

11.8.1 The Brú na Bóinne connection

The best comparison to Ballynahatty in Ireland in terms of individual sites and the wider landscape is Brú na Bóinne in Co. Meath. The east coast lowland locations set in a bend in the Boyne and a loop in the Lagan are easily comparable. The Knowth 'square-in-circle' timber enclosure, though reconstructed with a roof, has similar features to BNH6 and a lithic assemblage which utilised Antrim flint. The intensity of monumentality in Brú na Bóinne is breath-taking and drone photographs, taken in 2018, the same year as the Ballynahatty images, both sharpened views of existing monuments and discovered spectacular new ones in the Newgrange floodplain (Condit and Keegan 2018).

One of these, the Newgrange 'Four Poster Enclosure' (*ibid.*, 25), is identical to BNH6 (Plates 93 and 94). In Figure 11.17, the BNH6 posts are laid over the GIS outline of the Newgrange postholes (left). The proportional match is remarkable, even to the extent of hinting at an intermediate post south of the entrance. This figure also shows their actual sizes at a common scale (right) – Newgrange is Ballynahatty '× 2'. This confirms that a conceptualised plan had been developed that potentially could be reproduced anywhere and at any scale. The interpretation of the

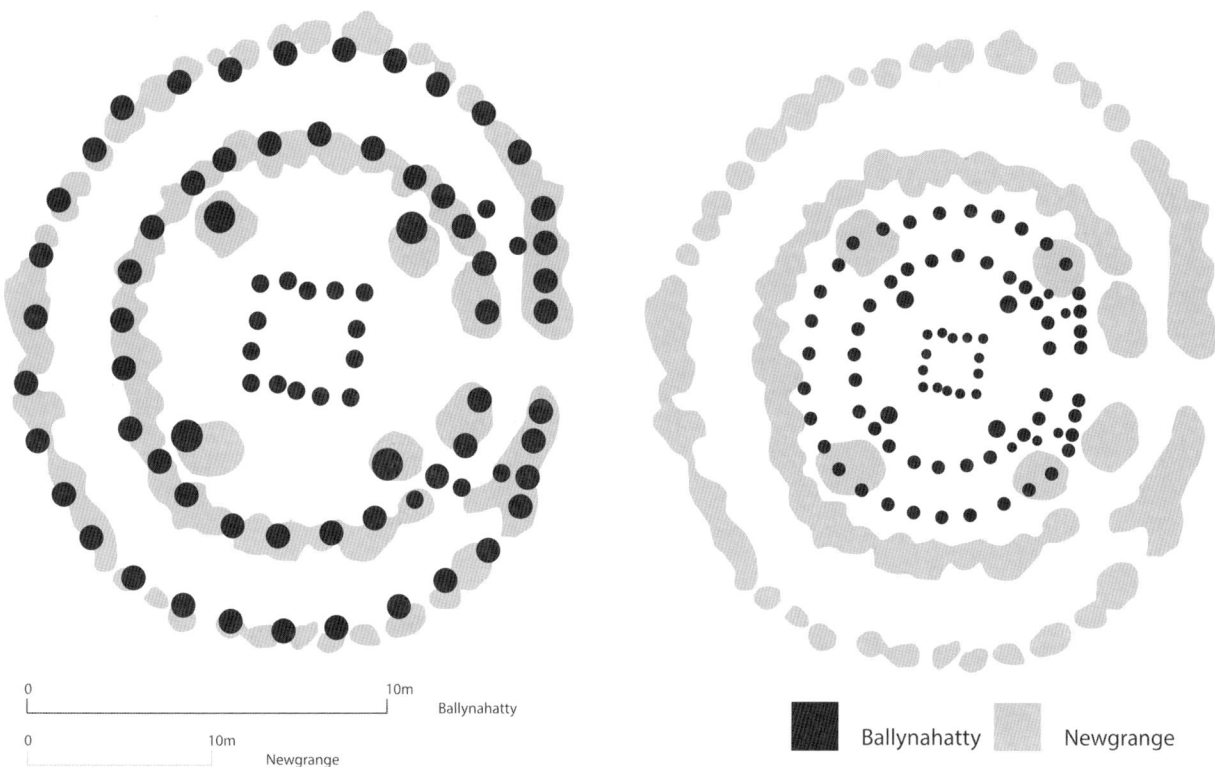

Figure 11.17 Comparison between the Newgrange Four Poster Enclosure (after Condit and Keegan 2018, Fig. 24) and BNH6: left: by proportion; right: by size.

Newgrange imagery suggests that the concentric circles contained a bank, so views of the interior were similarly restricted. There are more postholes in the longer Newgrange circuit, but the interval appears to be identical to that of BNH6, so a common unit of measurement was in use. Like BNH5, the Newgrange Four-Poster is set within a larger enclosure (Plate 94) of paired posts set on or containing a bank.

There are similarities with other sites in these two complexes. The Giant's Ring (BNH1) appears to be set out around the passage tomb (BNH2). BNH2 is centrally placed within the henge to north and south with a present diameter between bank tops of 201 m. East to west, the distances are 76 m and 125 m but combine to 201 m (Fig. 11.18). It seems that the intention was to create an embankment the equivalent of 200 m in diameter around the passage tomb. The limitation was the edge of the plateau and a steep slope 80 m to the east of BNH2 (Plate 95). This was solved by sliding the east–west diameter across the tomb to the west and creating a straight section at top break of slope in the east and a corresponding bulge on the west side. Therefore, the bank deliberately encircles the passage tomb and the diameter, and consequently, the circumference was predetermined and unalterable. The resultant shape of the Giant's Ring is, therefore, the product of the passage tomb position, the required size of the embankment and the distinctive local topography of Ballynahatty.

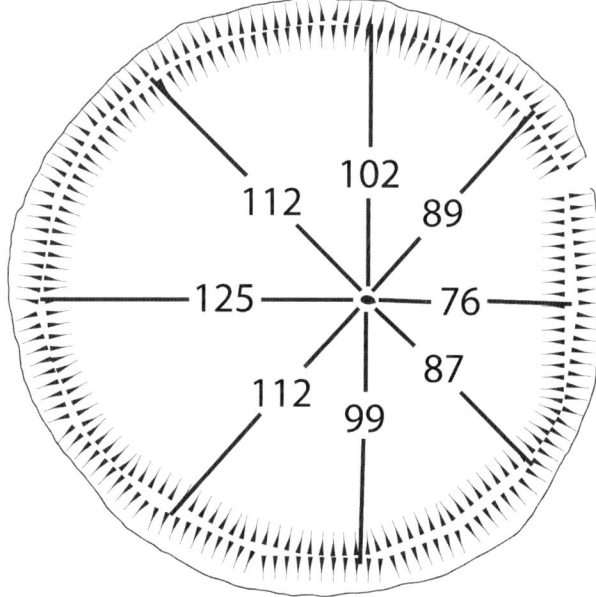

Figure 11.18 Giant's Ring, showing distance (m) between the passage tomb and the embankment as radii and diameters (Barrie Hartwell).

The rather unruly-looking Henge A at Brú na Bóinne (Plate 96) similarly surrounds a small passage tomb (Condit and Keegan, 2018, fig. 27). It has a flat side and an opposing bulge, and any diameter measured across the tomb is approximately 135 m. The tomb is centrally placed

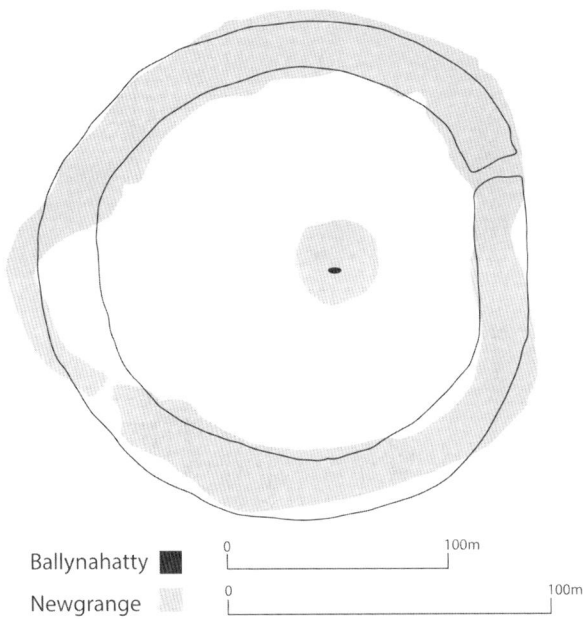

Figure 11.19 Giant's Ring and Newgrange Henge A: proportional relationship (Newgrange after Condit and Keegan 2018).

north to south, but along the axis is 45 m and 90 m – which still maintains a combined diameter of 135 m and a similar 1:2 ratio as the Giant's Ring. However, Henge A sits on flat ground with no constraints of topography to produce this shape. It is probable; therefore, that is a poorly executed imitation of the much larger Giant's Ring henge (Fig. 11.19). Its unrefined shape stands out in a landscape of deftly constructed monuments but could have been the first henge in the evolving later ritual landscape of Newgrange. Moving further east to Henge P one sees a regularisation of this shape and the development of the bulge into an annexe, and ultimately, further east again is the strict formality of the Geometric Henge. This may also account for the distorted shape of the Dowth Hall henge, the best surviving of the Brú na Bóinne henges. The Giant's Ring is larger by a considerable degree than any of the Boyne henges, and at a time when competitive aggrandisement was the order of the day, the Ballynahatty henge must have exerted a profound influence.

A third site, BNH43 (see Figs 2.10 and 4.1), is more ephemeral but is unusual on the Ballynahatty plateau because of its straight sides. It has an east–west orientation and appears to end in a circular site, recalling the layout of the Great Rectangular Palisaded Enclosure of the Boyne (Condit and Keegan 2018, fig. 93).

Both Ballynahatty and Brú na Bóinne have a common suite of ritual sites and the example of BNH6 and the Newgrange Four-Poster shows that these links and influences are not simply to be inferred from a common Grooved Ware tradition but are expressed architecturally and are tangible. The innovative development of novel architectural forms and concepts, the subtle use of height and slope to enhance the ritual experience and the emphasis on sheer size may plausibly have originated in the Ballynahatty landscape but were expanded and developed with verve and gusto in the greater expanse of Brú na Bóinne. We will never know whether this architectural movement was fed by a renewed religious fervour or the genius of a single individual, an Irish 'Imhotep'. What matters is a recognition that, following the end of the tradition of passage tomb construction and before the arrival of metal-using 'Beaker people' in Ireland, some groups were able to undertake large-scale, sophisticated projects of monument construction – and also went to extraordinary lengths to decommission some of these monuments at the end of their useful lives. The parallelism in monument development between Ballynahatty and Brú na Bóinne during the first half of the 3rd millennium not only demonstrates that the inhabitants of these areas were in contact with each other, but also suggests a degree of competitiveness in monument construction.

11.9 The Ballynahatty timber monument complex: what was it all for?

The key to this is in the interpretation of BNH6 as a mortuary enclosure and, in particular, the function of the central structure. It has been argued that BNH6 was an open, unroofed building and that the central structure was an open, unroofed platform. The high, enclosed nature of the walls, the narrow entrance and the limited available space within it precludes this platform from functioning as a stage for addressing an audience – the defined space is much too intimate, almost claustrophobic. The Four-Poster structure, towering over the rest of the 'Temple', marks this out as a position of great importance, open to the sky but not open to general view. From the outside, the location of an important act can be seen, but the actual process is hidden behind two impenetrable enclosure walls. The inner 'Temple' is the end of a ceremonial journey through the Ballynahatty landscape, but it is also the end of the journey through life and was responsible for the transformative process of becoming an ancestor. The function of the platform could have been as an altar which incorporated the exposure and processing of the dead – the process of excarnation. Such a mortuary function would fit with the nature of the Ballynahatty landscape and could be seen as the ultimate focus of community ceremony and death rituals (Hartwell 1998, 44). Similar interpretations for structures at Balfarg have been proposed (Barclay and Russell-White 1993). An alternative could be that BNH6 is the home of the ancestors and that the larger enclosure, BNH5, was for the exposure of corpses. It may be, however, that the treatment of the dead was only one of a number of aspects of ceremonial practices at Ballynahatty (Cooney 2000, 172). The elaborate nature of the monument complex also seems to have functioned to establish protocol for interaction between the living and the dead (Thomas 1999, 60).

The rituals and practices evident from the excavated remains would have been vital to maintaining the stability of the society which created the monuments, reinforcing the social order and ensuring continuity (Cooney and Grogan 1994, 90; Gibson 1998, 77). The importance of ceremonial procession is evident (Hartwell 2002, 530). The intricate layout of the design, the imposing façade and entrance way and the passageways were ultimately necessary to choreograph movement. In discussing enclosure elsewhere, Thomas states: 'A place which was already important was enclosed, and access to it at particular times regulated, while the precise experience of the location which a person could achieve was constrained' (1991, 60).

How, then, should we describe BNH5 and 6? 'Timber enclosure' is a bland expression of a defined space and gives no indication of function. The care, precision and public effort of the build displays high status. The development of the Grand Façade and the position on the ridge shows that this was meant to impress and possibly to intimidate. The control of access, the choreographing of movement without and within and the careful organisation of space show that this is a structure built for predetermined rituals. The placement within a known mortuary landscape and the link to the earlier passage tomb indicate that this is about the management of death and the referencing of an ancestral past. Although we cannot say for certain, this probably functioned in whole or in part as an excarnation site. Without further excavation, what happened in the larger enclosure remains a mystery but is a precinct set apart from the outside world. The organisation of BNH6 and the Eastern Settings gives a rare opportunity to witness a specific point at which ritual observance clearly happened. The soils of Ballynahatty do not preserve bone, and inhumation in soil would not be recognised, so the term 'mortuary house' in which a body is buried may not be appropriate. 'Charnel house', which encompasses excarnation, exhumation, cremation and storage of human skeletons, would certainly be an apt description. But what actually happened at that crucial point between the Eastern Settings, marked by a deposit of Knowth 1-style Grooved Ware pottery? Was this a glimpse of eternity – communing with the ancestors and making offerings? Was this the point beyond which only a priestly caste (or equivalent) could walk? A point where the simple observance of the natural order, as a human body decays, becomes a journey from death back to life but in a world of ancestors? Experiences such as these constitute a system of religious belief and ordered ritual tied to a specific building, so perhaps BNH6 is more than simply a warehouse for the dead or a shrine for the ancestors but is also a temple for the living.

12

Digitally recreating Ballynahatty and simulating astronomical alignments in Irish timber circles

R.P. Barrett

12.1 Introduction

With the advent of new digital technologies, archaeology possesses new tools for the presentation and investigation of archaeological contexts. Not only is it possible to produce realistic renders of the past but, with the addition of coding and simulations, archaeological queries can be investigated remotely and in a controlled environment. Many possibilities of research are now available and this project utilised two of these digital techniques to investigate areas of interest in Ballynahatty and surrounding sites: the recreation of the archaeological site using 3D reconstruction and a simulation of astronomical alignment patterns within the wider Irish landscape.

3D approximations, or 3D reconstructions, are user-generated virtual geometries based on existing or inferred archaeological features. Unlike other 3D digital techniques, 3D approximations do not directly measure standing features but rely on the interpretation of a modeller, who transliterates the current archaeological interpretation into a visual model. The primary advantage of this technique is the ability to present not only elements that are still present *in situ*, but also hypothetical elements, producing images that show the archaeological site at a time of flourishing rather than in ruins (Rua and Alvito 2011; Georgopoulos 2014). This follows the tradition of archaeological illustration championed by Sorrell (1981; Earl 2006).

In the case of Ballynahatty, 3D approximation can provide a fresh insight into life (and death) within the timber circle. Overall, the extent of the archaeological site is well preserved, with a clear indication of a plan provided by the excavated postholes. The shape of the structures helps determine movement around the site, with preserved entrances and focal points. Nonetheless, features above ground level are no longer present, meaning that some structural elements require inference. 3D approximation not only allows the presentation of these hypothetical components in an immediate and intuitive manner but also helps researchers visualise alternative theories providing new phenomenological observations that could not be obtained *in situ* (Patay-Horvath 2014).

Computer simulations are a form of experimentation that use virtual models to test theories in a non-destructive and interactive environment. Simulations have a variety of applications in every field but, for the purpose of this project, the question of alignment in Irish timber circles was selected. As with many archaeological sites, timber circles have been the subject of extensive speculation, with possible solar, lunar and sidereal alignments mentioned in the literature. Nonetheless, the intentionality of alignments in archaeology is a contentious issue, as a singular alignment is prone to be the result of random chance rather than design (Schaefer 2006; Polcaro and Polcaro 2009; Ruggles 2011). For this reason, the custom-built software *TarxienCore* was used to investigate alignment patterns for 22 different timber circles, providing new data regarding the accuracy of alignments and changes through space and time.

12.2 3D approximation: *in situ* elements

The 3D approximation of Ballynahatty combines *in situ* and hypothetical elements to showcase the current understanding of the archaeological remains (Plate 97). Because the images cannot convey the difference between what can be inferred directly from the archaeological remains and what has been extrapolated from them, a description of the approximation process is provided in this chapter. Overall, the 3D modelling of an archaeological site involves source collation, initial modelling of the standing features, the addition of hypothetical elements and verification of consistency between archaeological understanding and model.

The primary written sources used for the Ballynahatty model were the thesis published by Sarah Gormley (2004) and articles by Barrie Hartwell (1991; 1998; 2002). These provide an overview of the excavation and features uncovered, the chronology of the site and the wider landscape context. Pre-publication plans, sections and feature descriptions were also available, having recently been digitised for this publication. These data provided accurate measurements for the location and shape of postholes and helped clarify the stratigraphical relationship between areas of the site. Perhaps more significant was the guidance of the original site excavator Barrie Hartwell, who clarified conflicting reports and interpretations and shaped the model's overall look and feel.

The first elements that were reconstructed were those with direct evidence within the excavated remains. Primarily, these were the posts that originally stood within the excavated postholes. Each posthole was traced from the plan and extruded to the relative height. Gormley (2004) provides a graph of the estimated height of posts based on the depth of each hole (Fig. 5.6). Generally, the height of each post above the surface is given as three times the depth of the posthole. There are, however, some complications relating to the overall surface level, as an unspecified amount of original soil within the site has been lost through ploughing. During excavation, it was clear that the surface of the excavation was lower than the original ground level, meaning the postholes were originally deeper than measured. Nonetheless, the general application of the rule of posts being three times the posthole depth is applied to other sites where similar landscape changes occurred, meaning that the height of the posts in the Ballynahatty model is consistent with estimates from other sites.

While some postholes were directly recorded during excavations, others were inferred from aerial photography. In particular, the full extent of the external ring was obtained from cropmarks, which clearly show two rows of posts beyond the boundaries of excavation. These marks are mostly well defined, with occasional patches of uncertainty due to the nature of the soil. In particular, the circle portion just to the north of the excavated features seems to reside in a low part of the landscape, where pooling has occurred, making the features less clearly defined. Nonetheless, the post rows can be traced with some certainty. To the northwest of the circle, there is a section in which the posthole arrangement seems to differ. Although difficult to confirm with any certainty, it appears that the distance between the outer and inner postholes is reduced, suggesting that two of the posts may be recessed towards to centre of the circle. It is unclear whether this arrangement indicates the presence of an entrance or additional structure (Plate 1).

A further arrangement of posts visible in the aerial photography but not excavated is a mirror of the Eastern Setting. The Eastern Setting is present *in situ*, but the aerial photograph clearly shows a very similar arrangement to the west. Individual postholes can be identified in the aerial photograph and show that the settings were clearly a pair flanking a central walkway. As there is no discernible difference between the excavated and non-excavated features in the plans, these have been reconstructed identically.

Although most postholes are independent of one another and possess little stratigraphical relationship overall, it has been possible to distinguish different building phases and subsequent transformation of the site over its use. Several overlapping postholes suggest that an initial, smaller structure was later expanded with the addition of the annexes and an entranceway, as well as two phases of façade. (For a detailed description of the chronology, see Chapters 10 and 11.) To ensure the model would differentiate between different phases of activity, the posts were placed into different component groups, which allowed the modeller to toggle quickly certain structures from visible to invisible, hence producing different models that represent alternative time periods and highlight chronology (Plate 98).

The model itself was placed within its wider topographical context by making use of a 3D landscape obtained from drone footage captured in 2018. This footage was used to create a surface geometry using photogrammetry by David Craig, and the finished model was imported into *Sketchup*. To ensure the model was still workable given the large scale of the landscape model, *Meshlab* was used to decimate the points in the model, hence reducing the computational impact. This did result in a loss of definition; however, given that the model is used primarily for background rendering, it does not significantly affect accuracy. To ensure that the model fits within the landscape appropriately, the *DropGC* plugin by Smustard was used to ensure each posthole followed the landscape undulation. In terms of the viewers' experience, this also enhanced the accuracy of the model, especially as a prominent ridge runs through the site (Hartwell 1991).

12.3 3D approximation: hypothetical elements

Once the 3D model of the standing features was complete it was possible to add more hypothetical elements. These were based on the current understanding of the archaeology, as outlined in Chapter 5. Several elements are based on circumstantial archaeological evidence, such as panelling between the uprights and the presence of a central platform. In the model, panels have been added to the front of the outer circle of BNH6, as well as around the Entrance Chamber and the settings. The presence of panels is suggested by a small gully excavated between two of the postholes in BNH6, which contained traces of burning. The shape and fill of the gully seemed to indicate the presence of a horizontal plank of wood wedged between the two postholes to form a wall. It is possible that other gullies were present throughout the site but have been destroyed by subsequent soil activity. In the model, the panels are formed of roughly rectangular loose planks of wood, which would have slotted into grooves on the side of the posts. Their exact use is unknown, although they could have been used to obscure the view towards certain features, emphasise an entry or to carry some form of decoration. Given the highly ritualistic arrangement of space within the site, control of visibility could have played a significant role. As such, the panelling has also been used in the model in locations of high ritual meaning, such as the front of BNH6 and the Entrance Chamber, the latter to create separation from the annexes. It is worth noting that the arrangement of upright logs would have formed a natural barrier to visibility, and, in some cases (such as the façade), the presence of close-set posts would have obscured vision entirely in certain areas.

Panelling has also been used in the pair of Eastern Settings. Although there is no evidence of gullies, the arrangement of posts and the small size of the structures seem to suggest these were rooms or at least enclosed spaces. The shorter side, visible from a point between them, is formed of only two posts more widely spaced than the other posts in the structure. It is, therefore, likely that these formed an entranceway, with panelling occluding the view of the other three sides. The exact use of these areas is not discernible from the excavated remains; however, given the high visibility of these structures from the path leading from the entrance to BNH6, the current interpretation is that they could have been used as ossuaries. In the model, several skulls were placed to highlight this theory and contribute to the overall feel of the site (Plate 99).

In the centre of BNH6, a roughly square arrangement of small postholes has been interpreted as a platform (Plate 100). The closeness of the postholes makes it impossible for an individual to enter the enclosed area from any side. It is, therefore, plausible that rather than delimiting a space, the posts were used to support a wooden platform. The central location makes this structure a focal point of the timber circle, highly visible from the path between the entrance and the central circle. As for use, the current hypothesis is excarnation (Hartwell 2002; Chapter 11). Ballynahatty is seen as a crucial location for burial rites, and the arrangement of structures suggests that theatrical elements played an important role. Visitors would progressively enter the site and witness different stages of bodily decomposition, such as the bones within the ossuaries and decaying corpses on display at the centre. The platform would therefore allow the deceased to be placed in full view of anyone who entered.

In addition to hypothetical elements based on archaeological remains, other elements have been added based on speculation. In particular, the presence of lintels and of wooden infill of the lines of posts is not based on observed remains, but on assumptions of site use. The lintels created between posts in BNH6 and in the Entrance Chamber fit within a larger discussion of timber circle roofing (for a discussion of which see, for example, Eogan and Roche 1997). The absence of roofing due to site deterioration has led to uncertainty regarding the existence of a possible roof on parts of these structures. While roofing is certainly a possibility, in Ballynahatty, this has been discounted. The similarity in posthole size between the inner and outer central circle is inconsistent with a sloped roof, which would require larger inner posts and shorter outer posts. However, the absence of a roof does not preclude the presence of additional horizontal posts on top of the vertical ones. These would provide more stability to the structure and would let specific features stand out (Gibson and Simpson 1998). If lintels were not used, it would be difficult to identify individual parts of the timber circle as all posts would appear to be unconnected.

A similar suggestion involves the infill of the rows of posts. BNH5 and BNH6 are composed of two lines of vertical parallel postholes, which leave a central gap. While it is entirely possible that the intention was to have simple lines of posts to identify the confines of the site, these may have also been infilled to create walls. Walls would provide a physical barrier and force access through the entrance of the site, as well as limiting visibility from outside. While no infill has been found during excavation, materials would have been easily available in the form of wood offcuts from the creation of the posts. Branches, leaves and the top of trees would have been natural by-products of the building of Ballynahatty and could have been placed between the rows of posts. The absence of remains would be consistent as organic remains would not survive to the present day. Several methods of infill were tested during modelling (Plate 101).

12.4 Astronomical simulation: from sites to data

In addition to recreating the site in 3D, this project involved using simulations to investigate astronomical alignment at Ballynahatty and timber circles throughout Ireland. This investigation is based on methodologies developed within the field of archaeoastronomy which focuses on establishing an interest in astronomical events within past cultures. A primary emphasis in archaeoastronomy is alignment, the intentional orientation of sites and features to specific cosmological events. Many sites have been the subject of alignment studies, often identifying their relations with solar, lunar or sidereal events (see Ruggles 2015 for examples).

However, while single alignments frequently occur in archaeology, they are not sufficient to confirm intentionality, as it is statistically likely that a random orientation will by chance be aligned to a cosmological event (Polcaro and Polcaro 2009; Ruggles 2011). Intentionality is, therefore, determined by several factors, including closeness of alignments, evidence in the archaeological record and recurring patterns throughout multiple sites (Schaefer 2006). In the case of Ballynahatty and other timber circles in Ireland, the absence of literary sources or specific iconography make it difficult to determine whether there was an interest in astronomical events. Therefore, intentionality cannot be confirmed. Nonetheless, the presence of over 20 timber circles in Ireland allows for an investigation into patterns of alignment, which can provide new data for future research.

This study uses custom software *TarxienCore*, which has previously been used to identify alignments in Neolithic Malta (Barratt 2022). *TarxienCore* uses calculations by Meeus (1987) to measure the closeness of astronomical alignments throughout a year. The results are used to identify common alignments throughout sites at different time intervals, developing a chronology of alignments. While *TarxienCore* was originally written for *C#* in *Unity*, it has been rewritten for this project in *R*. This change has made the software more user-friendly and increased the accuracy of the results by removing 3D functionalities in favour of more robust mathematical models.

For this project, 22 sites were selected from the literature. While more sites were considered, those chosen had to follow guidelines to ensure the results were consistent. Primarily, sites had to have accessible site reports, with orientated plans and discernible centres and entrances. Unfortunately, it was not possible to visit the sites in person, so measurements had to be inferred from the plans. Table 12.1 presents a list of sites identified with measurements obtained. For each site, the primary axis was obtained by finding the centre of the structure and the midpoint of the entrance. As structures are not always entirely circular, and the reliability of plans is not always guaranteed, an undefined degree of error is introduced in this step.

For each site, *TarxienCore* uses the inputted measurements to calculate the position of astronomical bodies in relation to the sites at a determined point in time. To see changes of orientation throughout the Neolithic, this study chose the range 3000 BC to 1000 BC sampling alignments every 500 years. The software currently calculates the position of the sun and of the top 20 brightest stars in the sky. Therefore, for every minute of the selected years, the software determines the position of each astronomical body; it determines the angle between the centre of the site and the body in the celestial sphere (measured as ascension and declination) and returns the difference between this angle and the orientation of the site. To view this as a yearly trend, the software determines the average alignment when the star enters and exits the entrance area. The final value obtained is the accuracy of the alignment, or the difference between the average point of start and end of the alignment and the orientation of the site. The closer this value is to 0, the more precise is the orientation.

Table 12.1 List of sites used for this study, with orientation calculated from plans. Site reports used can be found in the reference list below.

Location	County	Site	Easting	Northing	Phase	Orientation
Armalughey	Tyrone	Site 20	264357	356840	LN	127°
Balgatheran 4 (1)	Louth	Structure A	302674	309977	–	105°
Balgatheran 4 (2)	Louth	Structure B	302674	309977	–	100°
Balgatheran 4 (3)	Louth	Structure C	302674	309977	–	108°
Ballynacarriga (1)	Cork	Structure 1	181470	102601	LN	146°
Ballynacarriga (2)	Cork	Structure 3	181470	102601	LN	152°
Ballynacarriga 2	Cork	Structure 7	181470	102601	LN	154°
Ballynahatty	Down	–	144473	523299	LN	115°
Bettystown	Meath	–	715526	773218	LN	163°
Cappagh Beg	Derry	Structure B	282959	436520	–	112°
Derrybeg	Derry	Site 12	307000	328350	MBA	122°
Glenshane Road Quarry	Derry	Structure 2	73821	567307	LN	113°
Kilbride	Mayo	–	527610	778927	LN	155°
Kilmainham 3	Meath	Structure 2	276100	274700	LN	116°
Knowth	Meath	–	699668	773447	LN	103°
Lagavooren	Meath	–	307320	273150	LN	133°
Lowpark	Mayo	–	147233	300643	LN	24°
Paulstown (A)	Kilkenny	Structure A	665600	658694	–	172°
Paulstown (B)	Kilkenny	Structure B	665600	658694	–	160°
Prumplestown	Carlow	–	276839	183174	–	121°
Scart	Kilkenny	Structure 1	656739	622635	LBA	148°
Whitewell	Westmeath	–	244708	244193	LN	143°

When an astronomical body or event is aligned to a high number of sites, the results become significant, resulting in a clear pattern of alignment. Nonetheless, it is important to note that even when close significant alignments occur, they can still be coincidental. It is unlikely that all astronomical bodies within an orientation would be the focus of alignment. But because the movement of the celestial sphere is overall consistent, aligning the site towards a singular astronomical body will result in false positive alignments with other bodies following similar trajectories. More importantly, external factors may affect site orientation. For example, orientating sites towards the winter sun out of a desire to maximise warmth or to be downwind would increase the chance of coincidental astronomical alignments, without involving a specific interest in astronomy.

Additionally, the software and methodology can result in significant errors that may change the results presented here. The orientation of the sites measured from plans can increase inaccuracies. Measurement of the angles *in situ* would aid the precision of the results, but with all of the sites, there is no standing archaeology, making it impossible to obtain better data. Further information was not available for this research but can be used in the future to aid interpretation. In particular, the height of the horizon plays a significant role in astronomy, as a high or low horizon significantly alters the point of rising of a cosmological body. Similarly, the maximum altitude of a cosmological body considered for alignments was a conservative 10° above the horizon, beyond which it is considered outside the entranceway. A study on the relationship between entrances to central circles and entrances to outer palisades would aid the precision of the results.

12.5 Astronomical simulation: results

The sites considered in Ireland have a clear preference in orientation towards the southeast, with only Lowpark having an orientation towards the north west (Fig 12.1). With 22 sites considered, the chances of this outcome are well below the r-value in a binomial distribution test (roughly $2.558 \times 10^{-11}\%$). Although dealing with a subset of all sites, these were not selected according to a variable that would affect this outcome. As such, statistics suggest that the site orientations are intentional, albeit not necessarily due to astronomical interest.

The astronomical simulations conducted with *TarxienCore* show significant alignments with several astronomical bodies (Table 12.2). Sirius, Rigel, Betelgeuse, Aldebaran, and Antares have a high incidence of alignments, with Rigel being aligned to the largest number of sites (13). Some of the alignments appear only late or

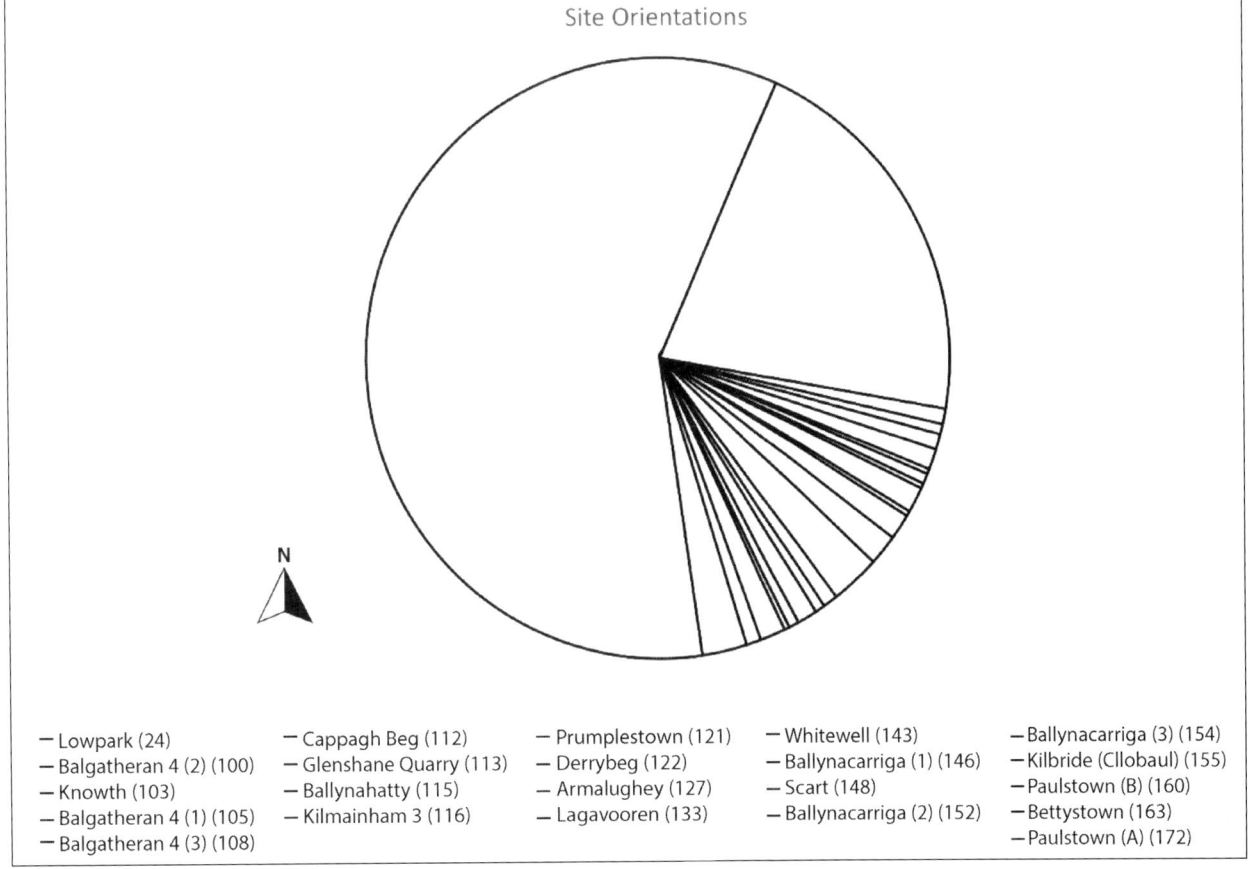

Figure 12.1 Orientation chart of Irish timber circles. Of all sites considered, only Lowpark is not orientated towards the SW. A binomial probability test shows that this arrangement is highly non-random.

Table 12.2 Results of alignment study using TarxienCore. Each table shows the error in orientation between the yearly average position of the cosmological body and the site orientation. On the left, the position of the body is given at alignment start (which may differ from rising if the body rises beyond the entrance), in the middle it is given at the end of alignment, and to the right the average of the two values is presented. Values in italics are <2° error, and as such are highlighted as significant, although they still may be coincidental.

Spring Equinox (Start)

	3000 BC	2500 BC	2000 BC	1500 BC	1000 BC
Balgatheran 4 (1)	4.8	4.86	4.94	4.91	4.83
Balgatheran 4 (2)	4.88	4.94	4.82	4.99	4.92
Balgatheran 4 (3)	4.92	4.98	4.85	4.82	4.95
Knowth	4.98	4.84	4.92	4.89	4.82

Spring Equinox (End)

	3000 BC	2500 BC	2000 BC	1500 BC	1000 BC
Balgatheran 4 (1)	*0.05*	*0.11*	*0.81*	*0.78*	*0.91*
Balgatheran 4 (2)	-4.94	-4.88	-4.18	-4.21	-4.08
Balgatheran 4 (3)	3.05	3.11	3.81	3.78	3.91
Knowth	*-1.99*	*-1.93*	*-1.23*	*-1.06*	*-1.13*

Spring Equinox (Average)

	3000 BC	2500 BC	2000 BC	1500 BC	1000 BC
Balgatheran 4 (1)	2.43	2.49	2.88	2.85	2.87
Balgatheran 4 (2)	*-0.03*	*0.03*	*0.32*	*0.39*	*0.42*
Balgatheran 4 (3)	3.99	4.05	4.33	4.3	4.43
Knowth	*1.5*	*1.46*	*1.85*	*1.92*	*1.85*

Autumn Equinox (Start)

	3000 BC	2500 BC	2000 BC	1500 BC	1000 BC
Balgatheran 4 (1)	4.85	4.8	4.85	4.86	4.91
Balgatheran 4 (2)	4.96	4.91	4.95	4.96	4.81
Knowth	4.85	4.8	4.84	4.85	4.91

Autumn Equinox (End)

	3000 BC	2500 BC	2000 BC	1500 BC	1000 BC
Balgatheran 4 (1)	2.17	2.53	2.16	2.59	2.84
Balgatheran 4 (2)	-2.82	-2.46	-2.83	-2.4	-2.15
Knowth	*0.32*	*0.68*	*0.32*	*0.53*	*0.79*

Autumn Equinox (Average)

	3000 BC	2500 BC	2000 BC	1500 BC	1000 BC
Balgatheran 4 (1)	3.51	3.67	3.51	3.73	3.88
Balgatheran 4 (2)	*1.07*	*1.23*	*1.06*	*1.28*	*1.33*
Knowth	2.59	2.74	2.58	2.69	2.85

Winter Solstice (Start)

	3000 BC	2500 BC	2000 BC	1500 BC	1000 BC
Ballynacarriga (1)	4.97	4.92	4.99	4.81	4.8
Ballynacarriga (2)	4.85	4.79	4.86	4.89	4.88
Ballynacarriga 2	4.91	4.85	4.92	4.95	4.93
Bettystown	4.77	4.93	4.99	4.8	4.99
Kilbride	4.87	4.81	4.87	4.91	4.89
Lagavooren	*0.61*	*0.76*	*0.83*	*0.87*	*1.06*
Paulstown (B)	4.94	4.88	4.94	-	-
Scart	4.96	4.91	4.97	4.8	4.99
Whitewell	4.97	4.92	4.99	4.82	4.8

Winter Solstice (End)

	3000 BC	2500 BC	2000 BC	1500 BC	1000 BC
Ballynacarriga (1)	-5.03	-5.01	-5.03	-5	-5.02
Ballynacarriga (2)	*-2.12*	*-1.96*	*-1.67*	*-1.64*	*-1.44*
Ballynacarriga 2	*-0.12*	*0.03*	*0.32*	*0.35*	*0.55*
Bettystown	3.65	3.8	3.87	4.12	4.54
Kilbride	-4.66	-4.5	-4.22	-3.96	-3.76
Lagavooren	-5.07	-5.12	-5.05	-5.01	-5.03
Paulstown (B)	4.28	4.43	4.72	-	-
Scart	-5.15	-5.21	-5.15	-5.12	-5.14
Whitewell	-5.13	-5.19	-5.13	-5.08	-5.11

Winter Solstice (Average)

	3000 BC	2500 BC	2000 BC	1500 BC	1000 BC
Ballynacarriga (1)	*-0.03*	*-0.04*	*-0.02*	*-0.1*	*-0.11*
Ballynacarriga (2)	*1.37*	*1.42*	*1.6*	*1.63*	*1.72*
Ballynacarriga 2	2.4	2.44	2.62	2.65	2.74
Bettystown	4.21	4.37	4.43	4.46	4.77
Kilbride	*0.11*	*0.16*	*0.33*	*0.48*	*0.57*
Lagavooren	-2.23	-2.18	-2.11	-2.07	-1.99
Paulstown (B)	4.61	4.66	4.83	-	-
Scart	*-0.1*	*-0.15*	*-0.09*	*-0.16*	*-0.07*
Whitewell	*-0.08*	*-0.14*	*-0.07*	*-0.13*	*-0.16*

(Continued)

Table 12.2 (Continued)

Sirius (Start)

	3000 BC	2500 BC	2000 BC	1500 BC	1000 BC
Armalughey	-3.16	*0.16*	*2.94*	4.65	4.67
Ballynacarriga (1)	4.73	4.87	-	-	-
Ballynacarriga (2)	4.88	-	-	-	-
Ballynacarriga 2	4.94	-	-	-	-
Ballynahatty	-	-	-	-	-4.67
Derrybeg	-	-4.48	*-1.68*	*0.62*	*2.42*
Kilbride	4.86	-	-	-	-
Kilmainhan 3	-	-	-	-4.9	-3.09
Lagavooren	3.49	4.65	4.63	4.69	4.67
Prumplestown	-	-4.24	*-1.55*	*0.65*	*2.35*
Scart	4.78	4.88	-	-	-
Whitewell	4.63	4.69	4.82	4.88	-

Sirius (End)

	3000 BC	2500 BC	2000 BC	1500 BC	1000 BC
Armalughey	-5.1	-5.1	-5.11	-5.1	-5.11
Ballynacarriga (1)	-3.23	*1.45*	-	-	-
Ballynacarriga (2)	2.77	-	-	-	-
Ballynacarriga 2	4.72	-	-	-	-
Ballynahatty	-	-	-	-	-5.07
Derrybeg	-	-5.1	-	-	-5.1
Kilbride	*1.36*	-	-	-	-
Kilmainhan 3	-	-	-	-5.1	-5.1
Lagavooren	-5.11	-5.1	-5.11	-5.1	-5.11
Prumplestown	-	-5.1	-5.1	-5.1	-5.1
Scart	*-1.68*	3.08	-	-	-
Whitewell	-5.11	-4.42	*-0.34*	*2.85*	-

Sirius (Average)

	3000 BC	2500 BC	2000 BC	1500 BC	1000 BC
Armalughey	-4.13	*-2.47*	*-1.09*	*-0.23*	*-0.22*
Ballynacarriga (1)	*0.75*	3.16	-	-	-
Ballynacarriga (2)	3.83	-	-	-	-
Ballynacarriga 2	4.83	-	-	-	-
Ballynahatty	-	-	-	-	-4.87
Derrybeg	-	-4.79	-3.39	*-2.24*	*-1.34*
Kilbride	3.11	-	-	-	-
Kilmainhan 3	-	-	-	-5	-4.1
Lagavooren	*-0.81*	*-0.23*	*-0.24*	*-0.21*	*-0.22*
Prumplestown	-	-4.67	-3.33	*-2.23*	*-1.38*
Scart	*1.55*	3.98	-	-	-
Whitewell	*-0.24*	*0.14*	*2.24*	3.87	-

Rigil Kent (Start)

	3000 BC	2500 BC	2000 BC	1500 BC	1000 BC
Ballynacarriga 2	-3.88	-	-	-	-
Bettystown	*-0.6*	-	-	-	-
Paulstown (A)	4.61	*-2.26*	-	-	-

Rigil Kent (End)

	3000 BC	2500 BC	2000 BC	1500 BC	1000 BC
Ballynacarriga 2	-5.1	-	-	-	-
Bettystown	-5.1	-	-	-	-
Paulstown (A)	-5.1	-5.1	-	-	-

Rigil Kent (Average)

	3000 BC	2500 BC	2000 BC	1500 BC	1000 BC
Ballynacarriga 2	-4.49	-	-	-	-
Bettystown	*-2.85*	-	-	-	-
Paulstown (A)	*-0.25*	-3.68	-	-	-

Arcturus (Start)

	3000 BC	2500 BC	2000 BC	1500 BC	1000 BC
Lowpark	-	-	4.9	4.63	4.63

Arcturus (End)

	3000 BC	2500 BC	2000 BC	1500 BC	1000 BC
Lowpark	-	-	2.54	-5.09	-5.1

Arcturus (Average)

	3000 BC	2500 BC	2000 BC	1500 BC	1000 BC
Lowpark	-	-	3.72	*-0.23*	*-0.24*

Vega (Start)

	3000 BC	2500 BC	2000 BC	1500 BC	1000 BC
Lowpark	-	4.89	4.76	4.61	4.61

Vega (End)

	3000 BC	2500 BC	2000 BC	1500 BC	1000 BC
Lowpark	-	*2.13*	-1.75	-4.69	-5.09

Vega (Average)

	3000 BC	2500 BC	2000 BC	1500 BC	1000 BC
Lowpark	-	3.51	*1.51*	*-0.04*	*-0.24*

(Continued)

Table 12.2 Results of alignment study using TarxienCore. Each table shows the error in orientation between the yearly average position of the cosmological body and the site orientation. On the left, the position of the body is given at alignment start (which may differ from rising if the body rises beyond the entrance), in the middle it is given at the end of alignment, and to the right the average of the two values is presented. Values in italics are <2° error, and as such are highlighted as significant, although they still may be coincidental. (Continued)

Capella (Start)

	3000 BC	2500 BC	2000 BC	1500 BC	1000 BC
Lowpark	-	-	*-1.92*	4.64	4.62

Capella (End)

	3000 BC	2500 BC	2000 BC	1500 BC	1000 BC
Lowpark	-	-	-5.1	-5.1	-5.1

Capella (Average)

	3000 BC	2500 BC	2000 BC	1500 BC	1000 BC
Lowpark	-	-	-3.51	*-0.23*	*-0.24*

Rigel (Start)

	3000 BC	2500 BC	2000 BC	1500 BC	1000 BC
Armalughey	-	-4.56	*0.38*	4.63	4.64
Ballynacarriga (1)	4.62	4.63	4.87	-	-
Ballynacarriga (2)	4.63	4.85	-	-	-
Ballynacarriga 2	4.64	4.88	-	-	-
Ballynahatty	-	-	-	-	*-2.97*
Bettystown	4.65	-	-	-	-
Derrybeg	-	-	-4.26	*0.17*	4.02
Kilbride	4.65	4.78	-	-	-
Kilmainhan 3	-	-	-	-	*-1.44*
Lagavooren	-3.21	*2.17*	4.61	4.63	4.75
Paulstown (B)	4.72	-	-	-	-
Prumplestown	-	-	-4.03	*0.2*	3.92
Scart	4.64	4.69	4.89	-	-
Whitewell	4.63	4.65	4.69	4.87	-

Rigel (End)

	3000 BC	2500 BC	2000 BC	1500 BC	1000 BC
Armalughey	-	-5.07	-5.1	-5.11	-5.11
Ballynacarriga (1)	-5.11	-5.11	*1.74*	-	-
Ballynacarriga (2)	-5.11	*0.64*	-	-	-
Ballynacarriga 2	-5.11	*2.64*	-	-	-
Ballynahatty	-	-	-	-	-5.06
Bettystown	-5.11	-	-	-	-
Derrybeg	-	-	-5.1	-5.1	-5.11
Kilbride	-5.11	*-1.22*	-	-	-
Kilmainhan 3	-	-	-	-	-5.1
Lagavooren	-5.1	-5.1	-5.1	-5.1	*-2.97*
Paulstown (B)	-3.33	-	-	-	-
Prumplestown	-	-	-5.1	-5.1	-5.11
Scart	-5.11	-3.85	3.37	-	-
Whitewell	-5.1	-5.11	-4.09	*2.22*	-

Rigel (Average)

	3000 BC	2500 BC	2000 BC	1500 BC	1000 BC
Armalughey	-	-4.82	-2.36	*-0.24*	*-0.24*
Ballynacarriga (1)	*-0.25*	*-0.24*	3.31	-	-
Ballynacarriga (2)	*-0.24*	*2.75*	-	-	-
Ballynacarriga 2	*-0.24*	3.76	-	-	-
Ballynahatty	-	-	-	-	-4.02
Bettystown	*-0.23*	-	-	-	-
Derrybeg	-	-	-4.68	-2.47	*-0.55*
Kilbride	*-0.23*	*1.78*	-	-	-
Kilmainhan 3	-	-	-	-	-3.27
Lagavooren	-4.16	*-1.47*	*-0.25*	*-0.24*	*0.89*
Paulstown (B)	*0.7*	-	-	-	-
Prumplestown	-	-	-4.57	-2.45	*-0.6*
Scart	*-0.24*	*0.42*	4.13	-	-
Whitewell	*-0.24*	*-0.23*	*0.3*	3.55	-

Procyon (Start)

	3000 BC	2500 BC	2000 BC	1500 BC	1000 BC
Balgatheran 4 (2)	4.86	4.89	-	-	-
Knowth	4.9	-	-	-	-

Procyon (End)

	3000 BC	2500 BC	2000 BC	1500 BC	1000 BC
Balgatheran 4 (2)	*1.36*	4.06	-	-	-
Knowth	4.38	-	-	-	-

Procyon (Average)

	3000 BC	2500 BC	2000 BC	1500 BC	1000 BC
Balgatheran 4 (2)	3.11	4.48	-	-	-
Knowth	4.64	-	-	-	-

(Continued)

12. Digitally recreating Ballynahatty and simulating astronomical alignments in Irish timber circles   195

Table 12.2 (Continued)

Betelgeuse (Start)

	3000 BC	2500 BC	2000 BC	1500 BC	1000 BC
Balgatheran 4 (1)	2.69	4.61	4.7	4.84	4.89
Balgatheran 4 (2)	-2.17	2.02	4.61	4.66	4.8
Balgatheran 4 (3)	4.63	4.64	4.81	4.89	-
Ballynahatty	4.71	4.87	-	-	-
Cappagh Beg	4.65	4.76	4.87	-	-
Derrybeg	4.89	-	-	-	-
Glenshane Road Quarry	4.64	4.81	4.89	-	-
Kilmainhan 3	4.79	4.88	-	-	-
Knowth	0.76	4.63	4.64	4.78	4.87
Prumplestown	4.89	-	-	-	-

Betelgeuse (End)

	3000 BC	2500 BC	2000 BC	1500 BC	1000 BC
Balgatheran 4 (1)	-5.1	-5.11	-3.72	0.17	3.62
Balgatheran 4 (2)	-5.1	-5.1	-5.1	-4.83	-1.38
Balgatheran 4 (3)	-5.11	-5.05	-0.72	3.17	-
Ballynahatty	-3.53	1.34	-	-	-
Cappagh Beg	-5.11	-2.18	2.33	-	-
Derrybeg	3.79	-	-	-	-
Glenshane Road Quarry	-5.11	-0.95	3.52	-	-
Kilmainhan 3	-1.76	2.99	-	-	-
Knowth	-5.1	-5.11	-5.1	-1.8	1.65
Prumplestown	3.97	-	-	-	-

Betelgeuse (Average)

	3000 BC	2500 BC	2000 BC	1500 BC	1000 BC
Balgatheran 4 (1)	-1.21	-0.25	0.49	2.51	4.26
Balgatheran 4 (2)	-3.64	-1.54	-0.25	-0.09	1.71
Balgatheran 4 (3)	-0.24	-0.21	2.05	4.03	-
Ballynahatty	0.59	3.11	-	-	-
Cappagh Beg	-0.23	1.29	3.6	-	-
Derrybeg	4.34	-	-	-	-
Glenshane Road Quarry	-0.24	1.93	4.21	-	-
Kilmainhan 3	1.52	3.94	-	-	-
Knowth	-2.17	-0.24	-0.23	1.49	3.26
Prumplestown	4.43	-	-	-	-

Beta Centauri (Start)

	3000 BC	2500 BC	2000 BC	1500 BC	1000 BC
Ballynacarriga (2)	-1.06	-	-	-	-
Ballynacarriga 2	0.87	-	-	-	-
Bettystown	4.63	-	-	-	-
Kilbride	-2.52	-	-	-	-
Paulstown (A)	4.64	4.62	-	-	-
Paulstown (B)	4.61	-3.51	-	-	-

Beta Centauri (End)

	3000 BC	2500 BC	2000 BC	1500 BC	1000 BC
Ballynacarriga (2)	-5.1	-	-	-	-
Ballynacarriga 2	-5.1	-	-	-	-
Bettystown	-5.1	-	-	-	-
Kilbride	-5.1	-	-	-	-
Paulstown (A)	-5.1	-5.1	-	-	-
Paulstown (B)	-5.1	-5.1	-	-	-

Beta Centauri (Average)

	3000 BC	2500 BC	2000 BC	1500 BC	1000 BC
Ballynacarriga (2)	-3.08	-	-	-	-
Ballynacarriga 2	-2.12	-	-	-	-
Bettystown	-0.24	-	-	-	-
Kilbride	-3.81	-	-	-	-
Paulstown (A)	-0.23	-0.24	-	-	-
Paulstown (B)	-0.25	-4.31	-	-	-

Acrux (Start)

	3000 BC	2500 BC	2000 BC	1500 BC	1000 BC
Paulstown (A)	-4.6	-	-	-	-

Acrux (End)

	3000 BC	2500 BC	2000 BC	1500 BC	1000 BC
Paulstown (A)	-5.1	-	-	-	-

Acrux (Average)

	3000 BC	2500 BC	2000 BC	1500 BC	1000 BC
Paulstown (A)	-4.85	-	-	-	-

(Continued)

Table 12.2 Results of alignment study using TarxienCore. Each table shows the error in orientation between the yearly average position of the cosmological body and the site orientation. On the left, the position of the body is given at alignment start (which may differ from rising if the body rises beyond the entrance), in the middle it is given at the end of alignment, and to the right the average of the two values is presented. Values in italics are <2° error, and as such are highlighted as significant, although they still may be coincidental. (Continued)

Aldebaran (Start)

	3000 BC	2500 BC	2000 BC	1500 BC	1000 BC
Balgatheran 4 (1)	4.64	4.74	4.87	-	-
Balgatheran 4 (2)	*2.04*	4.62	4.72	4.86	-
Balgatheran 4 (3)	4.63	4.84	-	-	-
Ballynahatty	4.86	-	-	-	-
Cappagh Beg	4.75	4.88	-	-	-
Glenshane Road Quarry	4.81	4.89	-	-	-
Kilmainhan 3	4.89	-	-	-	-
Knowth	4.62	4.65	4.83	4.9	-

Aldebaran (End)

	3000 BC	2500 BC	2000 BC	1500 BC	1000 BC
Balgatheran 4 (1)	-5.11	*-2.9*	*1.98*	-	-
Balgatheran 4 (2)	-5.1	-5.1	-3.02	*1.62*	-
Balgatheran 4 (3)	-5.02	*0.1*	-	-	-
Ballynahatty	*1.37*	-	-	-	-
Cappagh Beg	*-2.14*	3.19	-	-	-
Glenshane Road Quarry	*-0.91*	4.37	-	-	-
Kilmainhan 3	3.02	-	-	-	-
Knowth	-5.11	-4.87	*0.01*	4.64	-

Aldebaran (Average)

	3000 BC	2500 BC	2000 BC	1500 BC	1000 BC
Balgatheran 4 (1)	*-0.24*	*0.92*	3.43	-	-
Balgatheran 4 (2)	*-1.53*	*-0.24*	*0.85*	3.24	-
Balgatheran 4 (3)	*-0.2*	2.47	-	-	-
Ballynahatty	3.12	-	-	-	-
Cappagh Beg	*1.31*	4.04	-	-	-
Glenshane Road Quarry	*1.95*	4.63	-	-	-
Kilmainhan 3	3.96	-	-	-	-
Knowth	*-0.25*	*-0.11*	2.42	4.77	-

Antares (Start)

	3000 BC	2500 BC	2000 BC	1500 BC	1000 BC
Armalughey	-	-	4.9	4.8	4.64
Balgatheran 4 (1)	4.61	4.11	*-0.53*	-	-
Balgatheran 4 (2)	3.95	*-0.71*	-	-	-
Balgatheran 4 (3)	4.75	4.63	2.36	-2.27	-
Ballynahatty	4.89	4.77	4.65	4.11	*-0.49*
Cappagh Beg	4.84	4.64	4.63	*0.94*	-3.8
Derrybeg	-	-	4.84	4.66	4.62
Glenshane Road Quarry	4.85	4.68	4.63	*2.04*	-2.67
Kilmainhan 3	-	4.83	4.63	4.63	*1*
Knowth	4.63	*2.2*	-2.46	-	-
Lagavooren	-	-	-	-	4.83
Prumplestown	-	-	4.83	4.63	4.64

Antares (End)

	3000 BC	2500 BC	2000 BC	1500 BC	1000 BC
Armalughey	-	-	4.68	*-1.04*	-5.11
Balgatheran 4 (1)	-5.1	-5.1	-5.1	-	-
Balgatheran 4 (2)	-5.1	-5.1	-	-	-
Balgatheran 4 (3)	-2.85	-5.11	-5.1	-5.1	-5.1
Ballynahatty	3.58	*-1.82*	-5.11	-5.1	-5.1
Cappagh Beg	*0.11*	-5.11	-5.11	-5.1	-5.06
Derrybeg	-	-	*-0.03*	-5.11	-5.11
Glenshane Road Quarry	*1.32*	-4.15	-5.11	-5.1	-5.1
Kilmainhan 3	-	*-0.1*	-5.11	-5.11	-5.11
Knowth	-5.1	-5.1	-5.1	-	-
Lagavooren	-	-	-	-	*0.22*
Prumplestown	-	-	*0.3*	-5.1	-5.1

Antares (Average)

	3000 BC	2500 BC	2000 BC	1500 BC	1000 BC
Armalughey	-	-	4.79	*1.88*	*-0.24*
Balgatheran 4 (1)	*-0.25*	*-0.5*	-2.82	-	-
Balgatheran 4 (2)	*-0.58*	-2.91	-	-	-
Balgatheran 4 (3)	*0.95*	*-0.24*	*-1.37*	-3.69	-
Ballynahatty	4.24	*1.48*	*-0.23*	*-0.5*	-2.8
Cappagh Beg	2.48	*-0.24*	*-0.24*	-2.08	-4.43
Derrybeg	-	-	2.41	*-0.23*	*-0.25*
Glenshane Road Quarry	3.09	*0.27*	*-0.24*	*-1.53*	-3.89
Kilmainhan 3	-	2.37	*-0.24*	*-0.24*	-2.06
Knowth	*-0.24*	*-1.45*	-3.78	-	-
Lagavooren	-	-	-	-	*2.53*
Prumplestown	-	-	2.57	*-0.24*	*-0.23*

(Continued)

Table 12.2 (Continued)

	Beta Crux (Start)					Beta Crux (End)					Beta Crux (Average)				
	3000 BC	2500 BC	2000 BC	1500 BC	1000 BC	3000 BC	2500 BC	2000 BC	1500 BC	1000 BC	3000 BC	2500 BC	2000 BC	1500 BC	1000 BC
Ballynacarriga (2)	*-1.96*	-	-	-	-	-5.1	-	-	-	-	-3.53	-	-	-	-
Ballynacarriga 2	*0*	-	-	-	-	-5.1	-	-	-	-	*-2.55*	-	-	-	-
Bettystown	4.33	-	-	-	-	-5.1	-	-	-	-	*-0.39*	-	-	-	-
Kilbride	-3.64	-	-	-	-	-5.1	-	-	-	-	-4.37	-	-	-	-
Paulstown (A)	4.62	4.62	-	-	-	-5.1	-5.1	-	-	-	*-0.24*	*-0.24*	-	-	-

	Deneb (Start)					Deneb (End)					Deneb (Average)				
	3000 BC	2500 BC	2000 BC	1500 BC	1000 BC	3000 BC	2500 BC	2000 BC	1500 BC	1000 BC	3000 BC	2500 BC	2000 BC	1500 BC	1000 BC
Lowpark	4.6	4.62	4.6	4.63	4.62	-5.1	-5.1	-5.09	-5.09	-5.09	*-0.25*	*-0.24*	*-0.25*	*-0.23*	*-0.24*

early in the chronology and, therefore, are discounted but, overall, sites have a high incidence of alignment with these stars throughout the Neolithic. The most likely contender for sidereal significance is Sirius. Although some of the alignments are unlikely due to inconsistency with the site chronology, they have comparable alignment occurrences as other stars, and it is the brightest. Alternatively, Antares has a higher incidence of good alignments, with approximately ten sites aligned throughout the Neolithic.

However, data from the sun suggest that a winter solstice alignment may have been more important than sidereal alignments. Nine sites are oriented towards the winter solstice throughout the Neolithic. When removing alignments that occurred only at the start or the end of the period considered, the alignment with the sun is the most common and precise. It is, therefore, plausible that the sites were aligned to the position of the sun at the winter solstice and coincidentally also pointed towards Sirius, Rigel, Betelgeuse, Aldebaran and Antares. On the day of the winter solstice, the sun would rise to the left of the entranceway and transit through the entrance before disappearing to the right. Given the chance of error and the lack of horizon data, it is possible the sun would rise directly in line with the left extent of the entrance but, with the current information, the sunrise is beyond the entrance area.

However, while the winter solstice is often thought to be a significant event in prehistory, it is important to note that the alignment with the sun would not be limited to the winter solstice but would extend to a roughly 3 month period from November to February. Throughout this time, the sun would complete a similar pattern in the entranceway. It, therefore, may be more correct to say that the sites are oriented towards a winter sunrise, with the sun playing an important symbolic role throughout this period. The absence of a precise alignment with the position of the sun on the winter solstice specifically, intentional alignment towards this position is unlikely.

12.6 Conclusions

The study undertaken was an attempt at enhancing the understanding of Ballynahatty and surrounding timber circles through new digital means. The 3D approximation of the site helps reader and researcher alike to visualise the circle beyond just postholes. By adding hypothetical elements, the relationship between parts of the site can be ascertained, and theories can be evaluated and adjusted. The addition of components beyond the structural helps to bring new life into the site, showing active use. Nonetheless, the risk of misleading the public with a seemingly realistic model based on conjecture must be considered, and this chapter has attempted to explain the reconstruction process to aid understanding. By presenting the metadata and paradata here, the veracity of the model can be judged, and misrepresentation minimised.

Overall, the 3D approximation is a useful tool for the presentation of the archaeology of Ballynahatty, but only if it is presented within the context of the excavation and reconstruction report.

As for the alignment of the Irish timber circles, the main outcome is the confirmation that the orientations of the sites are statistically non-random, suggesting a common target for alignment. *TarxienCore* has been used to identify possible targets, and sunrise on the winter solstice has been noted as the most likely event, with nine sites out of 22 showing a significant alignment. Nonetheless, within the context of the sun's movement, it is more correct to identify the winter sunrise as the primary target. Some stars also have significant alignments, although these may be coincidental. Sirius is a possible candidate as it presents a high number of alignments and is the brightest star in the sky. What can be said with confidence is that other stars and solar events are unlikely to be the focus of alignments in Irish timber circles. Interestingly, Ballynahatty itself does not present an alignment with the winter sun. The most likely alignment is with Antares, which has a high incidence of alignments, but this is an unlikely target as that star is not very bright.

13

The Ballynahatty landscape – past, present and future

13.1 Introduction

This volume has demonstrated how the landscape of Ballynahatty was transformed by the actions of its prehistoric (especially Neolithic) occupants. In this concluding chapter, an attempt will first be made to counterbalance and complement the objective reporting of the results of our excavations by offering a subjective interpretation of the Late Neolithic timber monument complex, giving a sense of experiencing it in use (*cf.* Tilley 1994). The second part of this chapter will briefly review how the Ballynahatty prehistoric landscape has been changed over the last few centuries as a result of agricultural and antiquarian interventions, and the impacts of these changes on the visible and sub-surface archaeology. The fate of some of the items that were discovered during the 19th century is also discussed. In the final part, we shall suggest some of the outstanding research questions thrown up by our work.

13.2 A short walk in the Neolithic

Here we make various assumptions about Neolithic Ballynahatty and its inhabitants and attempt to understand its meaning. In the spirit of a phenomenological approach (Tilley 2006, 1–31), we ask: how would Ballynahatty have been experienced? What follows is a fictionalised, and certainly oversimplified, reconstruction of a final journey.

We start with boats drifting down a river and pulling up at a wooden jetty. Thousands of years later, the river would be called the Lagan and the place Edenderry. A corpse is lifted out of one boat, covered in skins, and laid on a bier. The group of friends and family follow a well-worn path around the southern flank of a low plateau. As they walk on, shouldering their burden, the path gradually climbs and looking over the landscape, wisps of smoke in every direction show the sites of scattered homesteads. The landscape is studded with trees but there is some order.

In clearings there are regular fields divided by hedges and wooden fences. The lowing of cattle can be heard as they freely graze in abandoned clearings and amongst the naturally regenerating understorey. Echoing across the valley is the steady hollow thud of a stone-bladed axe as it half cuts, half splinters its way through the branches of a tree felled the previous year. The branches are fuelling the fire around the burning stump. Looking upslope, a bank appears on the skyline and, at the end of a steep, natural ramp, they reach the top of the plateau. The land here has been cleared and kept open by grazing animals. Scattered across it into the distance are small cairns of stones marking burials. These are the people who worked the land before them and, although of a different tradition, are still respected. For them, death meant cremation, and their eternity was being sealed in a pot under the ground. The entrance to the henge is immediately ahead, on the edge of the plateau at its southwest corner and, moving through, the party suddenly emerges into the great walled expanse of the interior, revealing a central mound – the focal point of the arena. This is the passage tomb, the house of the first ancestor, and due homage is paid as the party traverses the space on a straight path to the exit on the far side. As they do so, they look to the north and see their destination, a great timber enclosure perched on the ridge beyond. The tops of two sets of higher post loom above the enclosure marking the destinations of the living and the dead (Plate 102). Towards the east end, a line of individual posts fronts the wall facing towards them, totems of the ancestors. They emerge from the henge and find, like the entrance, that they are on the edge of the plateau. A short walk leads them past a pond to the end of the ridge. Pigs scramble around in a stockade on one side but in front of them is the second big reveal. A great façade of posts, on the skyline cuts across the ridge at the top of the slope creating an awe-inspiring sight (Plate 103). As they toil up the slope, drawing ever closer, they can see that adorning these massive posts are strings of skulls. At the centre is a narrow entrance and, extending downslope from the end of the façade, a line of posts leads the eye back to the passage tomb and the link to the first, the supreme, ancestor. They cluster in the vestibule; the gate is opened, and they are ushered into a square room (Plate 104), the tall walls open to the skies (Plate 105). Moving from the lower land along the Lagan to Ballynahatty is an upward journey, and once inside the complex, with all reference to the earthly landscape removed, the focus becomes the sky.

The corpse on the bier is placed on the ground between the four big marker posts and the party moves into an annexe, closed to the outside world, for the ritual feast (Plate 106). Out of a bag is produced a flint nodule, and, resting on a stone anvil, flakes are struck off with a hammerstone. Meanwhile, a pig is brought in from the stockade, ritually sacrificed and butchered with the purposely made tools and joints roasted on an open fire. The feasting continues inside until, satiated; the residue is thrown onto the fire with the flint as a final offering. The people in charge of the rituals lift the corpse and carry it to the inner sanctum. From the corridor, the family can only see the inside of the great enclosure and the body being carried off to the right. Eventually, the head of the family is allowed through to the inner entrance where, for the first time, he can see the inner temple. It is eerily quiet, apart from the squawking of squabbling birds. He progresses halfway towards it until, on either side, he can see the complete inside of the Eastern Settings containing racks of bones (Plate 99). These are the houses of the ancestors, and he is now on the border of their realm and can go no further. From here, he can commune with the dead. Straight ahead, he now sees that the body has been placed on a pile of other corpses in various stages of decomposition (Plate 107). Disturbed, carrion birds cluster on the tops of the posts, white with their droppings, and drifting over everything is the nauseous smell of decaying human flesh (Plate 108). It is here that he bears witness to the rebirth of the dead as an ancestor. The ceremonial journey from the world of the living to the world of the ancestors has started, with rituals to ensure a successful passage. In the following weeks, he will return to observe that drift into anonymity with the identifiable essence of the person being taken into the sky through the agency of the birds. The bones, washed by the rain and bleached by the sun, eventually joining the ancestral family in the confines of the Ballynahatty 'Temple'.

13.3 Ballynahatty's archaeology in the more recent past

The deliberate burning down and dismantling of the timber monument complex, and the careful placement of stones and artefacts where its posts had stood, marked a significant transformation in the landscape and rendered the formerly imposing structures invisible but not forgotten. As argued in Chapter 11, with the passing of the timber temple, the major communal construction project of the Giant's Ring henge refocussed attention on the ancestral passage tomb and this has survived as an upstanding monument to the present day.

The landscape has, however, been under continuous change since the Giant's Ring was constructed, and this section of the chapter explores the impact of the last few centuries' change on the survival and condition of the archaeology.

13.3.1 No room for the ancients: the tenant farmer

The 18th and 19th centuries saw two opposing approaches to this landscape – antiquarian study and agricultural opportunity. With opportunity came destruction, and while it is easy to condemn this as cultural vandalism, life as a tenant farmer was hard and the land had to provide rent, sustenance and living expenses. Improvement to the land was an incentive if the benefit could be passed on to future generations of a family such as the Bodels, who discovered the 1855 tomb. Tenantry succession was as important to the Bodels as was title and estate succession to the landowners, the Dungannons. Ties between family and land were equally important to both landowner and tenant – and life could be unforgiving for both.

Shortly after 1819, the Giant's Ring came under sustained pressure from agricultural improvements (Chapter 3), which permanently changed its appearance. The agricultural aim was probably not so much to remove the bank as to flatten the interior and this was certainly achieved and may also explain the several low points on the bank, which would additionally facilitate the movement of animals from the interior to fields outside. At various times, these depressions have been considered – wrongly – as original entrances.

In 1833, the Ordnance Survey described the megalith as having been 'formerly planted with trees but has been ploughed up and nearly destroyed.' On Charles Ligar's 1837 OS plan (Fig. 1.3), the Ring is shown as a perfect circle – although it is not, but there is an additional annotation which gives an insight into the state of the Ring:

> The Ring had been injured by the occupying tenant of the farm; only a part of the parapet had escaped the plough, which had converted its inner abrupt sides into smooth slopes and filled the enclosed area with some of the soil that composed it. The interior is now under tillage. (Day 2014, 141)

1837 also saw the publication of Samuel Lewis' *Topographical Dictionary of County Down*, in which he gives a brief description of the Ring but adds ominously, '... the land is now let, and the earthwork is being removed for the purpose of cultivation' (Lewis 2003, 43–44). In the same year, on 9 December, the *Down Recorder* confirmed continued destruction:

> GIANT'S RING. – We are informed that the circular mound which bounds the field of the Giant's Ring, near Drumbo, is in progress of removal for agricultural purposes, but we hope that the proprietors of the soil will not permit the spoilation of an object which not even the hand of time has meddled with for more than a thousand years.

It is difficult to see how incidental ploughing alone could reduce the bank as there is a clear berm between the base of the inner slope and the start of the quarry trench. However, both the 1917 and 1954 excavations showed that there was considerable infilling of the quarry ditch and, together with the surviving lower stones of a revetment, confirms a deliberate rather than a natural process of degrading the steeper inner bank slope. This degradation would have taken a considerable effort, probably beyond the capabilities of a single tenant farmer, although this may have been happening over a period of 18 years from 1819 onwards.

If the passage tomb inside the Ring had ever been covered by a mound, then all traces of any such mound would have disappeared by the time it was seen by Harris in 1744. He describes how the ground was often used for horse racing, each heat comprising six circuits of the Ring (Harris 1744, 218), so the interior was then probably under pasture with the central area subsequently planted with trees – possibly as an estate feature when the Dungannons were still resident at Belvoir House. Pattison's map of 1819 (*A map of part of Ballinahatty held by John McKeown and James Thomson surveyed by Thomas Pattison, 1819* PRONI T872/1) shows eight trees on the bank and trees or scrub at the megalith. By 1823, when seen by Benn, the trees had gone, the interior was under cultivation, and close ploughing may have resulted in the removal of any kerb stones that may have existed. Later excavation was to show that the central chamber had been 'excavated' to a depth of 4 feet/1.2 m (Lawlor 1918).

The 1819 map shows that this part of Ballynahatty had been divided into a number of small fields held by tenant farmers. Three fields to the northeast of the bank were held by John McKeown and two to the north by James Thomson; the latter include the BNH5/6 site. William Bodel held land beyond Thomson's to the northwest. Downslope to the east, the lands were held by William McClure and James Mullan. To the west was the estate of John Russell Esquire and, to the south, that of Charles Dunlop Esquire. The interior of the Ring, the Bank, the Ringfield to the west and a thin strip between it and the bank are all listed separately but not attributed to a particular tenant. As the map is of specific fields held by Thomson and McKeown, the probability is that they both intended to take on these four fields, which would have doubled their combined holdings to just over 24 Cunningham Acres (*c.* 12.5 hectares). The increased rent required would necessitate making maximum use of the land, which probably is what precipitated the damage to the bank and the attempted levelling of the interior for cultivation. The surviving height difference in the early 19th century between the quarry ditch and the interior must still have been sufficient to present a real problem to ploughing and to necessitate pulling down the bank. Charles Ligar was the first to indicate any variation in bank height and his plan of 1837 shows the current, and probably original, northeast entrance and four further low points. However, the antiquary William Borlase (1897, I, 277) was the first person to comment on them in 1897, seeing five original and two recent gaps. As all record of the gaps post-dates the agricultural destruction of the 1820s, they are best attributable to this cause rather than the original construction.

Robert MacAdam's account of 1855 provides an interesting focus on attitudes to antiquity from the point of view of the tenant farmer and his family, the Bodels (MacAdam and Getty 1855). The earlier 19th century represented hard times in Ballynahatty and every inch of the Bodels' fields would have been necessary for subsistence. They had no time for a sentimental attachment to the past when the present was so precarious. For decades the Bodels had daily toiled and improved the land within sight of the Ring, inadvertently uncovering and deliberately removing a much older landscape. William Bodel, whose name appears on the 1819 map, and his wife had at least seven children. His wife and four daughters, all aged under 21, died over a period of eight months from October 1838. Her 9 and 11 year old daughters died just 3 weeks before she did. William died in 1843, leaving the remainder of his family with his unmarried brother, the enlightened David Bodel, who had discovered the 'subterranean chamber' tomb. In 1845, two more of the children died within a month. The last of William's family, John, who erected the family gravestone, died on 21 November 1855, the day that details of the prehistoric burial chamber first appeared in the *Belfast Newsletter*. None of William's children lived beyond the age of 20.

David Bodel may have moved back to the farm when his brother died to look after both the farm and the remaining family. With their deaths, there would have been plenty of room for another nephew – Hamilton, his wife Sarah and, from 1862, their young family. But of Hamilton's six children, three died before the age of 21, and Hamilton and his wife predeceased David, aged 44 and 45.

David Bodel appears unique amongst the Ballynahatty farmers in having a sympathetic attitude to the past. After the discovery of the tomb in 1855, he prohibited anyone else from accessing the site to protect it. He even removed the skulls from the tomb and took them back to his house for safekeeping. It was also David who allowed the Belfast Natural History and Philosophical Society (BNHPS) to record the site and remove the artefacts and

Figure 13.1 The house, formerly belonging to the Bodels, before demolition (photo: Dunlop family, Ballynahatty).

informed MacAdam of all the other antiquities found on his farm. From the number of sites listed, it is clear that the Bodels were responsible, as tenants, for the initial clearing and opening up of at least one of the fields in Ballynahatty from the mid-18th century, probably earlier. The Bodel's house at Ballynahatty has been replaced by a larger modern house, but photos exist of the original structure. It was '... not many perches [perch = 21 ft or 6.40 m] distant ...' (MacAdam and Getty 1855, 364) from the tomb and had been built on the site of a mound containing several short stone coffins. The photo (Fig. 13.1) shows a substantial longhouse with a small gable window indicating a first floor.

The opening up and farming of the land of Ballynahatty had been both a revelation and a great cultural loss. But for the humble potato, David Bodel's interest and MacAdam's record, all knowledge of this prehistoric legacy would have been lost.

13.3.2 The Dungannons: owning the land. The intervention of antiquaries

Arthur Young, writing in his *Tour in Ireland* of 1780 (vol. 1, chap. 7), does not mention the Giant's Ring, but does record that he passed by Lord Dungannon's estate at Belvoir. Dungannon's mansion was situated just 2.40 km northeast of the Giant's Ring. Young also comments on the productivity of the land, stating that nearly a third of all the revenue from Counties Down and Antrim is derived from the Lagan Valley.

In common with others of their class and status, family was an asset not to be squandered. A male heir was of crucial importance, as was their marriage to an heiress (preferably aristocratic) in order to accumulate land, wealth and influence and so to consolidate estates and establish and maintain the family lineage. The bottom line, after the failure of a direct male heir, was usually preservation of the family name by making this a condition of inheritance. By the end of the 18th century, the Dungannons had accumulated a substantial estate in North Down, including the townland of Ballynahatty, centred on their house at Belvoir.

The Dungannons were descended from the Hills of Hillsborough, Co. Down, and the Trevors of Brynkinalt, Co. Denbigh. Anne, Viscountess Midleton, started the building of Belvoir house on her Co. Down estate in the 1740s. It was probably about this time that the Bodel family became tenants. Anne died in 1747, and her second son, Arthur Hill, succeeded to the Brynkinalt estates in Wales and assumed the name Trevor in 1759, and it was Arthur Hill-Trevor who was created 1st Viscount Dungannon of Dungannon and Baron Hill of Olderfleet in 1765. He was Lord Lieutenant of Ireland, died in 1771 and was buried in the old Breda graveyard in the grounds of Belvoir House – although he no longer resides there. He was succeeded by his grandson, Arthur Hill-Trevor (the 2nd Viscount), whose sister Anne married the Earl of Mornington and became the mother of the Duke of Wellington. Arthur married Charlotte in 1795 but, by 1808, they were resident at their Welsh estate at Brynkinalt.

Their agent, Captain Cortland Skinner, lived at Belvoir until 1808 when the contents were auctioned and the house and most of the demesne were sold to three merchants and then to Robert Bateson of Orangefield in 1811. The 2nd Viscount Dungannon sold the remainder of the Belvoir land to Bateson in 1818 but retained the greater estate, which included Ballynahatty. This progressive dissociation from their Co. Down property and focus on Brynkinalt probably explains their apparent lack of interest in what was happening at Ballynahatty. However, it could be argued that agricultural improvement was encouraged to maximise the value of the land and rental income to support their other properties.

On the death of the 2nd Viscount in 1837, his son, Arthur Hill-Trevor, became 3rd Viscount Dungannon, and he married Sophia, 4th daughter of Col. George Irvine of Castle Irvine, Co. Fermanagh. His cultural credentials – MA, FSA, MRIA, MRSL – display a keen interest in the past and, no doubt stung by all the destructive activity on his land, one of his first acts was to initiate the construction of a protective stone wall around the Ring embankment which is commemorated by a plaque at the entrance (Fig. 13.2):

> This wall for the protection of The Giant's Ring was erected AD 1841 by Arthur 3rd Viscount Dungannon. On whose estate this singular relique of antiquity is situated and who earnestly recommends it to the care of his successors.

It is quite possible that Sophia, Viscountess Dungannon had read MacAdam and Getty's 1855 article in the *Ulster Journal of Archaeology* because the following year, she visited the Ring. The family was no longer resident in Ulster, so this was an event rare enough to be commemorated in true Dungannon tradition by yet another plaque, this time at the inside of the gate on her husband's wall:

Figure 13.2 The Dungannon plaque at the entrance to the Ring (photo: Barrie Hartwell)

This tablet is erected to commemorate the visit to the Giant's Ring of Sophia Viscountess Dungannon on Friday April IV MDCCCLVI [1856] as well as the cordial and affectionate manner in which she was received and welcomed by the adjacent tenantry on that occasion.

The real reason for this tablet was probably her thanks to the tenantry who had commissioned a portrait for Sophia of her husband (Fig. 13.3). The family must have been held in high esteem for such a valuable gift. Knox (1875) commented 35 years later that the Dungannon wall had become dilapidated in places and the entrance wicket thrown down.

When the 3rd Viscount died without issue in 1862, his honours became extinct and the Brynkinalt Estate devolved to Arthur Trumbull Hill, 2nd son of the 3rd Marquess of Downshire. He assumed the additional surname of Trevor and became Lord Arthur Edwin Hill-Trevor in 1862 and 1st Baron Trevor of Brynkinalt, Co. Denbigh in 1880, the year Sophia died.

13.3.3 Cultural awakening: an intellectual puzzle

The 18th and 19th centuries saw an increasing interest in the past and the establishment of learned societies, such as the Dublin Philosophical Society in 1683 and the Royal Irish Academy (RIA) in 1785 as well as provincial societies such as the BNHPS in 1842. Their study of antiquities was of variable quality and unsystematic or, as Waddell has described the earliest accounts: 'short, undisciplined but enthusiastic' (Waddell 2005, 45). Attempts were made to answer the three big questions of 'who built what, why and when?' In the absence of an independent prehistoric framework, reference was made to scriptural studies and the classical world. In the rest of Europe, the Roman occupation formed a fixed point in time and a history against which non-historic cultures could be measured. In Ireland, with an absence of Roman remains, this was problematic. One solution was the Nordic incursions which resulted in most circular earthworks (raths or ringforts) in Ireland being initially labelled as 'Danish Forts'.

Figure 13.3 Arthur Hill-Trevor, 3rd Viscount Dungannon (1798–1862) by Stephen Catterson Smith PRHA, 1856 (Copyright National Portrait Gallery, London).

Another was the description by Roman commentators of contemporary Iron Age cultures resulting in the attribution of stone circles and tombs to the 'Druids'. To this mix was added an increasingly potent layer of Irish mythology and numerous 'Dermot and Grania's beds'. Which of these elements informed the protagonists largely depended on political affiliation and religious persuasion. Thus, the Protestant ascendancy saw Irish prehistory and history as one of invasion, especially from Britain, and improvement because the barbarous Irish were incapable of progressing themselves. The Catholic Nationalist agenda promoted a romantic Gaelic past in which a glorious indigenous Celtic Ireland was diminished by Britain though having received influences directly from the European mainland. Such dichotomous views have ameliorated with time and the great expansion of archaeological knowledge to provide a more harmonious view of the Irish past.

Harris, writing in 1744, describes the passage tomb in the Ring as an ancient altar or 'cromlech', a term used in Britain at the time, and ascribes its introduction to Ireland with the Druids (Harris 1744, 200–1). The Ring is described as an 'artificial rath' though different from others in having no advantage of height (*ibid.*, 218). Anne Plumptre (1817) recorded the 'cromlech' in her diary after visiting in 1814. It is clear from Anne's account that she considers the Ring to be of considerable antiquarian interest and she is evidently surprised at the lack of local knowledge of, or interest in, the features here. During her visit, Anne stayed with Dr James MacDonell, a leading Belfast physician who owned a collection of antiquities. Despite his antiquarian interests, it is curious that he had apparently never visited the Giant's Ring. It perhaps emphasises that, although within 4 miles/6.5 km of Belfast, the Giant's Ring was in a rural backwater.

In 1823, Benn described some antiquities in the neighbourhood of Belfast which deserved to be rescued from neglect and included 'that stupendous work, called the Giant's Ring'. He termed the passage tomb a 'cromlech' and 'Druidical Altar', nearly 2000 years old. He thought that as the Ring bank excluded the view of the surrounding countryside, and that it was used for the:

> ... idolatrous adoration of the sun ... the glorious luminary himself whose beams they worshipped. It is a place which is calculated to inspire an uninformed Druid with additional superstition, or with the necessity of increased mortification; and they who formed it had a just conception of those human feelings which are extensive in their influence, powerful in their operation, and most deeply to be moved by external nature. (Benn 1823, 257)

As late as 1855, the popular press could talk of the passage tomb as being from the period of the ancient Druids (*Belfast Newsletter*, 21 November,1855, 1).

In 1833, the Ordnance Survey swept through the area. They recorded the Parish of Drumbo, in which Ballynahatty sits, though still rural in character, as having a relatively dense population of 1451 families in 1420 inhabited houses (1821). The associated Memoir, although it records the Ring as a 'very perfect piece of antiquity' measuring 200 yards (*c.* 183 m) in diameter, also noted the damage to the tomb and the embankment (Bordes and Scott 1833).

MacAdam and Getty were more circumspect when they speculated that the 1855 'subterranean chamber' was as old as the Giant's Ring and that it dated from 'a period long antecedent to Tara' and was the product of an 'advancing and considerable population', constructed to meet the needs of a family or tribe who would have been part of an agricultural or pastoral society (MacAdam and Getty 1855, 364). They felt that matters were made more difficult by the fact that there were three types of interment within the tomb: cremated remains with and without an 'urn' and unaccompanied, disarticulated remains. MacAdam and Getty's visit to the 'Ancient Sepulchral chamber' resulted in a good eight-page scientific description and analysis, which has been discussed in Section 1.2.2 and referred to throughout this volume.

The tomb contents subsequently had an interesting history. Two of the surviving crania were drawn and measured by a Grattan craniometer in the Anatomy Department of the newly opened Queen's College of Belfast (Evans 2021, 127). There they languished over the next century until the department moved to a new building in what had since become the Queen's University of Belfast. The anatomical collection had suffered from a lack of care in the old building and one of the skulls was lost; however, the other has clearly been identified and is now in the Archaeology Teaching Collection at Queen's. The remaining human and animal bones and pottery were given to the Belfast Natural History and Philosophical Society's (BNHPS) own museum, which later formed the nucleus of the Belfast Municipal Museum, which in turn became the Ulster Museum. At some stage, the 3rd Viscount Dungannon became involved. In January 1856, he presented the remains of at least three pots from the tomb to the Royal Irish Academy and this subsequently passed to the collection of the National Museum of Ireland in Dublin.

In 1882 the premier status of the Giant's Ring was recognised when it became one of the first monuments in Ulster to be placed under Guardianship of the State. The conflicting approaches to this land continued and at the end of the 19th century William Gray (Gray 1890) was able to report that the Ring had been a 'happy hunting ground for Irish antiquarian students for many years past' and that 'since [1855] many new discoveries have been made, the latest being the finding of a very curious circular burial chamber on the farm of Mr J M'Connell of Ballynahatty'. He describes something remarkably similar to the 1855 tomb but discovery, however, went hand in hand with destruction, and he adds: 'Owing to the uncompromising exigencies of Mr M'Connell's farming operations, the monument was rather rudely handled, and intelligent local observers had not the opportunity of making that careful investigation they desired ...'. O'Laverty (1880, 240) refers to a funeral mound containing urns, on the grounds of Edenderry House, the property of Mr Dunlop, to the southwest of the Ring.

At the end of the 19th century, Borlase detailed the Giant's Ring in his comprehensive survey *The Dolmens of Ireland* (1897). In his account, he put forward ideas about the character of the monument, dismissing any notion that it might have had a military or defensive function. He instead refers to the historical use of the site as a racecourse and connects this with the celebration of funeral games or the survival of the fair, and ultimately sees the Giant's Ring as a place of assembly associated in some way with religious or ritual activities (1897, 277). Meanwhile, periodic cultivation persisted and the interior was used for ploughing competitions into the 20th century, and plough lines can still be seen in some aerial photographs. In 1895, the Giant's Ring was the venue for a Grand Bazaar and Summer Fete lasting for 3 days from the 13–15 June. This was organised by Ballycairn

Presbyterian Church to raise money for the enlargement of the Manse and the erection of a lecture hall. It featured a talk by the Rev. Geo. R. Buick, MA, on the 'Archeology of the Ring'.

13.3.4 20th century: excavation

The various excavations and other archaeological research that took place over the 20th century have already been described in Section 1.3, so will not be repeated here. Suffice it to note that H.C. Lawlor – an enthusiastic amateur with no formal training in archaeology – dug with breath-taking speed, excavating 400 square yards (c. 335 m^2) of trenches and a section through the bank in only 1 week (Lawlor 1918). This speed, and the narrowness of his trenches (18 inches/c. 0.46 m), explain why his results were so disappointing (Evans 1943, 442): his trenches left little room for observation of this difficult subsoil and the recording and adequate interpretation of the bank was largely absent. Lawlor's excavation was followed, in 1929, by a similarly inconclusive dig by D.A. Chart (Section 1.3.2) and then by A.E.P. Collins' scientific and competent excavation of the Ring in 1954 (Section 1.3.3).

13.3.5 The 21st century: old problems, new solution

The changes in agricultural practices over the 20th century, from horse-drawn to motorised cultivation equipment and the use of increasingly heavy equipment, have led to further damage of the prehistoric landscape. The small fields and trackways of the 1819 Pattison map and the 1834 and 1858 Ordnance Survey maps, which Bodel would have used, were replaced by the larger fields of the 1955–71 survey (1:10000 sheet 147) to facilitate the movement of larger pieces of machinery (see Fig. 1.1). The old track, which ran southwest along the ridge, crossing over BNH5 in the process, was smoothed over.

At the end of the excavation in 2000, all the existing field walls, fences and trackways were removed from the area. This made good farming sense in that it allowed unrestricted access to the largest machines across the most productive land. Archaeologically, however, it was very damaging. As an example, Figure 13.4 shows a 38.00 m profile along the top of the ridge at the eastern end of the BNH5 annexe, which crosses a north–south field fence. Years of heavy ploughing had degraded the east field, which was 0.50 m lower than the land on the other side of the fence. After 2000, the levelling up of the surface between these two adjacent fields further degraded the subsoil remains and would have been sufficient to remove all traces of any Middle Neolithic occupation that may have accompanied the funerary evidence found during excavation at the site's eastern end.

At the standing stone, which once stood on an elevated section at the west end of the ridge, where the 'cart loads' of skeletons had been found, the effect is even more dramatic (Fig. 13.5). Another, more recent damaging activity has been the use of subsoiling equipment to relieve soil compaction. This can readily be seen in the 2018 drone photographs as a regular mass of diagonal lines covering the whole area (Fig. 13.6).

Whereas regular ploughing on a level area can do little further damage – and many shallow features probably now only exist as spreads of small artefacts within the plough zone – deeper, previously safe features have now been damaged in this sensitive landscape. The importance of the Giant's Ring in local recreation and tourism is well established but the recent increase in footfall has brought its own problems of erosion exacerbated by the riding of mountain bikes (Fig. 13.7).

The Giant's Ring has survived for over 4500 years and has proved remarkably resilient to millennia of natural degradation. However, farming has had a substantial impact on the site. The Dungannon wall was one response to this, and the monument's early incorporation into the Guardianship scheme was another. The recognition of the Giant's Ring as an important archaeological site of national importance was assured, but what of the wider prehistoric landscape, which includes Ballynahatty 5/6? Finally, in 2019, the 'Henge, Passage Grave and Pre-Historic Ritual Landscape' was Scheduled under Article 3 of *Historic Monuments and Archaeological Objects (NI)*, 1999. At last, this extensive landscape received the statutory protection it deserved and can now be managed for the benefit of both the farming community and the public

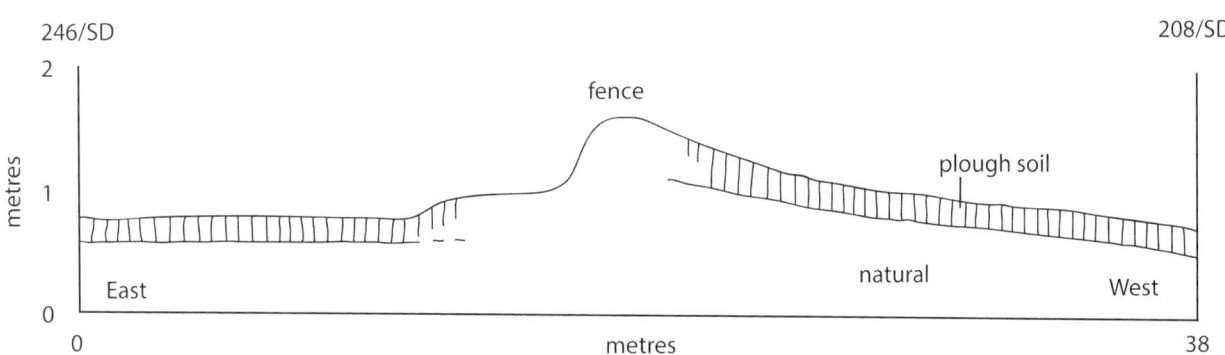

Figure 13.4 East–west section through the field fence on top of the ridge at the eastern end of the excavation.

Figure 13.5 The standing stone (BNH8) with the ground levelled around it and an accumulation of field stone (photo: Barrie Hartwell).

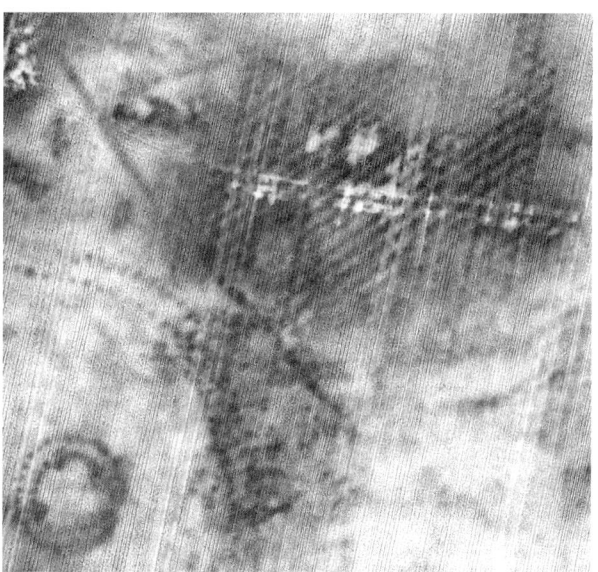

Figure 13.6 Diagonal lines show damage by subsoiling.

Figure 13.7 Erosion of the bank at the entrance to the Giant's Ring as a result of mountain bike use.

while preventing further degradation. Despite so many vicissitudes, farmers and landowners, antiquaries and archaeologists, the public and the State have somehow muddled through to exploit, experience, record, understand and preserve this remarkable landscape and which will allow future generations to unlock the enormous cultural value contained in the soils of Ballynahatty.

13.4 Ballynahatty, the future: outstanding research questions

The 1990–2000 excavations and the various remote sensing projects have served to reveal a wealth of archaeological material at Ballynahatty and have shed important light on a hitherto unknown complex of Late Neolithic timber monuments, as well as enhancing our understanding of non-monumental Middle Neolithic funerary practices and revealing evidence for activities in the area earlier in the 4th millennium BC and from the Early Bronze Age to the early medieval period. Other research undertaken independently of our project, including the DNA analysis of a woman's skull from the 1855 'subterranean chamber', has greatly enhanced our understanding of the individuals buried in the area.

Like all good research projects, however, ours has also thrown up new questions and challenges. The question of when the simple passage tomb inside the Giant's Ring was erected, and when the Ring itself was erected, remain to be resolved – although, given the degree of damage, prior excavation and 'hoking around' that has gone on, it is not guaranteed that any new excavation would succeed in obtaining definitive dating evidence. Other chronological issues include the need to date more of the calcined human remains from the timber monument complex to check whether many, or any, are of Late Neolithic date. The standing stone at the west end of the ridge (Fig. 13.5) remains a chronological conundrum: was it erected at the same time as the timber monument complex, or much later? Is the cursus-like monument that appears on aerial photographs a Neolithic cursus? The dating of this class of monument in Ireland leaves a great deal to be desired, so its investigation and dating should be accorded priority. There is much, too, that can be learned about the use of the landscape over time by 'ground-truthing' other features that have been identified from aerial and other remote surveying techniques. It would be particularly useful to learn more about where the users of the timber monument complex were living.

The Middle Neolithic funerary features associated with Coarse Ware raise the question of the broader nature of non-passage tomb Middle Neolithic funerary practices in Ireland, and it would be useful to revisit other funerary sites of that period (*e.g.* Ballynoe, Co. Down) to develop a more detailed picture of the range of practices and grave or tomb types.

Our project has demonstrated the huge potential of the area to shed significant new light on the evolution of a landscape that was both ceremonial and quotidian – as well as the serious loss of sites as a result of past farming and other practices. Our decade of fieldwork has scratched the surface and (to mix metaphors) revealed the tip of the iceberg; it is hoped that future research and fieldwork will reveal the rest of that fascinating iceberg. It would require Titanic effort but would be well worthwhile the effort (and would not end the same way as it did for the boat of the same name…).

Bibliography

Chapter 1: The landscape and historical research

Belfast Newsletter (1855) Discovery of an ancient tomb. 21 November.

Benn, G. (1823) *The History of the town of Belfast, with an Accurate Account of the Former and Present State; to which are added a Statistical Survey of the Parish of Belfast and a Description of Some Remarkable Antiquities in its Neighbourhood*. Belfast, Mackay.

Benn, G. (1834) Giant's Ring. *The Dublin Penny Journal* 3 (114), 77.

Bordes, Lt F.W. and Scott, G. (1833) *Parish of Drumbo*. Belfast, Ordnance Survey Memoirs.

Borlase, W.C. (1897) *The Dolmens of Ireland*. London, Chapman & Hall.

Collins, A.E.P. (1957) Excavations at the Giant's Ring, Ballynahatty. *Ulster Journal of Archaeology* 20, 44–50.

Day, A. (2014) *Glimpses of Ireland's Past: the Ordnance Survey memoir drawings: topography and technique*. Dublin, Royal Irish Academy.

Harris, W. (1744) *The Ancient and Present State of the County of Down*. Dublin, Reilly.

Hartwell, B. (1988) Air photography and fieldwork at the Giant's Ring. *Organisation of Irish Archaeologists Conference (Remote Sensing): Newsletter* 7, 22.

Hartwell, B. (1998) The Ballynahatty complex. In A.M. Gibson and D.D.A. Simpson (eds), *Prehistoric Ritual and Religion: essays in honour of Aubrey Burl*, 32–44. Stroud, Sutton.

Hinks, Rev T. (1825) *A Discourse read by the Rev. Thomas Hinks on the Anniversary of the Birth of Linnaeus where-in he Commemorates the Scientific Pursuits of the late John Templeton of Cranmore who died on the 15th December 1825*. Available online in The Templeton Journals at BNFC.org.uk.

Jope, M. (ed.) (1966) *An Archaeological Survey of County Down*. Belfast, HMSO.

Knox, A. (1875) *A History of the County of Down from the Most Remote Period Until the Present Day; Including an Account of its Early Colonisation, Ecclesiastical, Civil, and Military Polity, Geography, Topography, Antiquities and Natural History*. Dublin, Hodges, Foster & Co.

Lawlor, H.C. (1918) The Giant's Ring. *Proceedings of the Belfast Natural History and Philosophical Society 1918–1919*, 13–27.

MacAdam, R. and Getty, E. (1855) Discovery of an ancient sepulchral chamber. *Ulster Journal of Archaeology* 3, 358–365.

Macdonald, P. and Hartwell, B. (2009) Anne Plumptre and the Giant's Ring, County Down: an account of a possible bleach-green watch-tower. *Ulster Journal of Archaeology* 68, 152–157.

Nelis, E.L. (2003) *Lithics of the Northern Irish Neolithic* Unpublished PhD thesis, Queen's University Belfast.

Northern Ireland Land Act (1925) Estate of Elizabeth Dunlop (widow) County of Down. Record No. N.I. 789. Land Purchase Commission – Northern Ireland. Northern Ireland Land Act 1925, 2.

O'Reilly D. (2010) *Rivers of Belfast: a history*. Newtownards, Colourpoint.

Plumptre, A. (1817) *Narrative of a Residence in Ireland During the Summer of 1814, and that of 1815*. London, Henry Colborn.

Richmond, I.A. (1929) *An investigation of the Giant's Ring* Unpublished report, Northern Ireland Sites and Monuments Record at HERoNI@communities-ni.gov.uk.

Wilde, W.R. (1857) *A Descriptive Catalogue of the Antiquities of Stone, Earthen, Vegetable, and Animal Materials in the Museum of the Royal Irish Academy*. Dublin, M.H. Gill.

Chapter 2: Archaeological surveys

Davis, S. (2017) *Geophysical Survey at Ballynahatty, Co Down*. Preliminary report, University College Dublin, School of Archaeology.

Hartwell, B. (1988) Air photography and fieldwork at the Giant's Ring. *Organisation of Irish Archaeologists Conference (Remote Sensing): Newsletter* 7, 22.

Hartwell, B. (1998) The Ballynahatty complex. In A.M. Gibson and D.D.A. Simpson (eds) *Prehistoric Ritual and Religion. essays in honour of Aubrey Burl*, 32–44. Stroud, Sutton.

MacAdam, R. and Getty, E. (1855) Discovery of an ancient sepulchral chamber. *Ulster Journal of Archaeology* 3, 358–365.

Mercer, R.J. (1981) The excavation of a Late Neolithic henge-type enclosure at Balfarg, Markinch, Fife, Scotland, 1977–8. *Proceedings of the Society of Antiquaries of Scotland* 111, 63–171.

Nelis, E.L. (2003) *Lithics of the Northern Irish Neolithic* Unpublished PhD thesis, Queen's University Belfast.

Sheridan, J.A. (2004) Going round in circles? Understanding the Irish Grooved Ware 'complex' in its wider context. In H. Roche, E. Grogan, J. Bradley, J. Coles and B. Raftery (eds), *From Megaliths to Metals: essays in honour of George Eogan*, 26–37. Oxford, Oxbow Books.

Smith, I.F. and Keiller, A. (1965) *Windmill Hill and Avebury: excavations by Alexander Keiller, 1925–1939*. Oxford, Clarendon Press.

Thomas, J. (1999) *Understanding the Neolithic*. London, Routledge.

Woodman, P.C. (1994) Towards a definition of Irish Early Neolithic lithic assemblages. In N. Ashton and A. David (eds), *Stories in Stone*, 213–218. London, Lithic Studies Society Occasional Paper 14.

Chapter 3: Environmental history of the Ballynahatty area

Hirons, K.R. and Edwards, K.J. (1986) Events at and around the first and second *Ulmus* declines: palaeoecological investigations in Co. Tyrone, Northern Ireland. *New Phytologist* 104, 131–153.

O'Connell, M. and Molloy, K. (2019) Aran Islands, Western Ireland: farming history and environmental change, reconstructed from field surveys, historical sources, and pollen analyses. *Journal of the North Atlantic* (38), 1–27.

Pilcher, J., Hall, V.A. and McCormac, F.G. (1996) An outline tephrachronology for the Holocene of the North of Ireland. *Journal of Quaternary Science* 11, 485–494.

Plunkett, G., Carroll, F., Hartwell, B., Whitehouse, N.J. and Reimer, P.J. (2008) Vegetation history at the multi-period prehistoric complex at Ballynahatty, Co. Down, Northern Ireland. *Journal of Archaeological Science* 35, 181–190.

Smith, A.G. & Goddard, I.C. (1991) A 12500 year record of vegetational history at Sluggan Bog, Co. Antrim, N. Ireland (incorporating a pollen zone scheme for the non-specialist). *New Phytologist* 119, 167–187.

Torbenson, M.C.A, Plunkett, G., Brown, D.M., Pilcher, J.R. and Leuschner, H.H. (2015) Asynchrony on key Holocene chronologies: evidence from Irish bog pines. *Geology* 43, 799–802.

Trainor, P. (2011) *The Timing of the Holocene Alnus Rise in Ireland*. Unpublished MSc Dissertation, Queen's University Belfast, Belfast.

Chapter 5: Ballynahatty 5 and 6: excavating the enclosures

Connolly, M. and Coyne, F. (2005) *Underworld, Death and Burial in Cloghermore Cave, Co Kerry*. Bray, Wordwell.

Gibson, A.M. (2005) *Stonehenge and Timber Circles*. Stroud, Tempus.

Gleeson, P. and McLaughlin, R. (2021) Way of death – cremation and belief in 1st millennium Ireland. *Antiquity* 95 (380), 395–396.

Hartwell, B. (1991) Ballynahatty – a prehistoric ceremonial centre. *Archaeology Ireland* 18 (Winter), 12–15.

Hartwell, B. (1994) Late Neolithic ceremonies. *Archaeology Ireland* 30 (Winter), 10–13.

Hartwell, B. (1998) The Ballynahatty complex. In A.M. Gibson and D.D.A. Simpson (eds) *Prehistoric Ritual and Religion. essays in honour of Aubrey Burl*, 32–44. Stroud, Sutton.

Hartwell, B. (2002) A Neolithic ceremonial timber complex at Ballynahatty, Co. Down. *Antiquity* 76 (292), 526–532.

McClatchie, M., Bogaard, A., Colledge, S., Whitehouse, N.J., Schulting, R. and Barratt, P. (2016) The introduction of agriculture into Ireland: a review of the plant macro-remains evidence. *Antiquity* 350, 302–318.

McClatchie, M., Schulting, R., McLaughlin, R., Colledge, S., Bogaard, A., Barratt, P. and Whitehouse, N. (2022) Food production, processing and foodways in Neolithic Ireland. *Environmental Archaeology* 27, 80–92.

Monk, M.A. (1985/1986) Evidence from macroscopic plant remains for crop husbandry in prehistoric and Early Historic Ireland: a review. *Journal of Irish Archaeology* 3, 31–36.

Renfrew, J. 1973 *Palaeoethnobotany*. London, Methuen.

van der Veen, M. (1992) *Crop Husbandry Regimes*, Sheffield, Sheffield Archaeological Monographs 3, J.R. Collis Publications.

Chapter 6: The pottery

ApSimon, A. (1969) The earlier Bronze Age in the north of Ireland. *Ulster Journal of Archaeology* 32, 28–72.

Barkley, J. (2020) Appendix 1: ceramic report. In S. Nicol. and G. Donaghy *Archaeological Excavation Report: Glenshane Road Quarry, Claudy, Co Derry*. Unpublished excavation report for Northern Archaeological Consultancy, i–xiv. Belfast: Northern Archaeological Consultancy.

Brindley, A.L. (1999a) Irish Grooved Ware. In R. Cleal, and A. MacSween (eds), *Grooved Ware in Britain and Ireland*, 23–35. Oxford, Neolithic Studies Group Seminar Papers 3.

Brindley, A.L. (1999b) Sequence and dating in the Grooved Ware tradition. In R. Cleal and A. MacSween (eds), *Grooved Ware in Britain and Ireland*, 133–144. Oxford, Neolithic Studies Group Seminar Papers 3.

Brindley, A L (2007) *The Dating of Food Vessels and Urns in Ireland*. Galway, University of Ireland, Galway.

Brindley, A.L. and Lanting, J. (1990) Radiocarbon dates for Neolithic single burials. *Journal of Irish Archaeology* 5, 1–7.

Brindley, J.C. (1984) Appendix 3. Petrological examination of Beaker pottery from the Boyne Valley sites. In G. Eogan, *Excavations at Knowth 1*, 321–346. Dublin, Royal Irish Academy Monographs in Archaeology 1.

Carlin, N. (2011) *A Proper Place for Everything: the character and context of Beaker depositional practices in Ireland*. Unpublished PhD, University College Dublin.

Carlin, N. (2016) Discussion of the timber circle at Armalughey (sites 18 and 20) with a focus on the Grooved Ware and Beaker discoveries at the sites. In C. Dunlop and J. Barkley (eds), *Road to the West. A Road to the Past Volume 2. The Archaeology of the A4/A5 Road Improvements Scheme from Dungannon to Ballygawley*, 194–210. Belfast, Northern Archaeological Consultancy.

Carlin, N. (2017) Getting into the groove: exploring the relationship between Grooved Ware and developed passage tombs in Ireland *c.* 3000–2700 cal BC. *Proceedings of the Prehistoric Society* 83, 155–88.

Carlin, N. and Cooney, G. (2017) Transforming our understanding of Neolithic and Chalcolithic society (4000–2200 BC). In M. Stanley, R. Swan and A. O'Sullivan (eds), *Stories of*

Ireland's past. Knowledge Gained from NRA Roads Archaeology, 23–56. Dublin, Transport Infrastructure Ireland.

Cleary, R.M. (n.d.) Appendix II(a) Pottery from the N5 Charlestown bypass. In R. Gillespie, *Lowpark, Co. Mayo, multiperiod archaeological complex, Registration No. E3338, Ministerial Direction No. A020/012, final report*, 269–310. Unpublished report for Mayo County Council. [https://doi.org/10.7486/DRI.gf06vh467]

Collins, A.E.P. (1954) The excavation of a double horned cairn at Audleystown, Co. Down. *Ulster Journal of Archaeology* 17, 7–56.

Collins, A.E.P. (1959) Further Investigations in the Dundrum Sandhills. *Ulster Journal of Archaeology* 22, 5–20.

Collins, A.E.P. & Waterman, D.M. (1955) *Millin Bay: a Late Neolithic cairn in Co. Down*. Belfast, HMSO.

Cooney, G. (2000) *Landscapes of Neolithic Ireland*. London, Routledge.

Daniells, M.J. and Williams, B.B. (1977) Excavations at Kiltierney, Deerpark, Co. Fermanagh. *Ulster Journal of Archaeology* 40, 32–41.

Dunlop, C. and Barkley, J. (2016) *Road to the West. A Road to the Past Volume 2. The Archaeology of the A4/A5 Road Improvements Scheme from Dungannon to Ballygawley*. Belfast, Northern Archaeological Consultancy.

Eogan, G. and Roche, H. (1997) *Excavations at Knowth (2): Settlement and Ritual sites of the 4th & 3rd millennium BC*. Dublin, Royal Irish Academy Monographs in Archaeology.

Eogan, J. (2000) Bettystown, Co. Meath. In I. Bennett (ed.), *Excavations 1998: summary accounts of archaeological excavations in Ireland*, 161. Bray, Wordwell.

Garwood, P. (1999) Grooved Ware in southern Britain: chronology and interpretation. In R. Cleal and A. MacSween (eds), *Grooved Ware in Britain and Ireland*, 145–176. Oxford, Neolithic Studies Group Seminar Papers 3.

Gibson, A.M. (1982) *Beaker Domestic Sites: a study of the domestic pottery of the late 3rd & early 2nd millennia BC in the British Isles* Oxford, British Archaeological Report 107.

Gibson, A.M. (1999) Grooved Ware and timber circles. In R. Cleal and A. MacSween (eds), *Grooved Ware in Britain and Ireland*, 78–82. Oxford, Neolithic Studies Group Seminar Papers 3.

Gillespie, R. (n.d.) *Lowpark, Co. Mayo, multiperiod archaeological complex, Registration No. E3338, Ministerial Direction No. A020/012, final report*. Unpublished report for Mayo County Council. [https://doi.org/10.7486/DRI.gf06vh467]

Groenman-van Waateringe and Butler, J.J. (1976) The Ballynoe Stone Circle: excavations by A.E. Van Giffen 1937–1938. *Palaeohistoria* 18, 73–110.

Grogan, E. and Roche, H. (2010) Clay and fire: the development and distribution of pottery traditions.in Prehistoric Ireland. In M. Stanley, E. Danaher, E. and J. Eogan (eds), *Creative Minds: production, manufacturing and invention in ancient Ireland*, 27–46. Dublin, National Roads Authority.

Grogan, E. and Roche, H. (2011a) Prehistoric pottery. In P. Stevens, *Archaeological Excavations (A003/020, E3502) Ask, Sites 42–44, N11 Gorey to Arklow Link, Co. Wexford*, 155–178. Unpublished final report by Valerie J. Keeley Ltd. [https://doi.org/10.7486/DRI.h1291v592]

Grogan, E. and Roche, H. (2011b) Prehistoric pottery report. In Y. Whitty, *Archaeological Excavation Report E3144 Kilmainham 3, Co. Meath*, xli–xlvi. Unpublished report by Irish Archaeological Consultancy Ltd for Meath County Council. [https://doi.org/10.7486/DRI.pn89sn352]

Grogan, E. and Roche, H. (2012) The prehistoric pottery. In E. Stafford, *00E0914 Lagavooren 7 Final Report*, cxiii–cxxxvi. Unpublished report to the National Monuments Service, Department of the Environment, Heritage and Local Government. [https://www.europeana.eu/sv/item/255/_1j92vp21j]

Herity, M. (1974) *Irish Passage Graves: Neolithic tomb builders in Britain and Ireland 2500 BC*. Dublin, Irish University Press.

King, H.A. (1999) Excavations on the Fourknocks ridge, Co. Meath. *Proceedings of the Royal Irish Academy* 99C, 157–198.

Laidlaw, G. (2017) Excavations of the Late Neolithic Grooved Ware site at Scart, Co. Kilkenny. *Journal of Irish Archaeology* 26, 33–56.

Lehane, J., Johnston, P. and Leigh, D. (2010) Archaeological excavation report, E2412, Ballynacarriga 3, Co. Cork: prehistoric site with enclosure, structures, two ring-ditches and associated cist burials. *Eachtra* 10. http://eachtra.ie/index.php/journal/e2412-ballynacarriga3-co-cork/.

MacSween, A., Hunter, J., Sheridan, J.A., Bond, J., Bronk Ramsey, C., Reimer, P., Bayliss, A., Griffiths, S. and Whittle, A. (2015) Refining the chronology of the Neolithic settlement at Pool, Sanday, Orkney. *Proceedings of the Prehistoric Society* 81, 283–310.

Monteith, J. (2008) *Final Report on Excavations at Scart 1, Co. Kilkenny (E3001). N9/N10 Dunkitt to Sheepstown*. Unpublished report by Valerie J Keeley Ltd for Kilkenny County Council. Accessed online via: https://repository.dri.ie/.

Nicol, S. and Donaghy, G. (2020) *Archaeological Excavation Report: Glenshane Road Quarry, Claudy, Co Derry*. Unpublished excavation report for Northern Archaeological Consultancy Ltd. Belfast: Northern Archaeological Consultancy.

Ó Drisceoil, C. (2009) Archaeological excavation of a Late Neolithic Grooved Ware site at Balgatheran, County Louth. *County Louth Archaeological and Historical Journal* 27, 77–102.

O'Kelly, M.J. with Cleary, R. and Lehane, D. (1983) *Newgrange, Co. Meath, Ireland: the Late Neolithic/Beaker period settlement*. Oxford, British Archaeological Report S190.

Ó Ríordáin, S.P. (1954) Lough Gur excavations: Neolithic and Bronze Age houses on Knockadoon. *Proceedings of the Royal Irish Academy* 56C, 297–459.

Quinn, P.S. (2013) *Ceramic Petrography: the interpretation of archaeological pottery and related artefacts in thin section*. Oxford, Archaeopress.

Richards, C., Jones, A.M., MacSween, A., Sheridan, J.A., Dunbar, E., Reimer, P., Bayliss, A., Griffiths, S. and Whittle, A. (2016) Settlement duration and materiality: formal chronological models for the development of Barnhouse, a Grooved Ware settlement in Orkney. *Proceedings of the Prehistoric Society* 82, 193–225.

Roche, H. (1995) *Style and Context for Grooved Ware in Ireland: with special reference to the assemblage at Knowth, Co. Meath*. Unpublished M.A. thesis, National University of Ireland, Dublin.

Roche, H. and Grogan, E. (2008) Pottery identification in E. Cotter, *Kilbride, Co. Mayo*. Report submitted to Bord Gáis Eireann on the Mayo gas pipeline archaeological excavations.

Roche, H. and Grogan, E. (2010a) Prehistoric pottery report. In G. Laidlaw, *Final Report on Excavations at Scart North, Co. Kilkenny (E3021). N9/N10 Dunkitt to Sheepstown*. Unpublished report on behalf of V J Keeley Ltd for Kilkenny County Council, 52–60. Accessed online via: https://repository.dri.ie/.

Roche, H. and Grogan, E. (2010b) Appendix 7: Late Neolithic and Beaker Pottery. In Lehane, J., Johnston, P. and Leigh, D. Archaeological excavation report, E2412, Ballynacarriga 3, Co. Cork: prehistoric site with enclosure, structures, two ring-ditches and associated cist burials. *Eachtra* 10, 245–266

Schulting, R.J., Sheridan, A., Crozier, R. and Murphy, E. (2010) Revisiting Quanterness: new AMS dates and stable isotope data from an Orcadian chamber tomb. *Proceedings of the Society of Antiquaries of Scotland* 140, 1–50.

Schulting, R., Bronk Ramsey, C., Reimer, P., Eogan, G., Cleary, K., Cooney, G. and Sheridan, J.A. (2017) Dating Knowth. In G..Eogan with K. Cleary (eds), *Excavations at Knowth Volume 6: the passage tomb archaeology of the great mound at Knowth*, 331–379. Dublin, Royal Irish Academy.

Sheridan, J.A. (1995) Irish Neolithic pottery: the story in 1995. In I. Kinnes and G. Varndell (eds), *Unbaked Urns of Rudely Shape: essays on British and Irish pottery for Ian Longworth*, 3–21. Oxford, Oxbow Books.

Sheridan, J.A. (2007) From Picardie to Pickering and Pencraig Hill? New information on the 'Carinated Bowl Neolithic' in northern Britain. In A.W.R. Whittle & V. Cummings (eds), *Going Over: the Mesolithic–Neolithic transition in north–west Europe*, 441–492. Oxford, Proceedings of the British Academy 144.

Sheridan, J.A. (2016) Scottish Neolithic pottery in 2016: the big picture and some details of the narrative. In F.J. Hunter and J.A. Sheridan (eds), *Ancient Lives. Object, People and Place in Early Scotland. Essays for David V Clarke on his 70th Birthday*, 189–212. Leiden, Sidestone.

Sheridan, J.A. (2022) *Updated report on the pottery found at Eagle's Nest, Lambay Island*. Unpublished report produced for Professor Gabriel Cooney.

Unearthed. New discoveries in development-led archaeology in Northern Ireland 2015–2018 (2019) Belfast, Department for Communities, Historic Environment Department [https://www.communities-ni.gov.uk/publications/unearthed-new-discoveries-development-led-archaeology-northern-ireland-2015-2018]

Wainwright, G.J. and Longworth, I.H. (1971) *Durrington Walls: excavations 1966–1968.* London, Report of the Research Committee of the Society of Antiquaries of London 29.

Warren, S.H., Piggott, S., Clark, J.G.D., Burkitt, M.C. and Godwin, H. & Godwin, M.E. (1936) Archaeology of the submerged land surface of the Essex coast. *Proceedings of the Prehistoric Society* 2, 178–210.

Williams, J.Ll.W., & Jenkins, D.A. (2020) A petrographic study of pottery from the Late Neolithic/Beaker settlement at Newgrange, Co. Meath, Ireland. *Journal of Irish Archaeology* 29, 17–40.

Chapter 7: The lithic assemblage: chipped stone

Bailey, R. (2020) Appendix 2: lithic analysis and report. In S. Nichol and G. Donaghy, *Archaeological Excavation Report: Glenshane Road Quarry, Claudy, Co Derry*, xv–xxxv. Unpublished excavation report for Northern Archaeological Consultancy Ltd. Belfast: Northern Archaeological Consultancy.

Collins, A.E.P. (1981) The flint javelins heads of Ireland. In D. Ó Corráin (ed.), *Irish Antiquity, Festschrift to M.J. O'Kelly*, 111–133. Cork, Tower Books.

Cross, S. (1999) Analysis of the lithic collection. In H.A. King, Excavations on the Fourknocks ridge, Co. Meath, *Proceedings of the Royal Irish Academy* 99C, 183–187.

Dillon, F. (1997) The lithics. In G. Eogan and H. Roche, *Excavations at Knowth (2) Settlement and Ritual sites of the 4th & 3rd millennium BC*, 161–196. Dublin, Royal Irish Academy Monographs in Archaeology.

Gibson, A.M. (2015) Bridging the gap between typology and chronology. British Neolithic and Bronze Age Ceramics 3000–2000 BC. In Y. Tsetlin (ed.), *Proceedings of the International Symposium on Recent Approaches to Ancient Ceramics in Archaeology, 29–31 Oct 2013*,34–42. Moscow, Russian Academy of Sciences, Institute of Archaeology.

Gillespie, R. (n.d.) *Lowpark, Co. Mayo, multiperiod archaeological complex, Registration No. E3338, Ministerial Direction No. A020/012, final report*. Unpublished report for Mayo County Council. [https://doi.org/10.7486/DRI.gf06vh467]

Lehane, D. (1983) The flint work. In M.J. O'Kelly with R. Cleary, and D. Lehane, *Newgrange, Co. Meath, Ireland: The Late Neolithic/Beaker Period Settlement*, 118–167. Oxford, British Archaeological Report S190.

Mallory, J.P., Nelis, E. and Hartwell, B. (2011) *Excavations on Donegore Hill, Co. Antrim*. Bray, Wordwell.

Nelis, E.L. (2003) *Lithics of the Northern Irish Neolithic*. Unpublished Ph.D thesis, Queen's University Belfast.

Nelis, E. (2011) Appendix 2.3 lithic analysis report. In Y. Whitty, *Archaeological Excavation Report E3144 Kilmainham 3, Co. Meath*, xlix–lxxx. Unpublished report by Irish Archaeological Consultancy Ltd for Meath County Council. [https://doi.org/10.7486/DRI.pn89sn352]

Nichol, S. and Donaghy, G. (2020) *Archaeological Excavation Report: Glenshane Road Quarry, Claudy, Co Derry*. Unpublished excavation report for Northern Archaeological Consultancy Ltd. Belfast: Northern Archaeological Consultancy.

Ó Drisceoil, C. (2004) *Archaeological Excavation Report, 00E0905 Balgatheran 4, County Louth*. Unpublished report by Valerie J Keeley Ltd for the National Roads Authority [https://doi.org/10.7486/DRI.1c18sw53c]

Ó Drisceoil, C. (2009) Archaeological excavation of a Late Neolithic Grooved Ware site at Balgatheran, County Louth, *County Louth Archaeological and Historical Journal* 27, 77–102.

Sternke, F. (2012) Appendix 2.3 Lithic analysis report. In E. Stafford, *00E0914 Lagavooren 7 Final Report*, cxli–cc. Unpublished report to the National Monuments Service, Department of the Environment, Heritage and Local Government. [https://www.europeana.eu/sv/item/255/_1j92vp21j]

Chapter 8: Other artefacts from the excavation

Ballin, T.B. (2009) *Archaeological Pitchstone in Northern Britain: characterization and interpretation of an important prehistoric source.* Oxford, British Archaeological Report 476.

Eogan, G. and Roche, H. (1997) *Excavations at Knowth (2): Settlement and Ritual sites of the 4th & 3rd millennium BC.* Dublin, Royal Irish Academy Monographs in Archaeology.

Eogan, J. (2000) Bettystown, Co. Meath. In I. Bennett (ed.), *Excavations 1998: Summary accounts of archaeological excavations in Ireland*, 161. Bray, Wordwell.

Garrow, D. and Wilkin, N. (2022) *The World of Stonehenge*. London, British Museum Press.

Marshall, D. (1977) Carved stone balls. *Proceedings of the Society of Antiquaries of Scotland* 108, 40–72.

Marshall, D. (1984) Further notes on carved stone balls. *Proceedings of the Society of Antiquaries of Scotland* 113, 628–630.

Monteith, J. (2008) *Final Report on Excavations at Scart 1, Co. Kilkenny (E3001). N9/N10 Dunkitt to Sheepstown*. Unpublished report by Valerie J Keeley Ltd for Kilkenny County Council. Accessed online via: https://repository.dri.ie/.

Nicol, S. and Donaghy, G. (2020) *Archaeological Excavation Report: Glenshane Road Quarry, Claudy, Co Derry*. Unpublished excavation report for Northern Archaeological Consultancy Ltd. Belfast, Northern Archaeological Consultancy.

Ó Drisceoil, C (2009) Archaeological excavation of a Late Neolithic Grooved Ware site at Balgatheran, County Louth, *County Louth Archaeological and Historical Journal* 27, 77–102.

Sheridan, J.A. (1986) Porcellanite artefacts: a new survey, *Ulster Journal of Archaeology* 49, 19–32.

Sheridan, J.A. and Brophy, K. (eds) (2012) Special stone artefacts (e.g. carved stone balls and maceheads. *Neolithic Scotland: ScARF panel report*, 80–85. Edinburgh, Society of Antiquaries of Scotland.

Stafford, E. (2012) *00E0914 Lagavooren 7 Final Report*. Unpublished report to the National Monuments Service, Department of the Environment, Heritage and Local Government. [https://www.europeana.eu/sv/item/255/_1j92vp21j]

Stewart-Moffitt, C.L. (2022) *The Circular Archetype in Microcosm: the carved stone balls of Late Neolithic Scotland*. Oxford, Archaeopress.

Welsh, H. and Welsh, J. (2021) *The Prehistoric Artefacts of Northern Ireland*. Oxford, Archaeopress.

Chapter 9: Human remains from excavations at Ballynahatty

Brothwell, D.R. (1981) *Digging Up Bones*. London, British Museum of Natural History.

Cassidy, L.M., Martiniano, R., Murphy, E.M., Teasdale, M.D., Mallory, J., Hartwell, B. and Bradley, D.G. (2016) Neolithic and Bronze Age migration to Ireland and establishment of the insular Atlantic genome. *Proceedings of the National Academy of Sciences* 113, 368–373.

Cassidy, L.M., Maoldúin, R.Ó., Kador, T., Lynch, A., Jones, C., Woodman, P.C., Murphy, E., Ramsey, G., Dowd, M., Noonan, A., Campbell, C., Jones, E.R., Mattiangeli, V. and Bradley, D.G. (2020) A dynastic elite in monumental Neolithic society. *Nature* 582, 384–388. [doi: 10.1038/s41586-020-2378-6]

Ferembach, D., Schwidetzky, I. and Stloukal, M. (1980) Recommendations for age and sex diagnoses of skeletons. *Journal of Human Evolution* 9, 517–549.

Hartwell, B. (1994) Late Neolithic ceremonies. *Archaeology Ireland* 30 (Winter), 10–13.

MacAdam, R. and Getty, E. (1855) Discovery of an ancient sepulchral chamber. *Ulster Journal of Archaeology* 3, 358–365.

McCormick, F. (1997) The animal bones. In M.F. Hurley, O.M.B. Scully and S.W.J. McCutcheon (eds), *Late Viking Age and Medieval Waterford Excavations 1986–1992*, 819–853. Waterford, Waterford Corporation.

McKinley, J.I. (1994) *The Anglo-Saxon Cemetery at Spong Hill, North Elmham Part VIII: the cremations*. Gressenhall, East Anglian Archaeology 69.

Schaefer, M., Black, S. and Scheuer, L. (2009) *Juvenile Osteology: a laboratory and field manual*. London, Academic Press.

Schulting, R., Murphy, E., Jones, C. and Warren, G. (2012) A proposed chronology for Irish court tombs based on new dates from the north of the island. *Proceedings of the Royal Irish Academy* 112, 1–60.

Schwartz, J.H. (1995) *Skeleton Keys: an introduction to human skeletal morphology, development and analysis*. Oxford, Oxford University Press.

Silver, I.A. (1969) The ageing of domestic animals. In D. Brothwell, and E. Higgs (eds), *Science in Archaeology*, 283–302. London, Thames and Hudson.

Ubelaker, D.H. (1989) *Human Skeletal Remains* (2nd edn). Washington DC, Smithsonian Institution Manuals on Archaeology 2.

Chapter 10: Dating and chronology

Bolger, T., Moloney, C. and Shiels, D. (2015) *A Journey Along the Carlow Corridor. The Archaeology of the M9 Carlow Bypass*. Dublin, National Roads Authority.

Bronk Ramsey, C. (2009) Bayesian analysis of radiocarbon dates. *Radiocarbon* 51(1), 337–360.

Bronk Ramsey, C. (2017) Methods for summarizing radiocarbon datasets. *Radiocarbon* 59(2), 1809–1833.

Carlin, N. (2016) Discussion of the timber circle at Armalughey (sites 18 and 20) with a focus on the Grooved Ware and Beaker discoveries at the sites. In C. Dunlop and J. Barkley (eds), *Road to the West. A Road to the Past Volume 2. The Archaeology of the A4/A5 Road Improvements Scheme from Dungannon to Ballygawley*, 194–210. Belfast, Northern Archaeological Consultancy Ltd.

Coughlan, T. and Brick, M. (2009) Settlement, burial and ritual: a Bronze Age landscape on the N9/N10. *Senada* 4, 16–18

Chapple, R.M., McLaughlin, T.R. and Warren, G. (2022) '... Where they pass their unenterprising existence ...': change over time in the Mesolithic of Ireland as shown in radiocarbon dated activity. *Proceedings of the Royal Irish Academy* 122, 1–38.

Danaher, E. (2007) *Monumental Beginnings: the archaeology of the N4 Sligo inner relief road*. Dublin, National Roads Authority.

Dunlop, C. (2015) *Down The Road. A Road to the Past Volume 1. The Archaeology of the A1 Road Schemes between Lisburn and Newry*. Belfast, Northern Archaeological Consultancy.

Dunlop, C. and Barkley J. (2016) *Road to the West. A Road to the Past Volume 2. The Archaeology of the A4/A5 Road Improvements Scheme from Dungannon to Ballygawley*. Belfast, Northern Archaeological Consultancy.

Elder, S. (2009) *Report on the archaeological excavation of Raynestown 1, Co. Meath*. Unpublished Stratigraphic Report, ACS Ltd. Accessed online via: https://proxy.europeana.eu/255/_x633tf78s?view=https%3A%2F%-

2Frepository.dri.ie%2Fobjects%2Fx633tf78s%2Ffiles%2Fx-920vb624%3Fsurrogate%3Dpdf&disposition=inline&api_url=https%3A%2F%2Fapi.europeana.eu%2Fapi

Greaney, S., Hazell, Z., Barclay, A., Ramsey, C.B., Dunbar, E., Hajdas, I., Reimer, P., Pollard, J., Sharples, N. and Marshall, P. (2020) Tempo of a mega-henge: a new chronology for Mount Pleasant, Dorchester, Dorset. *Proceedings of the Prehistoric Society* 86, 199–236.

Grogan, E. (2007) *The Bronze Age Landscapes of the Pipeline to the West: an integrated archaeological and environmental assessment.* Bray, Wordwell.

Hannah, E. and McLaughlin R. (2019) Long-term archaeological perspectives on new genomic and environmental evidence from early medieval Ireland. *Journal of Archaeological Science* 106, 23–28.

Hartwell, B. (2002) A Neolithic ceremonial timber complex at Ballynahatty, Co. Down. Antiquity 76, 520–532.

IAC (n.d.) Website: www.iac.ie/a-late-neolithic-timber-circle-at-lagavooren-7-co-meath/; accessed November 2022

Kelly, B., Roycroft, N. and Stanley, M. (2012) Appendix 1 – radiocarbon dates from excavated archaeological sites described in these proceedings. In B. Kelly, N. Roycroft and M. Stanley, *Futures and Pasts: archaeological Science on Irish Road Schemes, Proceedings of a Public Seminar on Archaeological Discoveries on National Road Schemes, August 2012.* Dublin, Archaeology and the National Roads Authority, Monograph 10.

Lanting, J.N. and Brindley, A.L. (1999) Dating cremated bone: the scientific background. Trabajos de Prehistoria 56(2) 137–140. [https://www.ingentaconnect.com/content/doaj/00825638/1999/00000056/00000002/art00007]

Lehane, J., Johnston, P. and Leigh, D. (2011) Archaeological excavation report. E2412 – Ballynacarriga 3, Co. Cork. Prehistoric site with enclosure, structures, two ring ditches and associated cist burials. *Eachtra* 10.

McLaughlin, T.R. (2019) On applications of space-time modelling with open-source ^{14}C age calibration. *Journal of Archaeological Method and Theory* 26, 479–501.

McLaughlin, T.R. (2020) An archaeology of Ireland for the Information Age. Emania 25, 7–30.

McLaughlin, T.R., Whitehouse, N.J., Schulting, R.J., McClatchie, M., Barratt, P. and Bogaard, A. (2016) The changing face of Neolithic and Bronze Age Ireland: a big data approach to the settlement and burial archives. *Journal of World Prehistory* 29, 117–153.

NRA (n.d.) www.nra.ie/Archaeology/NRAArchaeologicalDatabase/ no longer available; accessed July 2010

O'Neill, N. (2013) Sacred places: Kilskeagh, Co. Galway and Neolithic earthen enclosures. *Proceedings of the Royal Irish Academy* 113C, 1–28.

O'Sullivan, M., Scarre, C. and Doyle, M. (2013) *Tara-- from the Past to the Future. Towards a New Research Agenda.* Dublin, Wordwell.

Plunkett, G., Carroll, F., Hartwell, B., Whitehouse, N.J. and Reimer, P.J. (2008) Vegetation history at the multi-period prehistoric complex at Ballynahatty, Co. Down, Northern Ireland. *Journal of Archaeological Science* 35, 181–190.

Reimer, P., Austin, W., Bard, E., Bayliss, A., Blackwell, P., Bronk Ramsey, C., Butzin, M., Cheng, H., Edwards, R., Friedrich, M., Grootes, P., Guilderson, T., Hajdas, I., Heaton, T., Hogg, A., Hughen, K., Kromer, B., Manning, S., Muscheler, R., Palmer, J., Pearson, C., van der Plicht, J., Reimer, R., Richards, D., Scott, E., Southon, J., Turney, C., Wacker, L., Adolphi, F., Büntgen, U., Capano, M., Fahrni, S., Fogtmann-Schulz, A., Friedrich, R., Köhler, P., Kudsk, S., Miyake, F., Olsen, J., Reinig, F., Sakamoto, M., Sookdeo, A. and Talamo, S. (2020) The IntCal20 Northern Hemisphere radiocarbon age calibration curve (0–55 cal kBP). *Radiocarbon* 62, 725–757.

Schulting, R.J., Murphy, E., Jones, C. and Warren, G. (2012) New dates from the north and a proposed chronology for Irish court tombs. *Proceedings of the Royal Irish Academy* 112C, 1–60.

Wainwright, G.J. (1989) *The Henge Monuments. Ceremony and Society in Prehistoric Britain.* London, Thames & Hudson.

Chapter 11: Interpreting the excavation results in the wider context of prehistoric Ballynahatty

Barclay, G. (1983) Sites of the third millennium BC to the first millennium AD at North Mains, Strathallan, Perthshire. *Proceedings of the Society of Antiquaries of Scotland* 113, 122–281.

Barclay, G.J. and Russell-White, C.J. (1993) Excavations in the ceremonial complex of the fourth to second millennium BC at Balfarg/Balbirnie, Glenrothes, Fife. *Proceedings of the Society of Antiquaries of Scotland* 123, 43–210.

Bayliss, A. and O'Sullivan, M. (2013) Interpreting chronologies for the Mound of the Hostages, Tara, and its contemporary context in Neolithic and Bronze Age Ireland. In M. O'Sullivan, C. Scarre and M. Doyle (eds), *Tara – from the Past to the Future*, 26–104. Dublin: Wordwell.

Bayliss, A., Marshall, P., Richards, C. and Whittle, A. (2017) Islands of history: the Late Neolithic timescape of Orkney. *Antiquity* 91, 1171–1188.

Bergh, S. and Hensey, R. (2013) Unpicking the chronology of Carrowmore. *Oxford Journal of Archaeology* 32, 343–366.

Bolger, T., Moloney, C. and Shiels, D. (2015) *A Journey Along the Carlow Corridor. The Archaeology of the M9 Carlow Bypass.* Dublin, National Roads Authority.

Bourke, E. (1997) Appendix 2: towards a reconstruction of the Grooved Ware circular wooden structure. In G. Eogan, and H. Roche *Excavations at Knowth (2): Settlement and Ritual sites of the 4th & 3rd millennium BC*, 283–294. Dublin, Royal Irish Academy Monographs in Archaeology.

Bradley, R (2003) A life less ordinary: the ritualization of the domestic sphere in later prehistoric Europe, *Cambridge Archaeological Journal* 13(1), 5–23.

Bradley, R.J. (2011) *Stages and Screens: an investigation of four henge monuments in northern and north-eastern Scotland.* Edinburgh, Society of Antiquaries of Scotland.

Brindley, A L (2007) *The Dating of Food Vessels and Urns in Ireland.* Galway, University of Ireland, Galway.

Brindley, A.L. and Lanting, J. (1990) Radiocarbon dates for Neolithic single burials. *Journal of Irish Archaeology* 5, 1–7.

Brophy, K. and Noble, G. (2020) *Prehistoric Forteviot.* York, Council for British Archaeology.

Carlin, N. (2011) *A Proper Place for Everything: the character and context of Beaker depositional practices in Ireland*. Unpublished PhD, University College Dublin.

Carlin, N. (2016) Discussion of the timber circle at Armalughey (sites 18 and 20) with a focus on the Grooved Ware and Beaker discoveries at the sites. In C. Dunlop and J. Barkley (eds), *Road to the West. A Road to the Past Volume 2. The Archaeology of the A4/A5 Road Improvements Scheme from Dungannon to Ballygawley*, 194–210. Belfast, Northern Archaeological Consultancy Ltd.

Carlin, N. (2017) Getting into the groove: exploring the relationship between Grooved Ware and developed passage tombs in Ireland, c. 3000–2700 cal BC. *Proceedings of the Prehistoric Society* 83, 155–188.

Carlin, N. and Cooney, G. (2017) Transforming our understanding of Neolithic and Chalcolithic society (4000–2200 BC). In M. Stanley, R. Swan and A. O'Sullivan (eds), *Stories of Ireland's Past. Knowledge Gained from NRA roads archaeology*, 23–56. Dublin, Transport Infrastructure Ireland.

Cassidy, L.M., Maoldúin, R.Ó., Kador, T., Lynch, A., Jones, C., Woodman, P.C., Murphy, E., Ramsey, G., Dowd, M., Noonan, A., Campbell, C., Jones, E.R., Mattiangeli, V. and Bradley, D.G. (2020) A dynastic elite in monumental Neolithic society. *Nature* 582, 384–388. [doi: 10.1038/s41586-020-2378-6]

Clark, G. (1936) The timber monument at Arminghall and its affinities. *Proceedings of the Prehistoric Society* 2, 1–51.

Collins, A.E.P. (1957) Excavations at the Giant's Ring, Ballynahatty. *Ulster Journal of Archaeology* 20, 44–50.

Collins, A.E.P. and Waterman, D.M. (1955) *Millin Bay: a Late Neolithic cairn in Co. Down*. Belfast, HMSO.

Condit, T. and Keegan, M. (2018) *Aerial Investigation and Mapping of the Newgrange landscape, Brú na Bóinne, Co. Meath: the archaeology of the Brú na Bóinne World Heritage site interim report, December 2018*. Belfast, Department of Culture, Heritage and Gaeltacht. [https://www.archaeology.ie/sites/default/files/files/bru-na-boinne-interim-report.pdf]

Cooney, G. (2000) *Landscapes of Neolithic Ireland*. London, Routledge.

Cooney, G. and Grogan, E. (1994) *Irish Prehistory. A Social Perspective*. Dublin, Wordwell.

Cotter, E. (2006) *Excavations at Kilbride, Co. Mayo*. Unpublished final report by Archaeological Consultancy Services Ltd.

Cunnington, M.E. (1929) *Woodhenge. A description of the site as revealed by excavations carried out there by Mr & Mrs B. H. Cunnington, 1926–7–8*. Devizes, George Simpson & Co.

Darvill, T. (2006) *Stonehenge: the biography of a landscape*. Stroud, Tempus.

Darvill. T. (in press) Magical rings: British stone and timber circles as ceremonial places. In P. Barnwell and T. Darvill (eds) *Places of worship in Britain and Ireland: Prehistoric and Roman* (Rowley House Studies in the Historic Environment 14). Doninton, Shaun Tyas.

De Paor, L. and Ó h-Eochaidhe, M. (1957) Unusual group of earthworks at Slieve Breagh, Co. Meath. *Journal of the Royal Society of Antiquaries of Ireland* 86, 97–10.

Dunlop, C. and Barkley, J. (2016) *Road to the West. A Road to the Past Volume 2. The Archaeology of the A4/A5 Road Improvements Scheme from Dungannon to Ballygawley*, Belfast, Northern Archaeological Consultancy Ltd.

Elliott, R. (2009) *Excavations at Paulstown, Co. Kilkenny*. Unpublished report by Irish Archaeological Consultancy on behalf of Kilkenny County Council and the National Roads Authority.

Eogan, G. (1991) Prehistoric and early historic culture change at Brughna Bóinne, *Proceedings of the Royal Irish Academy* 91C, 105–132.

Eogan, G. & Roche, H. (1997) *Excavations at Knowth (2): Settlement and Ritual sites of the 4th & 3rd millennium BC, RIA Monographs in Archaeology*. Dublin, Royal Irish Academy.

Eogan, J. (2000) *Excavations at Bettystown, Co. Meath [98E0072]*. Unpublished report for Archaeological Development Services Ltd.

Gibson, A.M. (1992), Possible timber circles at Dorchester-on-Thames. *Oxford Journal of Archaeology* 11, 85–91.

Gibson, A.M. (1994) Excavations at the Sarn-y-bryn-caled cursus complex, Welshpool, Powys and the timber circles of Great Britain and Ireland. *Proceedings of the Prehistoric Society* 60, 143–223.

Gibson, A.M. (1998) *Stonehenge and Timber Circles*. Stroud, Tempus.

Gibson, A.M. (2005) *Stonehenge and Timber Circles* (2nd edn). Stroud, Tempus.

Gillespie, R. (n.d.) *Lowpark, Co. Mayo, Multiperiod Archaeological Complex, Registration No. E3338, Ministerial Direction No. A020/012, final report*. Unpublished report for Mayo County Council. [https://doi.org/10.7486/DRI.gf06vh467]

Greaney, S., Hazell, Z., Barclay, A., Ramsey, C.B., Dunbar, E., Hajdas, I., Reimer, P., Pollard, J., Sharples, N. and Marshall, P. (2020) Tempo of a mega-henge: a new chronology for Mount Pleasant, Dorchester, Dorset. *Proceedings of the Prehistoric Society* 86, 199–236.

Green, M. (2000) *A Landscape Revealed: 10,000 years on a chalkland farm*. Stroud, Tempus.

Groenman-van Waateringe, W. and Butler, J.J. (1976) The Ballynoe stone circle: excavations by A.E. van Giffen 1937–8. *Palaeohistoria* 18, 73–104.

Haggarty, A. (1991) Machrie Moor, Arran: recent excavations at two stone circles. *Proceedings of the Society of Antiquaries of Scotland* 121, 51–94.

Harris, W. (1744) *The Ancient and Present State of the County of Down*. Dublin, Reilly.

Hartwell, B. (1998) The Ballynahatty complex. In A.M. Gibson and D.D.A. Simpson (eds), *Prehistoric Ritual and Religion: essays in honour of Aubrey Burl*, 32–44. Stroud, Sutton.

Hartwell, B. (2002) A Neolithic ceremonial timber complex at Ballynahatty, Co. Down. *Antiquity* 76, 520–532.

Herity, M. (1974) *Irish Passage Graves: Neolithic tomb builders in Britain and Ireland 2500 BC*. Dublin, Irish University Press.

Kador, T., Cassidy, L.M., Geber, J., Hensey, R., Meehan, P. and Moore, S. (2018) Rites of Passage: Mortuary Practice, Population Dynamics, and Chronology at the Carrowkeel Passage Tomb Complex, Co. Sligo, Ireland. *Proceedings of the Prehistoric Society* 84. 225–255.

Kilbride-Jones, H.E. (1950) The excavation of a composite early Iron Age monument with 'henge' features at Lugg, Co. Dublin. *Proceedings of the Royal Irish Academy* 53C, 311–332.

Laidlaw, G. (2017) Excavations of the Late Neolithic Grooved Ware site at Scart, Co. Kilkenny. *Journal of Irish Archaeology* 26, 33–56.

Lehane, J., Johnston, P. and Leigh, D. (2011) *Archaeological Excavation Report, E2412, Ballynacarriga 3, Co. Cork: prehistoric site with enclosure, structures, two ring-ditches and associated cist burials. Eachtra* 10.

Lehane, J., Johnston, P. and Leigh, D. (2019) 2.8 Ballynacarriga 3 – Multi-period prehistoric ceremonial site. In P. Johnston & J. Kiely (eds), *Hidden Voices: the archaeology of the M8 Fermoy–Mitchelstown motorway*, 40–51. Dublin, Transport Infrastructure Ireland.

Liversage, G.D. (1960) A Neolithic site at Townleyhall, Co. Louth. *Journal of the Royal Society of Antiquaries of Ireland* 90, 49–60.

MacAdam, R. and Getty, E. (1855) Discovery of an ancient sepulchral chamber. *Ulster Journal of Archaeology* 3, 358–365.

Mercer, R (1982) The excavation of a late Neolithic henge-type enclosure at Balfarg, Markinch, Fife, Scotland, 1977–78. *Proceedings of the Society of Antiquaries of Scotland* 111, 63–171.

Monteith, J. (2008) *Final Report on Excavations at Scart 1, Co. Kilkenny (E3001). N9/N10 Dunkitt to Sheepstown*. Unpublished report by Valerie J Keeley Ltd. for Kilkenny County Council. Accessed online via: https://repository.dri.ie/.

Musson, C.R. (1971) A study of possible building forms at Durrington Walls, Woodhenge and the Sanctuary. in G.J. Wainwright and I.H. Longworth, *Durrington Walls: excavations 1966–1968*, 363–377. London, Reports of the Research Committee of the Society of Antiquaries of London 29.

Noble, G., Greig, M., Millican, K., Anderson, S., Clarke, A., Johnson, M., McLaren, D. and Sheridan, J.A. (2012) Excavations at a multi-period site at Greenbogs, Aberdeenshire, Scotland and the four-post timber architecture tradition of Late Neolithic Britain and Ireland. *Proceedings of the Prehistoric Society* 78, 135–171.

Ó Drisceoil, C. (2009) Archaeological excavation of a Late Neolithic Grooved Ware site at Balgatheran, County Louth, *County Louth Archaeological and Historical Journal* 27, 77–102.

O'Sullivan, M. (2005) *Duma na nGiall. The Mound of the Hostages, Tara*. Bray: Wordwell.

Parker Pearson, M., Marshall, P., Pollard, J., Richards, C., Thomas, J. and Welham, K. (2013) Stonehenge. In H. Fokkens and J. Harding (eds), *The Oxford Handbook of the European Bronze Age*, 159–178. Oxford, Oxford University Press.

Parker Pearson, M., Pollard, J., Richards, C., Thomas, J., Tilley, C. and Welham, K. (2020) *Stonehenge for the Ancestors, Part 1: landscape and monuments*. Leiden, Sidestone.

Phelan, S. (2007) Whitewell. Grooved Ware timber circle. In E. Grogan, L. O'Donnell and P. Johnson (eds), *The Bronze Age Landscapes of the Pipeline to the West*, 349–350. Bray, Wordwell.

Pigière, F. and Smyth, J. (2023) First evidence for cattle traction in Middle Neolithic Ireland: a pivotal element for resource exploitation. *PLoS ONE* 18(1): e0279556. [https://doi.org/10.1371/journal.pone.0279556]

Pollard, J. (1992) The Sanctuary: new investigations on Overton Hill, Avebury. *Wiltshire Archaeological and Natural History Magazine* 94, 1–23.

Richards, C. and Thomas, J. (1984) Ritual activity and structured deposition in later Neolithic Wessex. In R. Bradley and J. Gardiner (eds), *Neolithic Studies: a review of some current research*, 189–218. Oxford, British Archaeological Report 133.

Ritchie, J.N.G. (1976) The Stones of Stenness, Orkney. *Proceedings of the Society of Antiquaries of Scotland* 107, 1–60.

Schulting, R., Murphy, E., Jones, C. and Warren, G. (2012) A proposed chronology for Irish court tombs based on new dates from the north of the island. *Proceedings of the Royal Irish Academy* 112, 1–60.

Sheridan, J.A. (1986) Megaliths and megalomania: an account, and interpretation, of the development of passage tombs in Ireland, *Journal of Irish Archaeology* 3, 17–30.

Sheridan, J.A. (2003) The National Museums of Scotland *Dating Cremated Bones Project*: dates obtained during 2002/3. *Discovery and Excavation in Scotland* 4, 167–9.

Sheridan, J.A. (2004) Going round in circles? Understanding the Irish Grooved Ware 'complex' in its wider context. In H. Roche, E. Grogan, J. Bradley, J. Coles and B. Raftery (eds), *From Megaliths to Metals: essays in honour of George Eogan*, 26–37. Oxford, Oxbow Books.

Sheridan, J.A. (2014) Little and large: the miniature 'carved stone ball' beads from the eastern tomb at Knowth, Ireland, and their broader significance. In R.M. Arbogast and A. Greffier-Richard (eds), *Entre archéologie et écologie, une préhistoire de tous les milieux. Mélanges offerts à Pierre Pétrequin*, 303–314. Besançon, Presses universitaires de Franche-Comté.

Smyth, J. (2011) The house and group identity in the Irish Neolithic. *Proceedings of the Royal Irish Academy. Section C: Archaeology, Celtic Studies, History, Linguistics, Literature* 111C, 1–31.

Smyth, J. (2013) Tides of change? The house through the Irish Neolithic. In D. Hofmann and J. Smyth, (eds), *Tracking the Neolithic House in Europe*, 301–327. New York, Springer One World Archaeology.

Smyth, J. (2014) *Settlement in the Irish Neolithic: new discoveries at the edge of Europe*. Oxford, Prehistoric Society Research Papers 6.

Stafford, E. (2012) *00E0914 Lagavooren 7 Final Report*. Unpublished report to the National Monuments Service, Department of the Environment, Heritage and Local Government. [https://www.europeana.eu/sv/item/255/_1j92vp21j]

Stout, G. (1991) Embanked enclosures of the Boyne region. *Proceedings of the Royal Irish Academy* 91 C, 245–264.

Sweetman, P.D. (1971) An earthen enclosure at Monknewtown, Slane, preliminary report. *Journal of the Royal Society of Antiquaries of Ireland* 101(2), 135–140.

Sweetman, P.D. (1976) An earthen enclosure at Monknewtown, Slane, Co. Meath. *Proceedings of the Royal Irish Academy* 76C, 25–72.

Thomas, J. (1991) *Rethinking the Neolithic*. Cambridge, Cambridge University Press.

Thomas. J. (1999) *Understanding the Neolithic*. London, Routledge.

Thomas, J. (2007a) *Place and Memory: excavations at the Pict's Knowe, Holywood and Holm Farm, Dumfries and Galloway, 1994–8*. Oxford: Oxbow Books.

Thomas, J. (2007b) The internal features at Durrington Walls: investigations in the Southern Circle and Western Enclosures 2005–6. In M. Larsson and M. Parker Pearson (eds), *From

Stonehenge to the Baltic: living with cultural diversity in the third millennium BC, 145–158. Oxford, British Archaeological Report S1692.

Thomas, J. (2010) The return of the Rinyo-Clacton folk? The cultural significance of the Grooved Ware Complex in later Neolithic Britain. *Cambridge Archaeological Journal* 20(1), 1–15.

Wainwright, G.J. and Longworth, I.H. (1971) *Durrington Walls: excavations 1966–1968.* London, Reports of the Research Committee of the Society of Antiquaries of London 29.

Whittle, A.W.R. (1997) *Sacred Mound Holy Rings. Silbury Hill and the West Kennet Palisade Enclosures: a later Neolithic complex in north Wiltshire.* Oxford, Oxbow Monograph 74.

Chapter 12: Digitally recreating Ballynahatty and simulating astronomical alignments in Irish timber circles

Barratt, R.P. (2022) The crux of astronomical alignment in Neolithic Malta: using 3D simulation to produce new data. *Digital Applications in Archaeology and Cultural Heritage* 26, 1–19.

Earl, G.P. (2006) At the edges of the lens. In T.L Evans and P. Daly (eds), *Digital Archaeology: bridging method and theory*, 173–188. London, Routledge.

Eogan, G. and Roche, H. (1997) *Excavations at Knowth (2): Settlement and Ritual sites of the 4th & 3rd millennium BC, RIA Monographs in Archaeology.* Dublin, Royal Irish Academy.

Georgopoulos, A. (2014) 3D virtual reconstruction of archaeological monuments. *Mediterranean Archaeology and Archaeometry* 14(4), 155–164.

Gibson, A. and Simpson, D.D.A. (1998) *Prehistoric Ritual and Religion: essays in honour of Audrey Burl.* Stroud, Sutton.

Gormley, S. (2004) *The Dating and Phasing of the Timber Circle Complex at Ballynahatty, Co. Down.* Unpublished thesis, Queen's University Belfast.

Hartwell, B. (1991) Ballynahatty – a prehistorical ceremonial centre. *Archaeology Ireland* 5(4), 12–15.

Hartwell, B. (1998) The Ballynahatty Complex. In A. Gibson, and D.D.A. Simpson (eds), *Prehistoric Ritual and Religion: essays in honour of Audrey Burl.*, 32–45. Stroud, Sutton.

Hartwell, B. (2002) A Neolithic ceremonial timber complex at Ballynahatty, Co. Down. *Antiquity* 76, 520–532.

Meeus, J. (1987) *Astronomical Algorithms.* Richmond, Willmann-Bell.

Patay-Horvátz, A. (2014) The virtual 3D reconstruction of the east pediment of the temple of Zeus at Olympia – an old puzzle of classical archaeology in the light of recent technologies. *Digital Applications in Archaeology and Cultural Heritage* 1, 12–22.

Polcaro, A. and Polcaro, V.F. (2009) Man and sky: problems and methods of archaeoastronomy. *Archeologia e Calcolatori* 20, 223–245.

Rua, H. and Alvito, P. (2011) Living the past: 3D models, virtual reality and game engines as tools for supporting archaeology and the reconstruction of cultural heritage – the case-study of the Roman villa of Casal de Freiria. *Journal of Archaeological Science* 38, 3298–3308.

Ruggles, C.L.N. (2011) Pushing back the frontiers or still running around the same circles? 'Interpretative archaeoastronomy' thirty years on. *Proceedings of the International Astronomical Union* 7, 1–18.

Ruggles, C.L.N. (2015) *Handbook of Archeoastronomy and Ethnoastronomy.* New York, Springer.

Schaefer, B.E. (2006) Case study of three of the most famous claimed archaeoastronomical alignments in North America. In T.W. Bostwick, and B. Bates (eds), *Viewing the Sky Through Past and Present Cultures*, 27–56. Phoenix AZ, Pueblo Grande Museum Anthropological Papers 15.

Sorrell, A. (1981) *Reconstructing the Past.* London, Batsford.

Chapter 13: The Ballynahatty landscape – past, present and future

A map of part of Ballinahatty held by John McKeown and James Thomson surveyed by Thomas Pattison, 1819 PRONI T872/1. Available at the Public Record Office Northern Ireland.

Belfast Newsletter (1855) Discovery of an ancient tomb, 21 November.

Benn, G. (1823) *The History of the Town of Belfast, with an Accurate Account of the Former and Present State; to which are added a Statistical Survey of the Parish of Belfast and a Description of some Remarkable Antiquities in its Neighbourhood.* Belfast, Mackay.

Bordes, Lt F.W. and Scott, G. (1833) *Parish of Drumbo.* Belfast, Ordnance Survey Memoirs.

Borlase, W.C. (1897) *The Dolmens of Ireland.* London, Chapman & Hall.

Day, A. (2014) *Glimpses of Ireland's Past: the Ordnance Survey memoir drawings: topography and technique.* Dublin, Royal Irish Academy.

Evans, A. (2021) John Grattan: pharmacist, phrenologist and physical anthropologist. *Ulster Journal of Archaeology* 76, 127.

Evans, E.E. (1943) Mr H. C. Lawlor. *Nature* 152, 441–442.

Gray, W. (1890) Discovery of an ancient sepulchre at the Giant's Ring, Belfast. *Journal of the Royal Society of Antiquaries of Ireland* 1(2), 164–165.

Harris, W. (1744) *The Ancient and Present State of the County of Down.* Dublin, Reilly.

Knox, A. (1875) *A History of the County of Down from the Most Remote Period until the Present Day; Including an Account of its Early Colonisation, Ecclesiastical, Civil, and Military Polity, Geography, Topography, Antiquities and Natural History.* Dublin, Hodges, Foster & Co.

Lawlor, H.C. (1918) The Giant's Ring. *Proceedings of the Belfast Natural History and Philosophical Society*, 13–27.

Lewis, S. (2003) *Topographical Dictionary of County Down.* Belfast, Friar's Bush Press.

MacAdam, R. and Getty, E. (1855) Discovery of an ancient sepulchral chamber. *Ulster Journal of Archaeology* 3, 358–365.

O'Laverty, J. (1880) *An Historical Account of the Diocese of Down and Connor, Ancient and Modern.* Dublin.

Plumptre, A. (1817) *Narrative of a Residence in Ireland During the Summer of 1814, and that of 1815.* London.

Tilley, C. Y. (1994) *A Phenomenology of Landscape: places paths and monuments.* Oxford, Berg.

Tilley, C. (2006) *The Materiality of Stone: explorations in landscape phenomenology: 1.* Oxford, Berg.

Waddell, J (2005) *Foundation Myths: the beginnings of Irish archaeology.* Bray, Wordwell.

Young, A. (1780) *A Tour in Ireland: with general observations on the present state of that kingdom. Made in the years 1776, 1777 and 1778. And brought to Down to the end of 1779.* London, H. Goldney.

Index

1855 tomb 5, 7, 8, 9, 18, 25–26, 31, 41, 90–91, 151, 153, 160–161, 201, 204, 206
    cremated bone 136, 139
    pottery 146
    unburnt bone 145, 146–148, 150

Acrux 195
aerial survey 14–16, 37, 180
agricultural damage 5, 13, 199, 201–202, 205
Aldebaran 191, 196, 197
animal bone 8, 9, 40, 103, 143, 145–146, 148, 164
Annaghilla, Tyrone 154
Anne, Viscountess Midleton 202
Annexe 13, 37–40, 58, 64–66, 68–69, 71, 75, 76, 85–87, 90, 97–98, 101–102, 123, 128, 132–133, 138–139, 144, 152, 155, 160, 162–165, 172–173, 176, 179, 185, 200, 205
Annexe entrance 140–141, 171–172, 174
Antares 191, 196–198
Antler 160, 174–175, 178
archaeobotanical remains 37, 78–81
Archaeology Teaching Collection, QUB 204
Arcturus 193
Armalughey, Tyrone 154, 170, 172, 175, 180, 183, 190, 193–194, 196
Arminghall, Norfolk 169
arrowhead, petit tranchet derivative 25–26, 114, 118–119, 123, 125–126, 129–130, 160, 162, 176
arrowheads, leaf-shaped 10, 12, 119–120, 159
artefacts 12, 25–26, 102, 107, 119, 121, 123, 125, 127, 131, 135, 175–177
Ask, Co. Wexford 89, 92, 105, 154
astronomical alignments 187, 189–191, 193, 195, 197
axehead 102, 117, 120, 127
    dolerite 133
    Mesolithic 26, 153
    porcellanite 132, 177
    rhyolite 132–133

Balfarg, Fife 26, 169, 185
Balgatheran, Co. Louth 196
Ballycairn Presbyterian Church Bazaar 204

Ballylesson xv, 17, 25
Ballynacarriga, Cork 92, 97, 154, 190, 192, 193–195, 197
Ballynahatty Bog 15, 20, 22, 28, 31, 58, 153, 169
Ballynahatty 3–4, 18, 26
Ballynoe, Co. Down 91, 162, 206
Bateson, Robert 202
bead
    glass 134–135
    faience 134
    lignite 131, 135, 177
    stone 135
Belfast Natural History and Philosophical Society (BNHPS) 7, 201, 204
Belvoir House 201–202
Benn, George 5–6, 8, 201, 204
Beta Centauri 195
Beta Crux 197
Betelgeuse 191, 195, 197
Bettystown, Co. Meath 131, 175, 183, 190, 192–195, 197
Bleasdale, Lancashire 171
BNH site numbers 14, 31–34
BNH3 12, 14, 16, 19, 25–26, 31, 78, 99, 164
BNH5 173, 176–77, 179–180, 184–186, 189, 201, 205
BNH5/6 d 15, 18, 21, 24, 26, 31, 33–34, 165, 172–173, 176–177, 179–180, 201
BNH6 3, 15, 18, 37–39, 43–44, 46–50, 52–28, 71, 85–87, 89–91, 97, 99, 100–102, 123–128, 131, 132–134, 138, 141–145, 150, 151–153, 155–156, 159–160, 162–177, 179–180, 183–186, 188–189
    construction 164–173
    planning 166
BNH25 18, 20, 31, 100
BNH36 18, 21, 31
BNH40 18, 21, 31
BNH42 18, 19, 31
BNH43 21, 22–23, 31, 185
Bodel family 8, 9, 17, 201
Bodel house 9, 12, 17, 202
bone
    animal, *see* animal bone
    burnt 12, 26, 41, 43, 65, 143, 159, 175, 183

cremated 8, 40, 41, 44, 52, 82, 90–91, 103, 105, 128, 136–137, 139–141, 144–145, 160–162, 164–166, 175, 178, 204; see also cremation deposit
   human 8–10, 52, 76, 78, 133, 145, 150, 161
   unburnt 102, 145–146, 160, 176
Borlase, William 6, 8, 10, 204
Brú na Bóinne, Co. Meath 180, 183–185
Brynkinalt, Co. Denbigh 202–203
Burning pit C1453 (early medieval) 78, 80–81, 134, 156

Capella 194
Cappagh Beg, Co. Derry 190, 195–196
Skinner, Captain Cortland 202
Carinated Bowl 82–83, 86, 88–90, 159–160
Carnbane, Co. Armagh 155
Carrowkeel cemetery, Co. Sligo 82, 90–91, 160–162, 178
Carrowkeel Ware, see Course Ware
Carrowmore cemetery, Co. Sligo 160, 162
Carrowreagh, Co. Down 155
cemetery mound 5, 8, 20
cereals 81, 156, 183
chamber C591 40, 42, 82–83, 86, 90–91, 101, 136, 160
   cremations 82–83, 136, 160
charcoal 10, 12, 29–30, 44, 46, 47–54, 57, 58, 60–65, 68–71, 76, 78, 89, 101–102, 134, 149–151, 152, 154, 159–160, 164–65, 169, 173–174, 180, 183
Chart, D.A. 10
chisel 10, 26, 134, 205
Coarse Ware 8, 9, 12, 25, 41, 43, 82–83, 85–86, 87, 90–91, 100–101, 123, 128, 136, 152, 160–162, 165, 178, 206
Collared Urn 25, 82, 105, 164
Collins, A.E.P. 'Pat' 5, 10, 12, 14, 90–91, 103, 130, 161–162, 177, 180, 205
control of movement 165
cord-impressed pottery 26
coring 28
cremation deposit
   C588 43, 44, 136, 141, 144, 160
   C943 43, 44, 128, 142, 144, 160
   C1012 43, 87, 128, 139–140, 144, 160
   C1014 43, 83, 87, 90, 128, 139, 141, 144, 160–161
   C1016 43, 139, 144, 160
   C1018 43, 139, 160
cremations 25, 26
cromlech 5, 9, 204
cursus 20, 31, 159, 206

dating
   14C 149–156
   Bayesian chronology 149, 151, 153, 155–156, 165
   Early Bronze Age 103, 144, 150, 152, 155–156
   early medieval 81, 135, 138, 150, 152, 156
   Early Neolithic 89, 134, 152–153, 156, 159–161, 179
   Kernel Density Estimate (KDE) 150, 153, 155
   Late Neolithic 153–155, 159
   Mesolithic 29, 150, 152–153, 156
   Middle Neolithic 90, 144, 147, 155, 159, 163
Deneb 197
Derrybeg, Co. Derry 190, 192, 194, 195
DNA analysis 145, 147, 161, 196, 206
Dorchester on Thames, Oxfordshire 175
draught animals 165
drone survey 15, 16, 56, 183, 188, 205
DropGC 188
Druids 7 203, 203, 204
Dublin Philosophical Society 203
Dungannon, Viscounts and family 5, 202–204
Dungannon wall 12, 15, 203, 205
Dunlop, Charles 203
Dunlop, Mrs 25
Durrington Walls, Wiltshire 102, 169, 172–173, 175, 179, 183

Early Neolithic 82, 88–89, 134, 152, 153, 156, 159, 160–161, 179–180
Eastern Setting 38, 44, 56, 57, 58, 78, 128, 162, 168, 169, 173, 188
Edenderry 4, 5, 20, 22, 26, 153, 199
Edenderry House 4, 25, 204
Entrance (BNH5) 57, 58, 86, 162, 165
Entrance (BNH6) 37, 38, 44, 162
Entrance Chamber posts (EC) 61
excarnation 162, 164, 176, 180, 185–186, 189
excavation grid 12, 26, 37–38
excavation licence 37
exclusion 170, 180

feature groups 39, 44, 64, 69–71, 86–87, 136, 163, 165–166, 171, 174
field collection of artefacts 13, 25–27
Four-Poster (BNH6), (EC) 39, 44, 52, 71, 86, 91, 100, 123–124, 127, 131–132, 138, 143–144, 152, 162, 163, 165–168, 174, 176–177, 183–185

Geology 3
geophysical survey 2, 31, 33, 58
Giant's Ring 3–12, 14–15, 17–18, 23–26, 200–206
   construction 9, 12, 21–22, 31, 39, 44, 128, 156, 164, 178, 180, 184–185, 200
   dating 25, 155–156, 160, 162, 177, 179–180
   function 34, 161
Glassdrummond, Down 154
Glenshane Road Quarry, Co. Derry 89
Globular Bowl 81–82, 85, 87–90, 160
Grange, Meath 103, 154
Gray, William 26, 204
Greenbogs, Aberdeenshire 183
Grooved Ware 65, 82–83, 85–87, 89–92, 94, 97–103, 105, 123, 128, 160, 165, 176, 183, 185–186

Knowth 1 82, 86–87, 95–97, 102–103, 176, 186
Knowth 2 82, 85–87, 91–92, 97, 102–103, 176
Knowth 3 86–87, 97, 100, 102–103, 176
Guardianship 204–205

Hammerstone 107, 109, 133–134, 200
Harris, Walter 5, 10, 12, 159, 201, 203
hazelnuts 78, 81, 103, 152, 164
Henge A, Brú na Bóinne 184–185
henge, *see* Giant's Ring
Hill-Trevor, *see* Dungannon

inhumations 186
Inner Façade 39–40, 69, 71, 76, 87, 90, 128, 132, 139, 145, 163, 173
Inner Ring (BNH5) 38–39, 44, 60–61, 64, 79, 101, 128, 132, 138, 163, 168
Inner Ring (BNH6) 38–39, 44, 46–48, 52–53, 56, 100, 123, 127–128, 132–133, 138, 141, 144, 152, 160, 163, 166–169, 176–177
intermediate postholes 56
iron pan 44, 46–47

javelin head, flint 119

Kilbride, Co. Mayo 92, 97, 190–195, 197
Kilmainham, Meath 92, 97, 127, 129, 154, 190
Kilshane, Co. Dublin 165
Kilskeagh, Co. Galway 155
Knowth, Co. Meath 91, 92, 94, 102–103, 126–127, 129–130, 155, 166, 168–169, 175–176, 180, 183, 185–187, 189
Knox, Alexander 203

Lagan 3–5, 15, 20, 22, 26, 153, 164–165, 183, 199–200, 202
Lagavooren, Co. Meath 89, 92, 97, 102, 105, 127, 131, 154, 183, 190, 192–194, 196
Late Mesolithic 25
Lawlor, Henry 5, 8–10, 12, 159, 205
Ligar, Charles 5, 201
linear features 20,
lintels 41, 58, 169, 189
lithic artefacts 123, 126–127, 129
lithic distribution 126, 176
Lowpark, Co. Mayo 89, 92, 97, 102, 105, 130, 175, 183, 190–191, 193–194, 197
Lugg, Co. Dublin 172

MacAdam, James 7–9, 25, 145–148, 160–161, 201–202, 204
MacDonell, Dr James 204
Machrie Moor, Arran 102, 171, 183
magnetic gradiometer, *Sensys* 22
magnetic susceptibility 13

marker cairn 41, 161, 174
McClure, William 201
McKeown, John 9, 201
Meshlab 188
Mesolithic 25, 26, 29–30, 134, 150, 152–153, 156
Middle Façade 39, 64, 69, 71, 87, 128, 163, 171–172, 180
Middle Neolithic 8, 12, 25, 30, 37, 43, 82, 85, 87, 89–90, 100–101, 133, 144, 147, 155, 159, 160, 161, 162, 163, 165, 176–177, 180, 205–206
Millin Bay, Co. Down 91, 162
Monknewtown, Co. Meath 162, 178
motte 5, 20
Mound of the Hostages, Tara, Co. Meath 91, 162
Mount Pleasant, Dorset 102, 155, 172, 179
Mullan, James 201

National Museum of Ireland 25, 204
Newgrange Four Poster 184–185
NISMR numbers 14, 31
North Mains, Perth and Kinross 171–172, 179
North–South posts (N–S) 39, 64–66, 79, 101, 128, 134, 159, 163, 165
Northern East–West posts (NEW) 39, 64–65, 68, 128, 163

Ordnance Survey 200, 204–205
orthophotograph 17
Outer Façade posts (OF) 38–39, 64, 69, 71–74, 85, 87, 90, 101, 128, 132, 138, 156, 163, 165, 171–173
Outer Ring (BNH5) 39, 58, 60, 64–65, 67, 76, 88, 101, 128, 132, 138, 163, 171, 173, 177
Outer Ring (BNH6) 44, 47, 49–50, 52, 56, 65, 100, 123, 125, 128, 132, 138, 144, 163, 165–166, 168, 177, 169

Parknahown, Laois 154
passage tomb 4–6, 8, 10, 12, 14–15, 17, 23, 24, 31, 89, 91, 133, 156, 159, 161–162, 172, 177–178, 179–180, 184–186, 199, 200, 201, 203–204, 206
Pattison Estate map, 1819 201, 205
Paulstown, Co. Kilkenny 105, 154, 175, 183, 190, 192–195, 197
Phase 1a 78, 79, 165–166, 176
Phase 1b 79, 165, 169, 176
Phase 2 162, 165, 171, 176
Phase 3 79, 163–164, 165, 171, 173, 176
Phase 4 80, 81, 164, 176
phenomenological description 188, 199
pitchstone, Arran 134, 159
planking 64, 168, 170–171
platform (excarnation) 15–16, 44, 48, 71, 162, 164, 167–169, 185, 187, 189
platform, lithic 26, 119, 108, 109, 110, 111, 117, 120–123, 128–130, 134
Plumptre, Anne 5, 204
post decay 41, 174
post extraction 60, 71, 102, 174–175, 177

post heights 45–48, 58, 65, 69, 71, 165, 167, 169–171, 174, 188
posthole excavation 38–40, 44, 46, 49, 50, 57–58, 64, 69, 71, 168, 171, 205
pottery distribution 87, 124
primary fill 41, 44, 46, 47, 50, 57, 58, 60, 65, 67, 69, 71, 84–85, 87, 97, 98, 100–101, 123, 134, 139, 144–145, 160, 167, 174, 176
Procyon 194
Prumplestown Lower, Kildare 154, 175, 190, 193–196
Purdy's Burn 3, 4, 20, 33,

Queen's University Belfast xv, 14, 26, 37, 78, 106, 165, 204

racecourse 204
ramp 32, 43, 44, 46–47, 50, 58, 60, 64, 69, 167–168, 174–75, 199, 171–172, 179–180, 186
Raynestown, Meath 154
resistivity survey 14, 22–23
ridge 12–15, 17, 25, 34, 37, 39, 40, 44, 58, 64–65, 68–69, 76, 78, 88, 92, 160–161, 164, 167, 169, 188, 199, 205–206
Rigel 191, 194, 197
Rigil Kent 193
ring ditches 17–18, 21, 31, 164
Ritchie, Gail 26
Romano-Germanic Commission, Frankfurt 22, 24
roof 7, 40, 58, 161, 167–169, 171, 183, 189
Royal Irish Academy (RIA) 8, 203, 204
Russell, John 5, 201

Sanctuary, Wiltshire 169, 172
Sarn-y-bryn-caled, Powys 169
Scart, Co. Kilkenny 89, 92, 97, 102, 131, 190, 192, 175, 183, 193–194
Schedule 205
secondary fill 44, 46, 48, 51, 53, 57–58, 60, 64–65, 67–69, 71, 76, 84–85, 91, 100–101, 123, 133–134, 136, 143–144, 168, 171, 174–175
sidereal significance 187, 189, 197
Sirius 191, 193, 197–198
Sketchfab 18
Sketchup 188
Skreen, Meath 154
sky burial 200
Slieve Breagh, Co. Meath 183

Southern East–West posts (SEW) 39, 65, 67, 71, 75–76, 138, 163, 165, 171, 176, 179
split stone C998 43–44
spot phosphate 12–13
standing stone 4, 12, 14, 16–17, 31, 164, 205–206
stone artefact distribution 106, 123–128, 131, 134
stone ball 133, 177
stratigraphy 12, 29, 39, 44, 47, 64, 69, 162, 165, 178

TarxienCore 188, 190–192, 194, 196, 198
temple 180, 186, 200
tenant 5, 8, 17, 200–201
tephra 28–29
Thompson, James xv, 37
timber circles, Ireland 81, 88–89, 92, 102–103, 105, 126–127, 130–131, 149, 153, 155, 168, 172, 175, 180, 182–183, 186–187, 189–191, 197–198
Tonafortes, Co. Sligo 155
topography 3, 184, 185
Townleyhall, Co. Louth 162
trackway 20, 22, 205

Ulster Archaeological Society xv, 37
Ulster Museum 12, 25, 126, 145
University College Dublin 82, 136

Vallency, General 5
Vega 193
vegetation history 28–29, 78, 81
    Early Bronze Age 39
    early post-glacial period 28–29
    Mesolithic 29–30
    Neolithic 30

walling 5, 9, 12, 15, 20, 41, 43, 46–47, 50, 58, 60, 65, 68, 71, 167, 169–171, 177–178, 185, 188–189, 199, 200, 202, 203, 205
weed seeds 80–81
West Kennet, Wiltshire 172
West–North–West posts (WNW) 39, 65–68, 71, 76, 103, 143, 163–165, 171
Whitewell, Co. Westmeath 154, 190, 192–194
winter solstice 197–198
winter sunrise 197–198
Woodhenge, Wiltshire 169, 171, 180
Wyke Down, Dorset 183